The Dynamic Ether of Cosmic Space

Correcting a Major Error in Modern Science

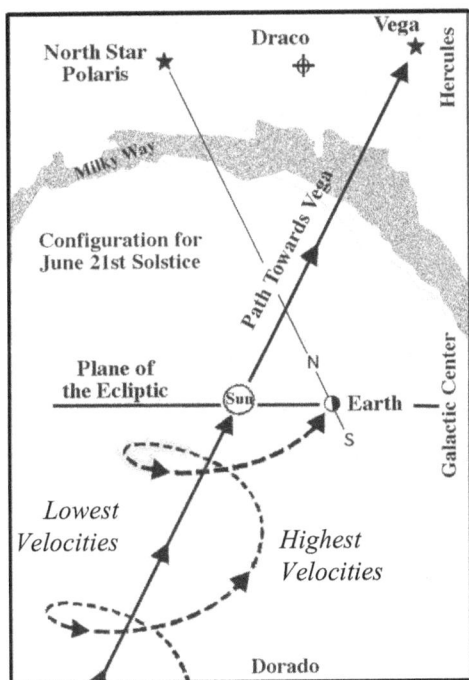

North Star Polaris

Draco

Vega

Hercules

Milky Way

Configuration for June 21st Solstice

Path Towards Vega

Plane of the Ecliptic

Sun

N

Earth

S

Galactic Center

Lowest Velocities

Highest Velocities

Dorado

James DeMeo, PhD

The Dynamic Ether of Cosmic Space

Correcting a Major Error in Modern Science

by

James DeMeo

2019
Natural Energy Works
Ashland, Oregon, USA
www.naturalenergyworks.net

Publication and worldwide distribution rights:

Natural Energy Works
PO Box 1148
Ashland, Oregon 97520
United States of America
http://www.naturalenergyworks.net
Email: info@naturalenergyworks.net

ISBN: 978-0-9974057-1-2

First Edition, 2019

200118

All photos and images have been identified as to their sources to the best extent humanly possible. Many are created by the author, and retain copyrights with this book. In a few cases where an item is mis-attributed or unattributed, when so informed of more exact sources or an error in permissions, we will be happy to either add the credits or remove the item from subsequent printings.

Front Cover: Author's graphic of Earth's spiral-form trajectory around the Sun, as the solar system moves towards the star Vega. Inset photo is Dayton Miller's light-beam interferometer, situated at Mount Wilson. **Rear Cover**: Authors graphic of 17 different independently determined axes of Earth's cosmic motion in space.
Cover Background: A portion of the blue-glowing halo of cosmic ether, surrounding Andromeda galaxy.

CONTENTS

Table of Figures

Table of Tables Page

ACKNOWLEDGMENTS

A work such as this could not have developed without the encourage-ments and assistance I received over a long period of my adult life. I wish to thank Professors Robert Nunley and Robert Haralick at the University of Kansas, who allowed and even encouraged me to think "outside the box". A thanks goes to John Chappell, who as a fellow KU PhD, went on to form the Natural Philosophy Alliance, and invited me to present my early findings on the ether-drift experiments to various NPA/AAAS conferences. My gratitude also to physicist Carolyn Th-ompson who, in many private discussions and along with Chappell, opened my eyes to serious problems in modern physics and astrophys-ics. I thank William Fickinger and the Archivists at Case Western Reserve University in Cleveland, Ohio, who granted access to their materials on Michelson-Morley and Dayton Miller. Likewise a thanks to the Archivists at Hebrew University in Jerusalem, who granted access to their Albert Einstein collections relevant to the ether question.

I am also grateful to Yuri Galaev and Hector Munera, for their ether-drift research and permissions to use their photos and graphs in this book. A thank-you goes to David Marett, Thanassis Mandafounis, Tom DiFerdinando and Gary Douglass, friends and colleagues already familiar with this line of research, for their helpful pre-publication proofreading and critique of this book. My deepest thanks and appre-ciations to my wife Daniela Brückner, who made numerous review-readings and helpful suggestions for this book, going back several years. She also translated several German-language documents reveal-ing Einstein's views on the ether subject. My appreciations to all who participated in long discussions and endured my many questions on ether-drift and astrophysical issues.

Finally, my profound gratitude to Dayton Miller and Wilhelm Reich, for their lifelong work on these subjects. When their discoveries are eventually taken seriously by the general public and scientific community, it will bring in a badly needed dose of *realism*, and change our small troubled world for the better.

James DeMeo, PhD
8 August 2019

"I believe that I have really found the relationship between gravitation and electricity, assuming that the Miller experiments are based on a fundamental error. Otherwise, the whole relativity theory collapses like a house of cards."
— Albert Einstein, letter to Robert Millikan
June 1921

"My opinion about Miller's experiments is the following. ... Should the positive result be confirmed, then the special theory of relativity and with it the general theory of relativity, in its current form, would be invalid. Experimentum summus judex. Only the equivalence of inertia and gravitation would remain, however, they would have to lead to a significantly different theory."
— Albert Einstein, letter to Edwin Slosson,
8 July 1925, Hebrew Univ. Archive
Jerusalem.

"The effect [of ether-drift] has persisted throughout. After considering all the possible sources of error, there always remained a positive effect."
— Dayton Miller, 1928, p.399

"You imagine that I look back on my life's work with calm satisfaction. But from nearby it looks quite different. There is not a single concept of which I am convinced that it will stand firm, and I feel uncertain whether I am in general on the right track."
— Albert Einstein, on his 70th birthday,
letter to Maurice Solovine,
28 March 1949

Author's Introduction

Intergalactic Medium! Interstellar Medium! Interstellar Wind!
Neutrino Sea! Neutrino Wind! Dark Matter! Dark Matter Wind!
Gravitational Waves! Higgs "God" Field! Cosmic Strings!
Cosmic Ray Anisotropy! CMBR Anisotropy!
Zero-Point Vacuum Fluctuation! Torsion Fields! Solitons!

Modern astrophysics and astronomy describe the cosmic space between the planets, stars and galaxies as an empty void, a hard vacuum lacking in inherent properties or substance. And yet, scientists working in these disciplines continue to discover "empty space" to be saturated with energy and particles, with turbulence and motion, as with the above concepts. Each is considered, by convention, to be a completely separate phenomenon from all the others, in spite of numerous points of similarities and agreement. Each term stands for its own presumed "soup" of discrete mystery particles. No matter how fantastically abundant, the space between them remains an empty void, save for scatterings of light and other electromagnetic waves. The scientists have identified all these specific "trees", but deny the existence of any "forest", whereby their basic nature could be more logically understood. As with the example of 10 blind men in a room with an elephant, each describes in exceedingly precise detail what they have individually grasped – the trunk, tusk, body, tail, legs – but the word "elephant" has become taboo. Like the proverbial naked emperor, nobody dares speak about a possible single ocean of cosmic energy, which offers a more unified and simpler understanding of all the diverse particles and "winds".

In a related manner, a casual look at images of deep space shows us billowing clouds of nebulae, of objects pushing through an unknown fluid and leaving behind a trail within a resisting transparent medium, all frozen in time. They appear more like something seen in the depths of the oceans or lakes. In some areas, a surrounding cosmic substance glows brilliantly with luminating stars, while elsewhere, everything appears darkened and dirty, as if smoke blanketed a patch of space.

1

The Dynamic Ether of Cosmic Space

There is a great amount of unexpected structure in these images. But only artists and poets, and not scientists, are permitted to speak about it in such a manner. Never mentioned is how a billowing cloud forms only within a resisting medium, as is the case with cumulus clouds in the atmosphere, or that an excitable medium is necessary to produce the sharply defined dark blue halos surrounding many galaxies, thousands of light years in thickness. Our lifetimes are but a pinpoint in time, so we do not get to see these vast cosmic events in motion, as they unfold.

Open cosmic space is nevertheless officially certified as empty and dead, in spite of the multiplicity of separate "particle winds" in a claimed metaphysical universe of never-observed big-bang explosions, black-hole myths, multi-universe unrealities, and relativistic space-time warps. While empirical astronomy struggles to keep on the path of observation and documentation, theoretical astrophysics increasingly appears more like the complex epicycles of the Ptolemaic astrologers. All the textbooks demand obedience from students and nonconforming professors, who risk expulsion and professional ruination if they stray from the orthodox catechism. Space is empty and dead. There is no cosmic medium for light waves. Nothing moves unless something else makes it move. The ultimate source of universal motion was a gigantic creation-event explosion 14 billion years ago. Before that, nothing existed whatsoever. Or so we are told to believe.

In this work, I will put forth the argument, with considerable evidence, that the open reaches of cosmic space are not empty, and the universe not so dead. There existed a robust theory of unitary cosmic functions, from planetary motions to light transmission, but prematurely discarded around 100 years ago. Then, open space was filled with a cosmic fluidic medium, somewhat similar to the above-mentioned modern "mediums", but lacking in any notions of an "empty space". There was movement, power and motion within cosmic space, much like an ocean of surging and swirling water. Something of an exceedingly thin and rarified nature, like a dynamic gas, but much less dense, filled all space. That theoretical cosmic ocean was called the *luminiferous ether*,[1] the word "luminiferous" meaning, the capacity to *transmit or produce light*. It was a potent theory developed over hundreds of years of sound logic, critical argument and optical experiments.

---

1. Throughout this book, the term *ether* will be used, as was the manner of science in the English-speaking world of the late 19th and early 20th Century. The context easily separates it from ether-gas as used by surgeons. This also removes it from the category of archaic irrelevancy, as with the "aether" spelling.

The cosmic luminiferous ether also provided a straightforward common-sense understanding which united the phenomenon of light waves with the similar wave-behavior of water and sound. The wave theory of light demanded a medium for transmission, just as sound waves required the air, and water waves the water. For light waves, the luminiferous cosmic ether provided the necessary medium. It was substantive and yet could penetrate matter such as crystals or glass, to allow light waves their passage. The cosmic ether was also deemed necessary for diverse physical phenomena, such as electrostatics, magnetism and gravitation. Rational debates proceeded as to how dense or material the ether was, how it moved, or if it had no material or motional properties whatsoever. New experiments were proposed and undertaken, not merely to better understand light waves, but to detect our planetary motion through the ether medium in which light waved.

The earliest of those efforts was the famous 1887 Michelson-Morley experiment. It is described in every physics textbook, but *always with one important, staggering error and falsification of science history: That their experiment produced a negative result, thereby "disproving the ether"*. However, **this is not true.** *Michelson-Morley in fact **did** detect an ether wind moving past their interferometer instrument at a velocity approaching 5 to 7.5 km/sec (kilometers per second)*. Additional ether-wind, or ether-drift experiments were undertaken in later years by Morley in association with Dayton Miller, and by Miller independently, using much better instrumentation and a far more ambitious and lengthy program of investigation. *Their results far outpaced the significance of the Michelson-Morley experiment, with a consistently detected ether wind of around 9 -11 km/sec.* And yet, most scientists either don't know about these later experiments, or if they do, have been badly misinformed about their results and significance.

An accurate presentation of these facts is the primary goal of this book, along with a thorough discussion of similar evidence, and the profound implications which logically follow. Twentieth-Century science was erroneously steered into a dead end cul-de-sac over the period of two major world wars, which took an immense toll on the human psyche, from which scientists and university professors were not immune. Science and medicine also adopted a combative and absolutist tone not seen since the times of Galileo and Copernicus. It is past time that old problems in science be exposed and reviewed with fresh eyes.

Background of My Interest in the Ether

During my undergraduate years, as a student of Aerospace and later Environmental Science in the early 1970s, I began exploring the various unorthodox ideas about cosmic energy, and similar concepts of biological energy, or life-energy. Foremost among the researchers I studied was Wilhelm Reich, a heretic whose findings from c.1930 through 1957 were so threatening to established institutions of power, that he was imprisoned and had his books burned by government decree. Aside from his writings on the origins of human violence, on how totalitarian governments developed, and about natural love and sexuality, Reich also clinically and experimentally documented the existence of a real *cosmic life energy.* He argued that the same energy of life was found in a free form in the atmosphere, and in the hard vacuum of space. His ubiquitous cosmic energy had ether-like properties, but of a far more dynamic nature than any of his predecessors. Reich called it the *orgone energy,* and wrote about its similarities with and differences from traditional concepts of cosmic ether.

I also came upon various studies of cosmic energy by other scientists, who gave it different names. Such was the case with chemist Giorgio Piccardi, who discovered a cosmic energy signature in his chemical-reaction studies; or biologist Frank Brown, who found a similar cosmic-energy phenomenon in his study of biological rhythms and cycles; also the Dean of American astronomy, Halton Arp, whose studies refuted redshifts as cosmological distance indicators, thereby demolishing the big-bang theory. Arp was treated miserably for his findings, and was banished from using the big American observatory telescopes he had helped to build. Then there were the many "free energy" investigators, and those experimenting with high vacuum and the "zero point" vacuum fluctuation. I identified a long list of such unorthodox 20th Century experimental findings, all of which appeared to identify the same basic phenomenon, of a singular cosmic energy.

By 1979, I was finishing a Master's degree in Geography with an Earth and Atmospheric Science specialization, at the University of Kansas (KU). By that time I had already read the research papers of Michelson-Morley, Morley-Miller and Dayton Miller, as well as the major writings and experimental protocols of Reich, Piccardi, Brown and others. Also informative was the 1972 book by ether-skeptic Lloyd Swenson, *The Ethereal Aether.* It was an educating work, except for the biased dismissal of Miller's positive ether-drift experiments, with unconvincing reasons being offered. This was quite strange, as Miller's original published papers showed confirming results verifying light speed variations and a real cosmic ether. I dug deeper for the facts.

While at KU working towards a doctorate, I attended a 1980 lecture by Arno Penzias, co-winner of a Nobel Prize in physics for discovery of the 3-degree K Cosmic Microwave Background Radiation (CMBR), the so-called "smoke left over from the big bang", or residual thermal energy. Penzias lectured on his findings and theory in the Physics Department lecture hall. After his presentation was concluded, questions were entertained from those attending. One of the students asked, "What existed before the big bang?" Penzias went silent for a few seconds, pondering, before delivering his reply: "We asked that question ourselves, and *as best as we can determine, space, time, matter and energy simply did not exist.*" Another long period of silence followed, upon which I broke out into a loud belly laugh, thinking he had made a big joke. The room then quickly filled with animated whispering, and heads swivelled around as if searching for the blasphemer in the darkened lecture hall who had dared to laugh at sacrament. I slunk down into my seat, to avoid being identified.

Penzias' response was exactly what the big-bang theory demands. *Before the big bang, absolutely nothing existed.* Nothing, including space and time itself! Modern empty-space physics had, by some cosmic comical tragedy, come full circle back to the sentiments of the Catholic priests at the time of Galileo. Their conclusions were hardly different from the *Book of Genesis,* except the timeline was now 14 billion years, rather than seven days. I found that disturbing, and also that the professors and other students were *not* disturbed about it. At a later KU physics lecture about Einstein's theory of relativity, under my questioning, the speaker admitted that, in the face of solid evidence for a cosmic ether and variable light speed, Einstein's theory would be fully invalidated. An honest physicist.

My graduate research at KU included experimental field trials of a truly heretical device, the *Reich cloudbuster.* It is a passive antenna used for stimulating cloud growth and rains during droughts or in deserts, where humidity is already very low, and at times when clouds and rains should not develop. That device posed as big a puzzle for modern atmospheric science as Galileo's telescope had been for astronomy in the 1600s. It required an ether-like and water-reactive energy substrate to function. My professors at KU, particularly Professors

Robert Haralick and Robert Nunley, showed considerable open-mindedness to allow that study to proceed. In spite of the controversy and headaches my work and ideas surely gave to the KU faculty, my research was honestly peer reviewed, advised and corrected where needed, and finally accepted. Reich's most controversial claim and invention, the cloudbuster, was shown to function just as he described. The theoretical implications were profound.

Even more controversial for physics, however, was my separate research paper, "Evidence for the Existence of a Principle of Atmospheric Continuity". That paper included a section on "The Luminiferous Cosmic Ether and Relativity", where I wrote about the Michelson-Morley and Miller experiments. I argued they really had measured the cosmic ether, and that it was probably the same phenomenon Reich, Brown, Piccardi and others had discovered in their independent research. I further elaborated on Halton Arp's findings, demonstrating that cosmological redshifts were not distance indicators, and therefore the big-bang theory of creation could not be correct. Also included was a discussion on solar-terrestrial influences, and other heresies of the 1970s which are still taboo today (ie., modern climate changes find a much better explanation in solar variability, rather than the mechanistic and flawed CO_2 theory). That controversial paper was updated and included as an appendix to my KU Thesis on the cloudbuster. Later I learned it had caused heads to explode within the KU Physics Department.

Nevertheless, at one time I was named as the top student within the Geography Department, where I worked as research and teaching assistant. Later I was appointed as a KU Instructor, proposed my own courses, and found grant money for the Department. I relate this as prelude to what happened thereafter, when my research into concepts of cosmic energy, or life-energy, became more widely known.

Some years later, one or more of the KU Physics Department professors unethically tried to derail the 1986 awarding of my PhD degree. My doctoral research was a 7-year project entitled *Saharasia*, on the role of severe global climate change at around 4000-3500 BC, wreaking havoc upon emerging human societies. KU Physics failed to block my PhD, fortunately, but similar intolerant reactions came from other conformity-demanding academics in universities where I subsequently served as professor, alongside a decade of slander attacks from the "skeptic clubs". Such irrational attacks always blocked my attaining of tenure, forcing a nomadic existence. After serving for a few years in short term contracts at universities in Illinois, Florida and Iowa,

I finally had a belly-full of academic intolerance, and began working independently towards building up my own private non-profit research institute. Nevertheless, for many years thereafter, the malicious and destructive attacks continued.

For example, in 1990 I was invited to present my findings on cosmic energy to the Piccardi Group, at the 12th Conference of the International Society for Biometeorology (ISB), in Vienna, Austria. That group was named after Giorgio Piccardi, professor of Chemistry at the University of Florence, Italy, an ISB founder along with geologist Solco Tromp, known internationally for his research affirming water dowsing. Piccardi identified a cosmic energy signature in his controlled chemical tests, and documented his findings over a lifetime of investigations. I write more about Piccardi in Part III of this book. The ISB Piccardi lecture sessions were organized by Eric Wedler, professor of Environmental Science at the Freie Universität Berlin.

After Piccardi's death in 1972, the ISB drifted away from acceptance of anything so controversial as a cosmic energy. For the Vienna Conference, the new dogmatic ISB leadership *ordered the "cleansing" and removal of the entire Piccardi Group of scholars, 20 international scientists, myself included, from that conference.* Wedler, who had devoted his life to the subject of cosmic energy phenomena, was devastated by this scandalous anti-scientific attack upon new research and discovery. Nobody in the censored group had been contacted or consulted beforehand. While attending the Vienna ISB conference, Wedler collapsed and died shortly thereafter in the hospital, a result of the ISB's new dictatorial arrogance, and Wedler's already aged and fragile condition. Upon learning of all this, which happened in quick sequence, I wrote a strong letter of protest to the top ISB leaders and resigned from the organization. I never got a reply back from anyone, which is all-too typical on how modern academic science frequently behaves. Over the years, I learned first-hand how this kind of academic censorship and back-stabbing is rather commonplace. Today, the subject of cosmic energy, even when articulated within conventional physics and chemistry, or using the theories and language of quantum physics, is still a hotly "taboo" subject.

Moving to California, and later to Oregon where my institute[2] now resides, I continued my research and wrote papers for different scientific conferences and journals on such subjects as cosmic ether and Reich's life-energy. In California I came into contact with John Chappell, a

---

2. The Orgone Biophysical Research Laboratory, www.orgonelab.org

fellow KU Geography PhD who, like myself, had become critical of conventional cosmology, and suffered professionally because of it. He had organized the *Natural Philosophy Alliance* (NPA), which included numerous physicists, engineers and astronomers, many of them well-known within top mainstream institutions. Chappell encouraged me to speak at his organized conferences. I also became an advisor to the NPA, meeting and learning from an amazing variety of brilliant scientists who dissented from modern astrophysical theory.

In 1994 I presented a paper on "Energy in Space: Empirical Evidence and Implications for Orthodox Theory" to a meeting of the American Association for the Advancement of Science (AAAS), for a special session on *Challenges to Contemporary Views in Physics and Astronomy* held at San Francisco State University. The works of Miller, Piccardi, Burr, Reich and a few others were discussed as refuting the concept of "empty space". In 1996, I presented two similar papers, on "Dayton Miller's Discovery of the Dynamic Ether Drift" and "Discovery of a Dynamic Bio-Cosmic Energy in Space and in the Atmosphere", to an AAAS conference held at Northern Arizona University. Both of these events had been organized by Chappell and the NPA, with cooperation of the AAAS. By 2000 I had further investigated the historical ether-drift experiments, finding evidence of serious academic bias and erasure of their positive evidence. This was presented to a California NPA conference in Berkeley, a "Critical Review of the Shankland, et al, Analysis of Dayton Miller's Ether-Drift Experiments". This paper is revised and included in this book as a chapter in Part II.

In 2001, I visited the Case Western Reserve University (CWRU), where many of the original ether-drift experiments were undertaken. At the CWRU Archive, I reviewed the original documents and publications by the central figures of the historical ether-drift experiments, including the correspondence of Michelson-Morley, Dayton Miller, Robert Shankland and others. I was given a tour of CWRU Physics and campus by William Fickinger, who was most gracious and helpful, even though he and I fully disagreed on Einstein's relativity theory, and about cosmic ether. At my urging, he later located a long-lost set of Dayton Miller's original data sheets and notebooks, where Miller recorded the results of his various experiments. Additional archive materials were obtained from other universities, on Michelson-Morley, Miller, and Einstein, now preserved in my own institute's archive.

From that background came a major article written a few years later on Dayton Miller's work, which achieved widespread review on internet, "Dayton Miller's Ether-Drift Experiments: A Fresh Look". Versions of that paper were also published in the *Journal of Scientific Exploration*, and as a chapter in the book *Should the Laws of Gravitation Be Reconsidered?* (Munera 2011), recognizing the research of Nobel Prize winner Maurice Allais. Allais graciously permitted reproduction of one of his articles in my institute's research journal, *Pulse of the Planet*, discussing his finding of newly uncovered patterns in Miller's ether-research data (Allais 2002). Those articles further stimulated a correspondence with other serious ether theorists and experimenters around the world. These included Yuri Galaev, who made independent ether-drift experiments in his position as engineer at the Institute of Radiophysics and Electronics, in Kharkiv, Ukraine. I encouraged him to come and lecture on his findings in the USA, but as often happens, funding for such an event was never obtained. We lost contact in 2014, after the Russian invasion of Ukraine.

From the above accounts, one will gain an appreciation for both the open-mindedness and support that exists for unorthodox research in some parts of the academic world, along with the regrettably ruthless reactions in other parts. This schizophrenic situation continues.

For example, one of my professional associates in Europe had his aspirations for the PhD in physics crushed, when he dared to write critically of Einstein's theory of relativity, including a discussion of Dayton Miller's work, and citing my publications supporting Miller. After a disturbing battle with one horrible professor acting like a Grand Inquisitor, he was allowed to finish the MS degree by redrafting his work to be more in line with conventional thinking. However, the PhD was then out of the question. He'd get no letters of recommendation, case closed, a better future denied. In another similar European case, an undergraduate university student, having what he thought was an open-minded discussion with his professor, mentioned my name in association with criticisms of Einstein's theory, after which he was promptly expelled from the university! Similar things have happened in the American universities, and it is not uncommon for young students and even professors without tenure, to write on such subjects under pseudonym.

In running my own private laboratory and institute, I no longer have a university position to protect, and so am free to speak and write openly. However, the university students interested in these subjects must carefully pick and choose the professors who might have command or control over their graduate research efforts. Freedom of inquiry, as well as freedom of speech, has been even more severely corroded in recent years, by the rise of politically motivated junk-science and intellectual intolerance. Fact and truth are today frequently defined by how "agreeable" they are with "politically correct" *scientism*, as aggressively promoted by "activists". A new form of group-think Lysenkoism threatens society and the Academy, across the disciplines, infecting all the professions. In this process, fact, truth and authentic science have not been well served.

A Preliminary Outline of the Major Theories of Cosmic Ether

Let's briefly review the various motions of Earth in space, as determined by conventional astronomy, and for which the ether-drift experiments were aiming to measure. There is the rotational velocity of the Earth on its axis, which at the equator is around 0.5 km/sec. Then there is the orbit of Earth around the Sun, producing an average 30 km/ sec velocity. Then the velocity of the Sun and solar system towards the star Vega, within the local cluster of stars at around 20 km/sec. Add to that, another ~230 km/sec velocity of the local cluster of stars aiming towards the center of the Milky Way Galaxy. Further to this there are motions of the Milky Way within the local cluster of galaxies, and of that cluster of galaxies towards other directions, and then the claimed expansion of the universe as a consequence of the big-bang theory. These latter motions range from many hundreds to hundreds of thousands of kilometers per second. If a scientist wanted to determine the Earth's absolute velocity through space, using the methods of light-beam interferometry, what exact motion would they look for? What would they expect to find? Would motion towards some distant galactic cluster be just as easily detected as the Earth's motion rotating on its axis, or orbiting the Sun?

Newtonian Static Ether, or "Absolute Space"
Some scientists, starting with Newton, visualized the ether as a fully static, immobilized and immaterial thing, lacking in substance and only playing a role in the transmission of light waves or particles. Such a static ether was granted the properties of a stiff gel, like the commercial Jello, by which it could vibrate as the carrier of light waves, but otherwise had no other motions or identifiable properties. This is the Newtonian "absolute space" static ether, through which the Earth and Sun moved without any resistance whatsoever.

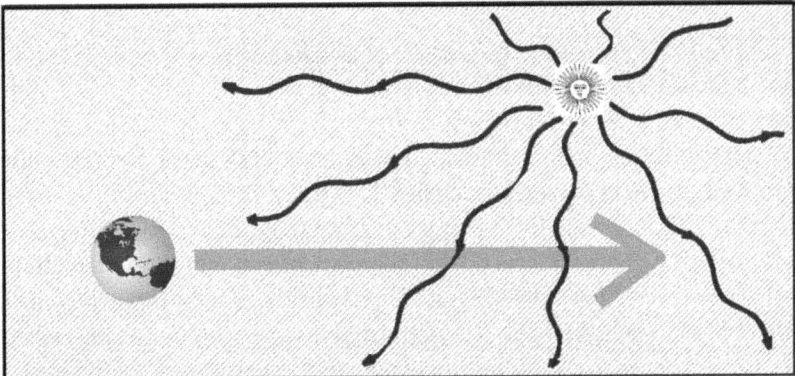

Figure 1. Newtonian Static Ether or "Absolute Space".
Ether exists throughout space as a static, unmoving substrate. Only subtle vibrations may occur, in association with light-wave transmission through the static ether. Planets and stars can move through this non-material ether with frictionless ease, never interacting or slowing down. The relative velocity between Earth and such a static ether is expected to be many hundreds or even thousands of kilometers per second.

Newton's static and immobilized cosmic ether was predicted to show a very high ether wind in any experiments designed to detect it, as a product of the Earth's and solar-system's higher-speed motions through it. Such a Newtonian static ether would blow straight through the Earth, down through the crust and through the planetary core, *without any interaction or reduction in Earth's velocity.* The ether wind created by Earth's motion would be the same at the surface of Earth as it was in nearby open space, at hundreds or thousands of km/sec. Figure 1 conceptualizes such an ether, of an exceedingly fine but immobile "something" which permeates everything, but has no role other than allowing for the transmission of light.

Newton is rightly credited for numerous important discoveries in physics, mathematics and optics. His ideas on the cosmic ether are not to be counted among them. While his ideas of a static and immobile cosmic ether were embraced by many scientists of his day and thereafter, others rejected it. I will provide details about the sources of Newton's ideas on the static ether, and those of other astronomers and optical scientists investigating the subject, in the next chapter.

A Static Ether with Material, Entrainable Properties

Newton's immobilized and static ether persisted into the late 19th and early 20th centuries, along with competing concepts of a material or motional ether, which interacted with the Earth's mass. Evidence for *light waves* also persisted alongside Newton's corpuscular or *particle theory* of light. Aside from being the accepted medium for light-wave transmission, many embraced the ether as having substance and matter-affecting properties. It was invoked as a causal agent for many physical phenomena, such as electrostatics and magnetism, which were described as *strains* or *tensions within the ether,* or due to motions of the ether. Such an active-dynamic ether was conceptualized as having sufficient material substance by which it could "touch" and interact with matter and energy, to affect and be affected by the material bodies of stars and planets which moved through it. And this being so, then it was quite logical that *a layer of slowed-down or entrained ether might develop and adhere close to the Earth's surface,* much as a viscous fluid clings to the vessel in which it is stored, or as water flowing in a pipe is frictionally slowed along the inside walls of the pipe.

Figure 2. A *Static Dragged* Ether of Slight Material Substance, slowed down by interaction with matter as the Earth pushes through it. The static-dragged material ether concept postulates the Earth moving through a motionless ether at high velocity, creating an ether flow opposite to the Earth's motion in the universe. An ether-entrainment effect would also occur (below), to *slow down ether-velocity at lower altitudes.* The speed of a material ether wind was thereby partly dependent upon altitude. Ether velocity can also be blocked by dense material obstructions such as stone buildings, basement locations, or dense wood or metal shields surrounding the interferometer devices used to measure it.

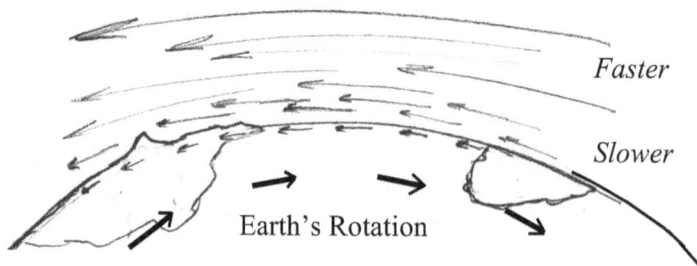

Faster

Slower

Earth's Rotation

The theory of a slightly material ether layer of increased density, adhering to the rotating Earth, came into discussion, along with the expectation that, would the ether eventually be measured, its velocity would be considerably slower than Newtonian expectations. The slowest ether velocity was thereby expected at the lowest altitudes, close to sea level, and in basement locations or deep in mines. The highest ether velocities would then be anticipated at higher altitudes. A slightly material and entrained ether, still fixed in the universe as per Newton's static concept, but with some unknown amount of velocity related to Earth's orbital motion of ~30 km/sec, is what the late 19th and early 20th Century optical scientists initially set out to investigate.

A material ether could cling to stars and planets as they moved through the cosmos, forming a substantial layer of compressed and slowed-down ether around them. This became an early dominant theory of how starlight could be bent and refracted to create such phenomena as stellar aberration. Before the appropriate experiments were undertaken, however, they could not definitively decide between a partially-dragged, or a fully-dragged and stagnant ether layer. Such determinations had to wait for the invention of new optical instruments, specifically the Michelson interferometer, to be described in a forthcoming chapter on *The Positive Results of the Michelson-Morley Experiment*. In the end, a partially-dragged ether, moving faster at higher altitudes, slowed down by heavy stone buildings or in basement locations, was in fact detected by nearly everyone who paid attention to these factors. Those who did not mostly got negative results, as I will detail. The diagram in Figure 2 gives an impression of such a fast cosmic ether being slowed and compressed at lower altitudes.

While the interferometer experiments did not detect the higher velocities of Newton's static ether theory, for reasons to be discussed, the experiments verifying a slower but definitive cosmic ether velocity of from 5 to 10 km/sec were ignored, due to a growing anti-ether bias.

Also in the background was a serious philosophical dilemma. If the Earth and all the other planets were moving through a tangible ether with slight substance, sufficient to create a dragged layer of ether at their surfaces, *this would apply a subtle braking force against all cosmic motion.* We must ask, if the ether was material, how could planetary motions get started against its resistance in the first instance, or later to be sustained over aeons of time? The solution offered by modern physics and astronomy, of a gigantic "big bang" explosion, was unconvincing in that regard. This is discussed in Part III.

Figure 3. Competing Theories of a Material Cosmic Ether: Static and Dragged? Or Dynamic and Motional?

The *static-dragged ether* concept postulated the Earth moving through a motionless ether, much like a cannon ball (above) or bullet moving at high velocity through a frictionally dragging atmosphere. Such an ether would interact with the Earth's atmosphere and crustal material, *reducing ether velocity at lower altitudes*. However, this implied an eventual slowing down of all planetary motions, unless there was a separate *prime mover* to keep the universe going. An Earth-entrained *dynamic and motional ether* (below) *provides such a prime mover*, with gravitational effects that push or float the planets and their suns along in orderly and lawful motions. *The ether wind moves in the same direction as the Earth, planets and Sun, dynamically moving and pushing or floating them along on their pathways* (Reich 1949, 1951a). Ether velocity could nevertheless be slowed and blocked by dense material obstructions such as stone buildings, basement locations, or dense wood or metal shields surrounding the interferometer devices. Both the static and dynamic ether theories imply variation in the velocity of light depending upon direction, but *only a dynamic moving and substantive ether solves the riddle of where cosmic motions come from, and what sustains them.*

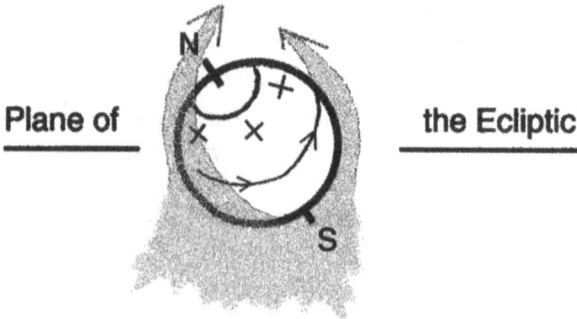

Plane of   **the Ecliptic**

Ether in motion, moves and carries the planets on their paths like balls floating on ocean waves or currents.

A Motional and Substantive Dynamic Ether

Open questions remain with us today, for which a motional, material and dynamic cosmic ether provides a solution. If Newton's laws of motion are universal, and nothing moves unless something else *makes it move*, then what is the *prime mover* that got the universe going and kept it moving in the first instance? How could any planetary or stellar motion get started against the braking force of a ubiquitous static and material, dragged ether, and keep moving in such a lawful manner? And how could ether be merely a "dragged" phenomenon, without ultimately slowing down the entire universe to a standstill? And what is the essence of gravitational force? How does matter emerge and build up to heavier-weight elements? Are they really created in the interior furnace of the stars, given how extreme temperatures tend to break matter down and apart into its basic elements and ionic components, and not build things up to greater complexity? And where does all the energy go from light and other electromagnetic frequencies, and from radioactivity, as emitted by all the stars and planets since the beginning of time, assuming time and the universe had a beginning?

Over my entire professional life, contemporary science has emphasized the principle of *Occam's Razor*, that simple explanations with fewer assumptions are most likely to be the more correct ones. But scientists generally ignore this important principle except for historical examples as with the Heliocentrism of Copernicus and Galileo's confirming observations. Instead, ever since the premature rejection of the cosmic ether, there has been an historical and psychological imperative to formulate non-intuitive and mystical theories with more, and not less complication, with more, and not fewer unproven assumptions. The most popular theories embrace convoluted mathematical "proofs" which lack empirical, real-world foundations towards which the maths are being applied. "Artist's renditions" have thereby increased in science journals and media, to "show" hypothetical things that nobody has ever seen or photographed.

A breakthrough out of this stagnant condition is summarized herein, firstly from an historical reappraisal, secondly from various heretic astronomers, and thirdly from unexpected sources, outside of astronomy or astrophysics. Scientists, physicians and naturalists as diverse as Wilhelm Reich, Harold Burr, Frank Brown and Giorgio Piccardi, discovered an ether-similar cosmic energetic force affecting the Earth's life and weather, as well as the properties and motions of the

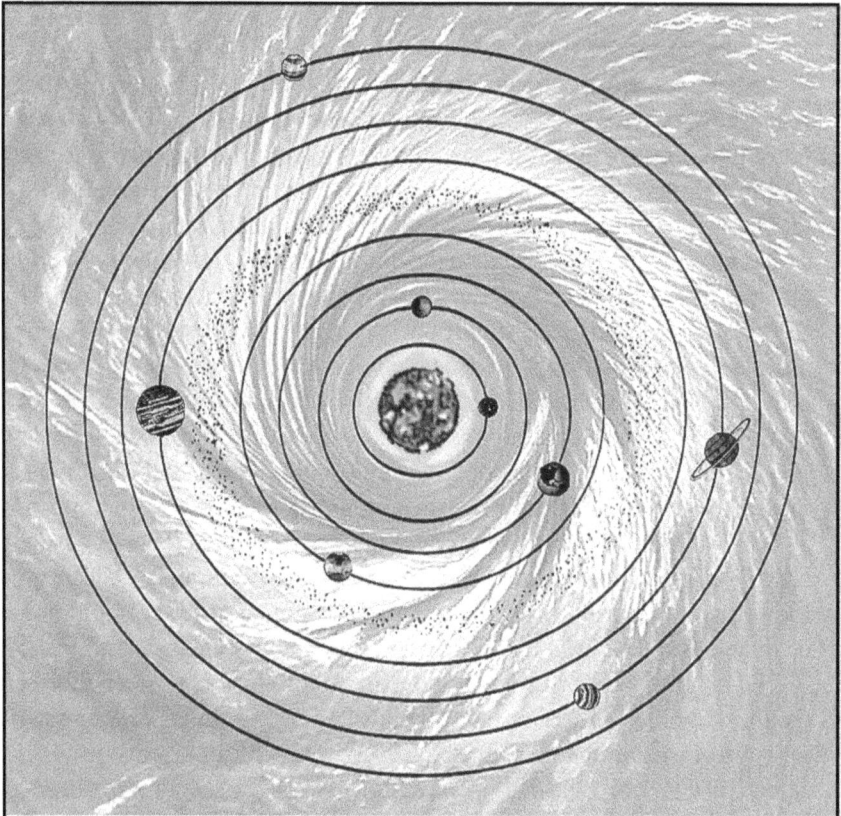

Figure 4. A Motional-Material Dynamic and Gravitational Ether, moving in a spiral vortex manner, propels the planets around the Sun. Separate vortices form around the planets. Slight variations in the cosmic ether wind are averaged out over time, due to the great mass of the Sun and planets. The inward sweeping of cosmic ether towards the Sun and around planets is gravitational, counter-balanced by centrifugal forces. The two forces together set the planets into regular lawful orbits, and likewise the moons into orbits around planets. (not to scale)

planets and stars. Among this notable group, Reich provided the most exacting clarifications on planetary motions, in a new theory he called *cosmic superimposition*, where his objectively demonstrated cosmic life-energy was attracted to matter, creating a negative entropy, to form and to coalesce matter into larger and more complex forms, as well as *to put matter into dynamic, vortexing and spiral-form motions.* I will detail these new contributions to the question of cosmic ether in Part III of this book.

By these new and often biologically-based determinations, some of the older theories of a vortex ether, first postulated by Descartes but ignored by nearly everyone else, find affirmation. *The cosmic ether is not merely a passive medium through which planets race and push, but is in motion, moving in large cosmic vortices of ether energy with a slight material property. It is a dynamic force in nature, which propels the planets on their paths around the Sun, captured in its swirling and merging motions.* Figure 4 gives a graphical representation of such a large vortex moving the planets of our solar system. Moons orbit around planets in similar but smaller vortices of cosmic energy, while the many stars and solar systems are swept along in even larger galactic vortices. Stable orbits then appear as a balance between the outward-pressing centrifugal forces, and the inward-pressing ether vortex forces. These are *not* "curved Einstein space-time" vortices, which were *declared into existence by a theory which demands no cosmic ether of*

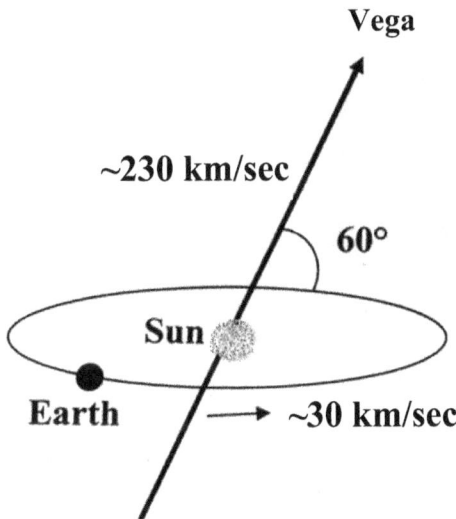

Figure 5. Earth's Net Motion Through the Galaxy?
Above is a simplified and half-accurate diagram, typical of what might be found in astronomy textbooks today. The Earth moves around the Sun, while the Sun moves towards the star Vega. However, the off-center spiral motion is generally misrepresented and not considered of any importance, given how modern astrophysical theory demands that space be empty, without anything like a substantive and motional cosmic ether.

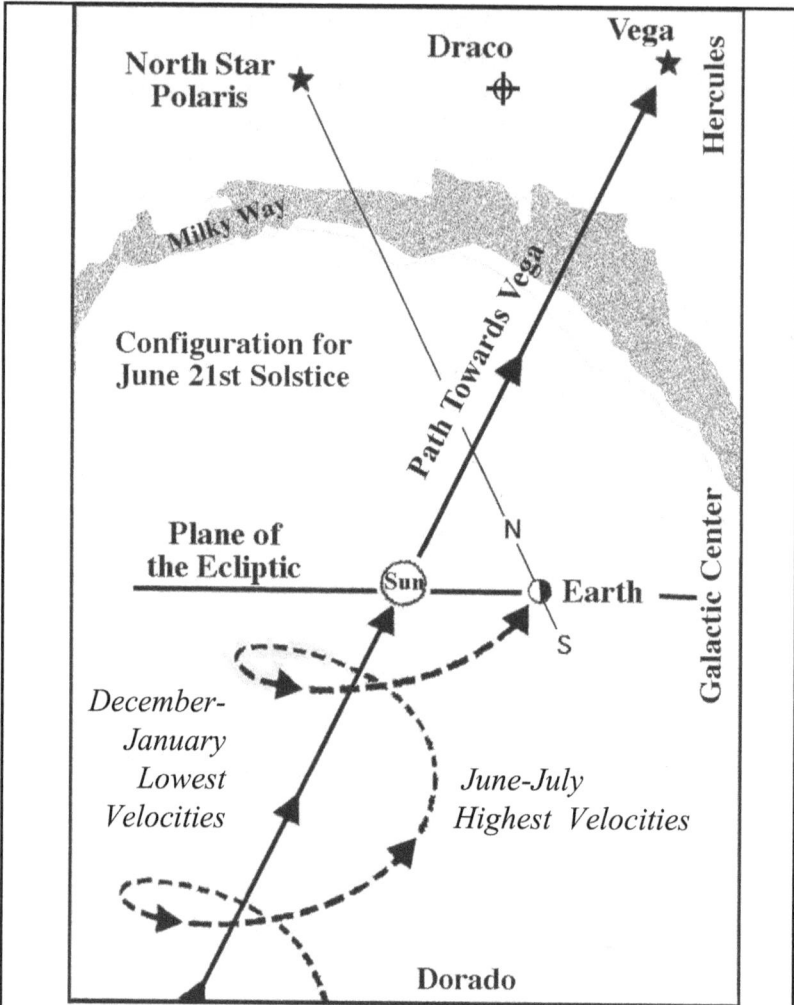

Figure 6. The Earth's Spiral Path Through the Cosmos.
Neither a flat circle nor an ellipse, nor even a symmetrical spiral as in a screw thread, the solar system's motion is a cambered, off-center spiral, which imparts *variable* velocities to the Earth and other planets over their annual orbits around the Sun. The northern pole of the ecliptic plane is identified at the "⊕" mark at the top-right center, while the Earth moves more closely towards the star Vega. The diagram shows the Earth at the June solstice position, a time of maximal spiral velocity through the cosmos. Actual ether velocity determinations are given in Part III, as well as in the chapter on Dayton Miller's work.

substance, and no variations in light speed. The ether-drift experiments provided clear evidence for light-speed variation, and should have shattered Einstein's theory of relativity even before it was written. I respect Einstein as a humanitarian, and for his E=mc² formula *as an approximation,* but not his relativity theory. More on this in Part II.

In Part III, and occasionally in hints elsewhere, I will present evidence that the cosmic ether vortex is, more exactly, an open-ended elongated spiral. Modern astronomy acknowledges the solar system's motion through the Milky Way Galaxy is a spiral. But the implications are ignored, as the current definitions give "outer space" no tangible substance or properties whereby such motions would be important to consider. Figure 5 presents such a half-accurate diagram, while Figure 6 shows a more precise image of the actual motions and vectors, which will be gradually introduced as we proceed.

The dynamic ether also has specific material properties as determined by new experiments, with influences upon the atmosphere and biology, in ways "empty-space" theory never predicted. The ether has a variable density that can range from a mass-free condition, condensing towards a material quality around planetary and stellar objects. This process also appears to form new particulate matter as a precipitate from its own concentrating primordial substance. It is also excitable, by which streaming motions are induced, and it can expand and contract in *pulsation* – a factor that may explain our own Sun's ~5 minute pulsation of expansion-contraction. This and other amazing solar phenomena (ie. solar flares following magnetic lines rather than ballistic trajectories) have no solution within Newtonian or Einsteinian theory, given their violations of physical laws governing mass and inertia.

From Reich we anticipate cosmic energetic streaming and pulsation, and together with ether theory, a better understanding has emerged. The dynamic ether is excitable and can flow and gather to higher potentials in the atmosphere and in living matter. It is *luminiferous*, not only as the carrier of light, but able to glow when sufficiently excited. Ether also becomes a motional gravitational force, with *life-energetic and negatively-entropic* properties. The dynamic ether concept as enhanced by Reich's findings, provides a satisfactory solution to *the long-sought self-organizing principle in the universe.* The ether is not something static or stagnant, dead and immobilized, through which the planets somehow push themselves, requiring a second mystery cosmic force to put the universe into motion. A simpler and more comprehensive understanding is possible. This book will present that evidence.

The Dynamic Ether of Cosmic Space

A motional vortexing gravitational ether was originally proposed by Descartes. This idea horrified Newton, as detailed in the next chapter. Michelson once mentioned such a cosmic ether vortex in passing, as a young man. Walter Russell and a few others did likewise in a philosophical context. However, the best scientific discussion favoring a dynamic theory of cosmic energy with spiral-form motions belongs to the 20th Century's most heretical scientist, Wilhelm Reich.

Public Slander and Destruction of New Discovery

Beyond the examples of unethical lapses in the universities as mentioned above, I must expose what is perhaps the most egregious example of medical and academic crushing down of unorthodox ideas in modern history: the public slandering and destruction of Wilhelm Reich, with government-ordered banning and burning of his books and research journals, and his death in a federal prison. My book *In Defense of Wilhelm Reich* covers that episode in detail, exposing the slander, lies and legal terrorism directed at him on two continents, along with a biographical sketch, and references to his medical and scientific discoveries, which were successfully reproduced and verified by others, myself included. (DeMeo 2013, Greenfield 1974, Martin 1999)

Reich's experimental work and discoveries are hardly known today, at least not in accurate presentations. The public slandering and lying about him began in Nazi and Communist newspapers in Europe of the 1930s, and followed him to America as he fled their terrorism. The lies have persisted into modern times, spread mostly by Marxist and Catholic writers, some of which has also spread, all too gleefully, into the mainstream of science and medicine. Having personally investigated Reich's biography and scientific claims, in depth, and having successfully replicated his most central experimental proofs, I find his work to be sound, good science, with amazing new inventions. His published experimental results with the *orgone energy accumulator*, for one example, are not merely important and eye-opening for science and the public health, but are *replicable and falsifiable* in the Popperian sense (see Part III, WebRef.1), assuming his protocols are followed.

More specific to astronomy, Reich's theory of *cosmic superimposition* is of paramount importance. He was the first, so far as I know, to have identified and emphasized the theoretical importance of Earth's *open-ended spiral-form motion* around the moving Sun. He reasoned how Kepler's equations for orbital velocities, which work fine for planetary motions along a flat 2-dimensional plane, become inexact and incomplete when applied to planetary spiral-form motions. *When planetary motions are viewed as an open-ended spiral, moving in 3-dimensions, Kepler fails, while Reich and ether-scientist Dayton Miller together provide a proof for cosmic ether of central importance.* Reich's books *Cosmic Superimposition* and *Ether, God and Devil* were so revolutionary and threatening, they were among those condemned to government flames in the 1950s. Today, they are reprinted. (Reich 1949, 1951a)

The censorship, hatred and violence directed towards him by nearly all the mainstream media, and by top leaders in European and American science and medicine, is stunning in its scope and viciousness. His research evidence is routinely erased, or distorted to create a better target for ridicule. My own experimental investigations nearly always confirmed Reich's claims and findings, often with new discovery in the process. (DeMeo 2011, 2014). For doing so, some of the same professionally-destructive slander, threats and hate-mail were, and continue to be hurled in my direction. The supporting studies undertaken in more recent years by scientific and medical professionals, my work included, is regularly erased in the falsified media and academic narratives. This trend of spreading lies and erasing evidence has continued today in just about all of the "top" science journals, in pop-psychology books, and now on internet. When alive, Reich wrote much about these attacks, and he appealed the legal persecutions all the way up to the US Supreme Court, whose judges basically rubber-stamped the book-burning and imprisonment. (Baker 1972, 1973, Blasband 1972, DeMeo 1989, 2013, Greenfield 1973, Martin 1999)

If the history of science tells us anything, it is that *only important books get burned, only important scientists get hysterically slandered in public media, and are then sent off to die in prison for technical violations of obscure laws.* Reich's mistreatment was worse than what Galileo was subjected to, but for similar reasons of being a threat to powerful institutions. He made significant new scientific discoveries that both the lay-public and professionals were ill-prepared to consider, *discoveries which threaten to up-turn nearly every major scientific theory of our time.* Reich's scientific works and findings are of Galilean stature and importance. He got it right, and was destroyed for it.

While copies of his books and research journals survive for a modern reconsideration, the distortions and slanders have continued. The so-called "skeptic clubs", Wikipedia and mainstream news organizations have been central in this process of public lying and falsification of history. (DeMeo 2013, WebRef.2) Added to this distortion of Reich's findings, is the new destructive mini-industry of eBay and internet hawkers selling all kinds of pendants, pyramids and gizmos, abusing Reich's name and orgone energy terms to sell trinkets with wildly exaggerated claims. This latter trend further muddies the water.

I make the above extended notation as a preventive, *to counter the prevailing trends of deliberate media and skeptic-club scientific-medical **lying***. *The widespread misinformation and distortions about Reich have no validity for him whatsoever*, no more than modern physics would wish to be defined by "quantum vitamin pills", or modern astronomy by "neo-geocentrism", or modern geology by the "flat earth society". (DeMeo et al. 2012, 2013, WebRef.3)

That having been said, in this work I will present the reader with evidence on the historical ether-drift experiments that most will not know, that will utterly refute the falsified opinions which today litter the textbooks. A review of this evidence has a powerful clarifying effect, sweeping aside mystically-inclined astrophysical confusions. It also leaves us with a calmer, more common-sense view of the cosmos. Mystic "black holes", relativistic "space-time distortions", "quantum magic", "multiple universes", "cosmic strings", "big-bang creationism" and similar mysticisms all came into being *only after early 20th Century science prematurely discarded the ubiquitous and interconnecting cosmic ether.*

Also included in this book are some of my own scientific findings, further supporting and clarifying cosmic ether and cosmic life-energy. An entirely different experimentally-developed, empirical and non-mystical way of viewing and understanding the universe is presented. Only minimal maths are included or necessary for this new understanding. The book is written in ordinary language, for the educated layperson and young student, as well as for the professionals, in hopes that a new generation will get the facts prior to being subjected to dogmatic indoctrination by the modern priesthood, worshiping "empty space" and a "dead universe", which "sprang into existence, from nothing".

— James DeMeo, PhD, August 2019
Greensprings, Ashland, Oregon

Part I:

Cosmic Ether as Theory

and Experimentally

Confirmed Fact

The Matter of Space,
Light Waves and Motion

"...when primordia are being carried downwards
straight through the void by their own weight, at times
quite undetermined and at undetermined spots they
push a little from their path, but only just so much as
you could call a change of trend. But if they were not
used to swerve, all things would fall downwards through
the deep void like drops of rain, nor could collision
come to be, nor a blow brought to pass for the primordia. So nature would never have brought anything into
existence."

– Lucretius, Roman Poet, c.75 BC
De Rerum Natura, Book II

Lucretius' primordial "swerve", quoted
above, was a reference to curved or circular
motion in the Great Void of the Cosmic
Heavens, an early concept of creation in
motion, resting upon ideas that ranged back
to Greek philosophers such as Aristotle, and
the Roman Epicureans. For those ancient
philosophers, creation was a role played out
by the gods, but they also put reasoned explanations to the physical world they could
touch and see. The nature of cosmic mo-

Lucretius (c.75 BC)

tions, the passage of the Sun, Moon, stars and
"wandering" planets, was always a central human interest, but only
dimly understood, and set apart from the confined material existence of
humankind on the Earth's surface.

Aristotle divided the material world into four elements, of fire, air,
water and earth, but the heavens were composed of a fifth element, a

weightless, unchanging and boundless "quintessence", which also was given the name of "Aether". Greek theology conceived of *Aether* as a primordial deity of the upper atmosphere. It was the pure essence of what the gods breathed, "heavenly air", as opposed to normal air, breathed by mortals. By his thinking, the cosmos was put into motion by a prime mover, related to the aether concept, which was also the godly "stuff" from which the planetary and stellar spheres were formed.

Aristotle
(384-322 BC)

Aristotle's philosophy, on matters of logic and the cosmos, captivated Western thinking and was even incorporated into the hierarchical astronomy as dictated by the Church of Rome, remaining so for centuries until new discoveries began to force their changes.

As more was learned about nature and the sky above, the mysteries of the gods were challenged. Aether came down to Earth as well, in the concept of a less theological and more physical *cosmic ether*, filling all the empty spaces of the Great Void. Slowly but surely, humanity limped towards a better understanding of the universe. Later still, in the modern era, the cosmic ether was firstly documented as a real thing, but later banished, prematurely discarded as the facts of science history show.

Waves in the Cosmic Ether of Space

The *luminiferous ether*, able to transmit and also to produce light, rose to dominance especially after the wave theory of light became more widely accepted in the 17th Century. If light expressed such easily demonstrated wave actions, and could even travel from the Sun, Moon, planets and stars down to the Earth, it must have a medium in the cosmos which fills all of space, and is present throughout the atmosphere and water, by which light waves could freely move through them. And since this cosmic ether could allow for passage of light waves through solid glass and other transparent materials, as well as through vacuum, air and water, then ether surely must be something of an exceedingly fine material density. It was thereby considered by early natural philosophy as a fourth phase of matter, of lesser density than solids, liquids or gasses.

Nicolaus Copernicus developed the heliocentric theory of the solar system, published in 1532, thereby eliminating the need for complex epicycle motions of the planets as required by geocentrism. His models also abandoned the old Aristotelian view of planetary spheres composed of ether, as the Earth was also a planet, but not composed of ether. He nevertheless retained aspects of both the spheres and ether in other contexts, merely placing the Sun in the center position of the solar system.

Copernicus (1473-1543)

Giordano Bruno incorporated the concepts of an ether medium (also termed *Spiritus*) into his philosophy and astronomy. He viewed the planets as independently moving, not fixed to celestial spheres, and abandoned the hierarchal astronomy embraced by the Church, which burned him alive in 1600 for multiple heresies.

Galileo Galilei mentioned the ether of space several times in his *Sidereal Messenger* (1610):

> "...there is round the body of the Moon, just as round the Earth, an envelope of some substance denser than the rest of the ether..." – Galileo, *Sidereal Messenger*, p.26.

Bruno (1548-1600)

> "...it seems to be by no means an untenable opinion to place round Jupiter also an atmosphere denser than the rest of the ether..." – Galileo, *Sidereal Messenger*, p.71

For Galileo, the ether not only existed, but condensed into an ether-layer surrounding the Moon and Jupiter, more dense than the ether found in open cosmic space. Gali-

Galileo (1564-1642)

leo also made, by modern standards, a crude attempt to measure the speed of light, where he and an assistant stood on two separate hills some distance from each other. Galileo held a lantern which would be opened to be seen by his assistant, who would then open his own lantern, allowing Galileo to approximate the time-lag. He computed the speed of light was something at least ten times the speed of sound.

Johannes Kepler, to whom we owe the mathematical laws of elliptical planetary motions, mentioned the cosmic ether in his work supporting Galileo, *Commentary on the Starry Messenger* (1610), wherein he referenced the Earth orbiting "among the planets through the ethereal plains", crossing the "free fields of ether", and of "Mercury crossing the liquid ether" of space. Kepler's *Dioptrice* (1611) and *Epitome of the Copernican Astronomy* (1617), also mention the ether. Reflecting the theology of his day, he viewed ether as the Holy Ghost and divine prime mover, a cosmic essence that brought light down from the heavens, and moved the planets in their orbits.

Kepler (1571-1630)

Wave phenomena preoccupied the emerging sciences of astronomy and physics, which relied upon analogies to observation of moving nature. Water waves, sound waves, light waves, and much later electromagnetic waves were studied and identified in their specific properties. Waves were a dominant characteristic of matter and energy, whereby influences could be transmitted over distances of space, often by invisible methods. An ocean wave moving through water could be seen, and its effects immediately understood as it crashed against a shoreline or seawall. Cause and effect were clear. The wave was not the same thing as the water, which moved only slightly as the waves passed through it, and water could also lay completely still at times without waves being present. Individual water waves could be additive or subtractive with other waves to yield standing waves, or oscillating waves at specific locations, such as where rivers poured out into the oceans, or during changes in tides, where current met countercurrent.

Sound and light waves had properties similar to water waves, in that they spread out and diminished over distance, could reflect against walls and diffuse through small openings. Sound waves also could bounce off walls, or be amplified by cupping the hand behind the ear. Just as with water waves, sound required a set time to travel from one hill to another, and the air was not observably moved by the sound passing through it.

The refraction of light, a more complicated matter, was also apparent to everyone who rowed a boat, seeing how their oar, when placed into the water, gained an angular distortion. It was also known to every successful spear- or bow-fisherman, who learned to compensate for this distortion by aiming below where a fish appeared to be in the water. Light would reflect off shiny or mirror surfaces, and like water waves and sound waves, light could bend around corners, further suggesting they all shared a wave nature.

Rocks thrown into a pond, or a loud crash of hands or metals clearly indicated simple forces that could create waves in the water or air. Sound and light were also episodic, and they had distant influences, as with the Sun's warming rays. A musical bow, composed of animal hair or hide, could create vibrational wave patterns on drumheads with a scattering of sand, and similar wave patterns in bowls of water.

However, in all these examples and unlike water waves, the waves of sound and light could not be seen directly, and no clarity existed as to how exactly they were transmitted. How could they move from here to there? Proofs were a long time in coming.

It was not until 1612 that Martin Mersenne measured the speed of sound in the open air. By the mid-1600s, further investigations by many others demonstrated that sound waves would diminish and eventually fail to transmit within an increasing vacuum, thereby confirming the role of air as the necessary medium for sound.

Light wave behaviors when reflected or refracted were generally described as early as the Second Century AD by Claudius Ptolemy. More than a thousand years would pass, however, before they were clarified more exactly by Rene Descartes, the "father of modern philosophy", who computed their first accurate geometry and maths.

Francesco Grimaldi described the diffraction of light, which clearly showed the relationship between water, sound and light waves. They all spread outwards in a circular wave-front, moving away from their point of creation, and they all could bend around corners, to regain a new circular wave-front, as when passing through a narrow opening.

Such observations on the wave nature of light led to new questions of just how light waves could be transmitted over the greatest distances, especially from the Sun, stars and planets down to the Earth's surface. A vacuum might block sound waves, but not light waves. That, and the identified decrease of atmospheric pressure with increasing altitude, as

Grimaldi (1618-1663) **Light Difraction** Light waves bend around corners and present a circular wave-front after passing through a narrow opening.

demonstrated when traversing from sea-level to high mountain tops where sunlight intensity increased, firmly indicated that air itself played little role in light transmission. Light intensity could be diminished by constituents in the air, such as colored vapors or particulate smokes or fogs, and the border between air and water created variations in its angle of refraction, but air was not the medium of light waves.

Transparent solids such as glass objects or panes, invented by the Alexandrian Romans around 100 AD, or transparent natural amber and mineral crystals, could also transmit light, and change light into a splay of colors. But, how exactly did the Sun and stars transmit their light through cosmic space, down through the atmosphere, and how, even with the humble candle, could light beams penetrate into and through

Descartes (1596-1650)

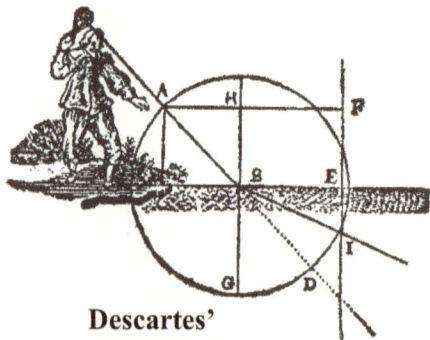

Descartes' Water Refraction

transparent solids? While charged with mystery, such questions continued to be generally answered in the postulate of a cosmic ether, not only as the medium for light waves, but also as the causal factor for gravity and planetary motions.

Descartes in his *Principles of Philosophy* (1644) proposed a continuous ether-fluid constituting a "second matter" which filled all space. His ether transmitted light, but also was divided up into large cosmic cells, each with whirlpools of ether motion providing the gravitational force that brought matter together, to create and put into motion the various stars, planets and moons. Sun and Earth were captured in one such vortex of motional ether, as was the Earth and the Moon. His theory in part developed from simple observations, such as how small pebbles in a stirred vessel would accumulate at the bottom center of the vortex.

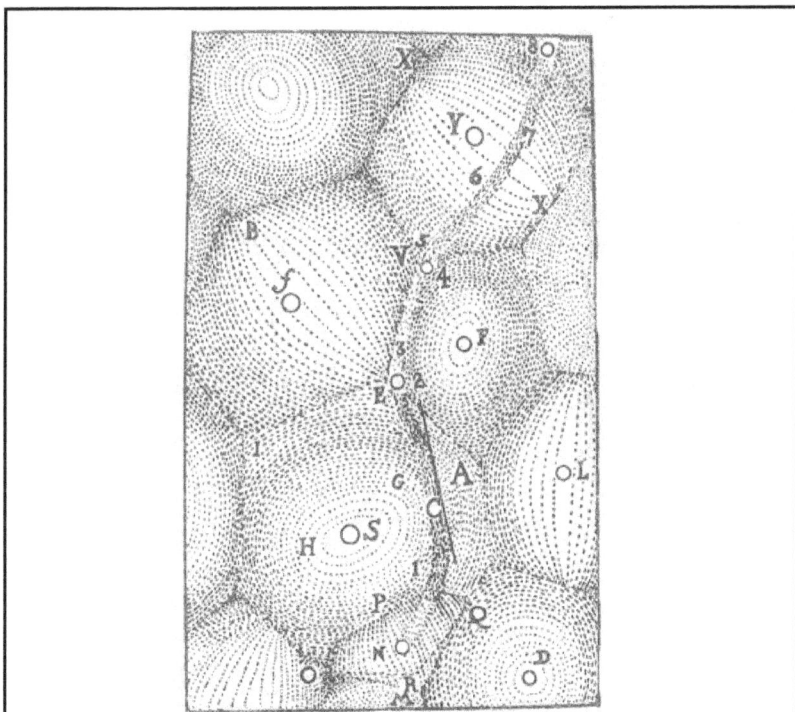

Rene Descartes' Vortex-Whirlpool Ether Theory
The universe was filled with an ether-fluid divided up into large cosmic cells, each of which contained a whirlpool of ether, of variable orientation. Ether whirlpools gathered matter together in their cores, creating and putting into motion the various stars, planets and moons.

The Dynamic Ether of Cosmic Space

While Descartes' spiral vortex theory was never fully carried forward by other scientists of his day, neither was it forgotten. His concepts of a spiral cosmic prime mover forming whirlpools and ether vortexes made a comeback some 130 years later, when astronomer Charles Messier listed over 100 "spiral nebulae" in his 1771 *Catalogue des Nébuleuses*. Messier's "nebulae" were in the 1920s identified as independent spiral galaxies by the astronomer Edwin Hubble, who measured the great distance of Andromeda and other "spiral nebulae", placing them well outside the region of closer stars forming our own Milky Way Galaxy. Albert Michelson, of earlier Michelson-Morley fame, also postulated a similar ether-vortex theory in his 1899 lecture series at the University of Chicago, later published in a 1903 book *Light Waves and Their Uses* (p.163). The newer discoveries of spiral nebulae, or galaxies, and ideas such as those of Michelson, provided Descartes some vindication, albeit several centuries after his death.

Astronomers today still puzzle over just how spiral galaxies form and maintain themselves, but rarely consider there might be some tangible "prime mover" cosmic ether with sufficient material composition to create a spiral-turning gravitational force acting ("swerving") over extremely large distances to produce such motions. While Descartes' name has nearly been forgotten in this context, the later discoveries of spiral nebulae/galaxies gave rise to another notable, but almost equally forgotten 20th Century scientist-heretic. In the 1950s, the physician and scientist Wilhelm Reich advocated a spiral theory of *cosmic superimposition*, of a matter-creating, gravitating, light-transmitting and luminating cosmic energy, at work in living and non-living matter, to be discussed later on.

Newton Kills the Motional Ether, Empties Space, and Diminishes Light Waves

Isaac Newton also embraced the luminiferous ether, asking the important question, "What is there in places empty of matter?" In 1679, at age 37, Newton wrote a letter to Robert Boyle (see *Appendix 2*), demonstrating agreement with ether concepts. He embraced a ponderable, moving and luminiferous ether, which was dynamically attracted to and penetrated into matter, exerting a "gravitational pressure" based upon variable ether density. He also argued for a residual denser blanket of ether surrounding planets, the Sun and smaller objects, by which the refractory effects of light could be understood. Newton wrote:

"... there is diffused through all places an aetherial substance, capable of contraction and dilatation, strongly elastic, and, in a word, much like air in all respects, but far more subtile. I suppose this aether pervades all gross bodies, but yet so as to stand rarer in their pores than in free spaces, and so much the rarer, as their pores are less; and this I suppose (with others) to be the cause why light incident on those bodies is refracted towards

Newton (1643-1727)

the perpendicular; why two well-polished metals cohere in a receiver exhausted of air; why mercury stands sometimes up to the top of a glass pipe, though much higher than thirty inches; and one of the main causes why the parts of all bodies cohere; also the cause of filtration, and of the rising of water in small glass pipes above the surface of the stagnating water they are dipped into; for I suspect the aether may stand rarer, not only in the insensible pores of bodies; but even in the very sensible cavities of those pipes; and the same principle may cause menstruums [solvents] to pervade with violence the pores of the bodies they dissolve, the surrounding aether, as well as the atmosphere, pressing them together." (Newton to Boyle, 1679. in *Appendix 2*)

This letter demonstrates the young Newton had a firm belief and working grasp of the ether of space as a thing of energy, substance and "ponderability". He embraced cosmic ether as a working force in optics, chemistry, electricity, magnetism, and gravitation, including in the gravitational motions of the planets. In this, the young Newton echoed in some measure the conceptual ideas of Descartes, Galileo and Kepler, all of whom had been an irritant to the Vatican bishops, who in the end would tolerate no possibility of a motional or gravitational "prime moving" force in nature other than *God*. The idea that ether and God might be identical in philosophical descriptions for some kind of creative self-organizing natural force, or even a "Holy Ghost ether" as Kepler proposed, or as a "cosmic prime mover", were equally intolerable to the Church. By their dictates, one could scientifically investigate and know the ether, but one could not measure or know "the divine". That was the purview of the Church.

However, 25 years later, Newton presented a different concept of the ether. In the 1704 edition of his *Opticks*, he listed 31 different "Queries" or Questions, wherein he noted established facts about light, heat, fire, optical perception and other subjects. He then expressed a *changed view of the ether medium as decidedly immaterial, lacking in properties he previously granted to it*:

> "Query 22: May not Planets and Comets, and all gross Bodies, perform their Motions more freely, and with less resistance in this Aethereal Medium than in any Fluid, which fills all Space adequately without leaving any Pores...? And may not its resistance be so small, as to be inconsiderable? ... And so small a resistance would scarce make any sensible alteration in the Motions of the Planets in ten thousand years." (*Opticks* 1704)

> "Query 28: ...against filling the Heavens with fluid Mediums, unless they be exceedingly rare, a great Objection arises from the regular and very lasting Motions of the Planets and Comets in all manner of Courses through the Heavens. For thence it is manifest, that the *Heavens are void of all sensible Resistance, and by consequence of all sensible Matter. ... it's necessary to empty the Heavens of all Matter...A dense Fluid can be of no use for explaining the Phaenomena of Nature, the Motions of the Planets and Comets being better explain'd without it.* It serves only to disturb and retard the Motions of those great Bodies... there is no evidence for its Existence and therefore it ought to be rejected. And if it be rejected, the Hypotheses that Light consists in Pression or Motion, propagated through such a Medium, are rejected with it." (*Opticks* 1704. Emphasis added)

From such statements one can see how the *older* Newton rejected the idea of a cosmic "ether-fluid" with any kind of slight mass able to push or retard the motions of the planets or stars. And from that he deduced wave theory was equally problematic. But there was another reason behind these particular Queries. After expressing his wonderment at the great order and beauty in the world, the marvel of the eye and the ear, and of animal instinct and senses, he turned to theology for explanations. He ended his Queries by making theological arguments that only God could be the prime mover, warding off any competition from a motional material ether, as if it would constitute heresy:

"Query 28: ...does it not appear from Phaenomena that there is a Being incorporeal, living, intelligent, omnipresent, who in infinite Space, as it were in his Sensory, sees the things themselves intimately, and thoroughly perceives them, and comprehends them wholly by their immediate presence to himself." (*Opticks* 1704)

Query 31: "Now by the help of these Principles, all material Things seem to have been composed of the hard and solid Particles, above-mention'd, variously associated in the first Creation by the Counsel of an intelligent Agent. For it became him who created them to set them in order. And if he did so, it's unphilosophical to seek for any other Origin of the World, or to pretend that it might arise out of a Chaos by the mere Laws of Nature; though being once formed, it may continue by those Laws for many ages..." (*Opticks* 1704)

While the older Newton's Queries were filled with good observations and brilliant insights, when grappling with the question of *original causation*, and specifically a cosmic ether which acted as the medium for the transmission of light waves, or motions of the planets, he expressed serious contradictions. He alternatively viewed ether as exceedingly rarified and unable to affect planetary matter moving through it, while at the same time positing a denser cosmic ether far away from the planets, out in open space, which could exert a serious gravitational pressure to push those same planets around. This was an effort to put a mechanism to his earlier law of gravitation, which was mathematically accurate, but in the end rested upon contradictory premises. Cosmic ether as a lawful motional force was theologically objectionable, and so he rendered it, ad-hoc, into a static or dead thing. *His static ether, however, contained numerous contradictions which remained imbedded within natural philosophy and science all the way up into the modern era.* Unable to go any farther in his conceptions, in the end he deferred to the Church. Scientific inquiry into the first origins of orderly cosmic motions, and by extension the origins of the universe, nature, life, etc., were considered "unphilosophical," or "taboo".

The older Newton further negated the possibility that the universe might have primary "laws of nature" that opposed "chaos", as with a cosmic *self-organizing principle* of some discoverable nature which might be scientifically identified. He did so, even while granting that

35

"once formed" matter and the universe "may continue by those laws for many ages." For Newton, God was a cosmic clock-maker who created the universe and set everything into motion, but then went on extended vacation. And it was "unphilosophical" to inquire further about it.

From that foundational conception, it is important to point out that Newton's laws of motion applied only to *existing* motional conditions, not to any primal cause in the sense of a first origin, and not to anything that might exert continuous, on-going motional influences, such as a cosmic ether with both slight mass and orderly motional properties. Even while Newton's equations for gravitation still provide the bedrock guide for our rockets to land on the Moon and Mars, he reduced the universe into a giant game of billiard-balls within a hard vacuum, colliding with each other as they moved around in a frictionless static ether, *ad-infinitum.*

Newton's answer to how gravitation, magnetism or electricity could affect objects at a distance was likewise abstracted. With erasure of an ether with more than static properties, one could assert the fact that actions took place over distances, but this gave no clarity as to *how* such actions were transmitted from one place to another. The mystical "action at a distance" non-explanation is today deeply imbedded within modern physics, which also forbids the taboo concept of a cosmic ether. Newton's redefined cosmic ether allowed for no such role in "actions", other than by contradictory ad-hoc speculations which never obtained a larger support, even among those who embraced the cosmic ether.

Newton also mostly abandoned the wave-theory of light in favor of a mechanistic "shower of particles" or "particle-rays", and described the luminiferous ether as somewhat identical to the stillness of a hard vacuum, except that it was sufficiently elastic to be vibrated as particles of light passed through it. This was completely different from the more fluid ether of earlier centuries, which transmitted light waves with minimal or no friction, but also interacted with matter sufficiently to produce cosmic motions.

Newton's ether of his later years was a dramatic departure from the motional and gravitational cosmic medium articulated by Galileo and Descartes. The cosmic ether of the elder Newton was rendered *static, stationary and generally immobilized.* All the planets, Sun, stars and comets could race through it without the slightest inhibition. The motion of the universe was the domain of the divine Hand of God, not of any motional ether with substance. Newtonian static ether could not even play a subordinated role within his theology, as with the deeply religious Kepler's "Holy Ghost" ether. The static ether could vibrate when shot through by light particles, but that was all. While Newton's theory of light particles was quickly dropped by most astronomers after his death, his conception of the cosmic ether as something static, immobile and *dead* would persist. A Newtonian static immobilized ether dominated subsequent scientific discussions, ultimately to be taken up by most of the scientists seeking to measure an ether drift, including Michelson and Morley some 200 years later.

Return to Light Waves, but the Ether Remains Dead, Static

While Newton was alive, he was strongly challenged by astronomer Christiaan Huygens, who successfully persuaded much of the science of his day in favor of light waves. After Newton's death in 1727, additional discoveries were made that fully resurrected the wave theory of light, primarily due to better measuring instruments and telescopes.

James Bradley discovered stellar aberration in 1728, whereby a star's location *appears* to be slightly different from it's actual location, based upon the changing direction of observation-angles of stars, as made at different times of year. It was then considered a proof of the ether, and later a result of wave-refraction according to increased optical density of an entrained or condensed layer of ether close to the Earth's surface.

Huygens (1629-1695)

The wave theory of light, along with the role of ether in its transmission, was further established by physician Thomas Young, who had earlier clarified how the human eye worked to focus light onto the retina. Young presented clear evidence for the *interference* of light waves when passing through two pinholes, which were then projected on a screen. Only light *waves* could produce such an effect, which was already known from the study of interfering water waves passing through two separate openings. He also at-

Bradley (1693-1762)

37

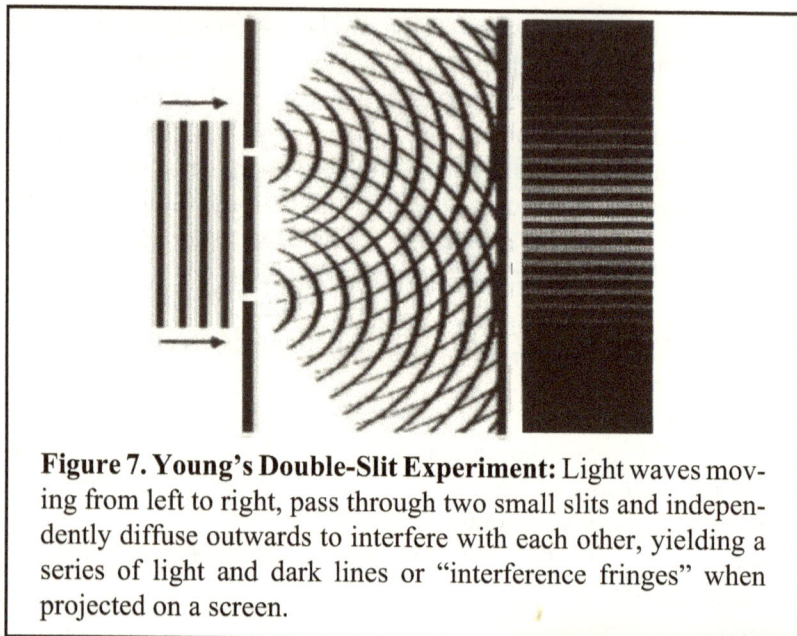

Figure 7. Young's Double-Slit Experiment: Light waves moving from left to right, pass through two small slits and independently diffuse outwards to interfere with each other, yielding a series of light and dark lines or "interference fringes" when projected on a screen.

tributed the different colors of light to different wavelengths, and argued for a gaseous type of cosmic ether.

August-Jean Fresnel continued investigations on light waves, discovering methods for light polarization and the transverse wave properties of light, vibrating at right-angles to the direction of its motion. His work further reinforced the reality of light waves, basically ending serious discussion on the Newtonian "shower of particles" corpuscular theory. Fresnel also advocated for a static but partially entrained ether, as necessary for light wave propagation and stellar aberration, but through which planets could move with ease and without significant obstruction. In the process, however, planets, stars and comets would drag along a layer of entrained ether-stuff, close to their surfaces, as they moved along in their orbits. From this he postulated an ether-drag coefficient based upon the index of refraction of trans-

Young (1773-1829)

Fresnel (1788-1827)

parent media, such as air or water. In later years, experiments would be undertaken to generally confirm Fresnel's ideas of a matter-dragged ether.

George Stokes went farther than Fresnel, arguing for a more significantly entrained or dragged ether, where the Earth carried a layer of fully stagnated ether along with it, much like a ship whose hull is coated with barnacles and weed drags a layer of water. *Stoke's idea was significant in how the results of later ether-drift experiments would be interpreted, or misinterpreted.* By his ideas, one could never detect such a fully-dragged ether at the Earth's surface, as there would be no motion within it to measure.

Stokes (1819-1903)

James Clerk Maxwell also embraced the cosmic ether, stating "It is inconceivable that a wave motion should propagate in empty space". He viewed ether as a necessary cosmic medium with dielectric properties, and considered his magnetic and electric "lines of force" as rotating tubes of ether within an otherwise stationary ether. By the time of Stokes and Maxwell, most scientific discussion about a possible Galilean-Keplerian-Descartes *ether-in-motion,* streaming or vortexing within the cosmos, had ended.

Such were the major ideas of the scientists of that period, who invoked cosmic

Maxwell(1831-1879)

ether as a primary causal factor in the world of light waves, even though at that time it remained invisible, lacking in *direct* evidence for its existence. Its reality and properties were inferred by observations of stellar aberration and other wave properties and behaviors of light, as previously noted. New experiments were being proposed and carried out in the 1800s, however, which would answer many open questions, while leaving others unanswered. Increasingly elaborate instruments were soon to be devised, using greatly improved optical-mechanical apparatus.

Ether Experiments and Theory
Prior to the Michelson-Morley Experiment

By the mid to late 1800s, the concept of light waves within a cosmic ether was widely accepted, but a static or stagnant ether theory dominated discussions. The idea of a material *ether wind,* in motion as a force to push the planets around, had receded. Arguments focused upon to what extent the ether was static and immobilized, and how the Earth and planets could pass through it without obstruction, even while ether was carrying light waves across great distances. Did the ether have some kind of mass or other interaction with matter, to create a dragged ether layer or atmosphere surrounding the planets, as Fresnel and Stokes had argued? Or, was the ether fully static and lacking in properties by which "ponderable substance" might be affected, merely being the medium for propagation of light waves? So the discussions went.

The wave-nature of light was well-demonstrated in numerous experiments, but the ether was only inferred to exist, and had yet to be experimentally proven by more direct evidence, with its properties firmly determined. Over a period of five years, however, new optical experiments allowed determinations of the speed of light, as well as detections of light-speed variations as it passed through air, water, and eventually through the cosmic ether itself.

In 1848, Hippolyte Fizeau undertook experiments to evaluate the speed of light over an 8.6 kilometer distance between two hills near Paris, for a total 17 kilometers of light path. He was testing out, with much better equipment, the hilltop lantern experiment firstly carried out by Galileo more than 200 years earlier. Fizeau constructed a rapidly rotating 720-tooth gear-wheel driven to high rotational speed by clockworks. With his apparatus set upon one hill, a beam of light was projected through the rotating gear teeth, which sequentially allowed or blocked the light beam towards the second, distant hill. A large mirror on the second hill reflected the light beam back to his apparatus on the first hill. The projected light beam went out and returned with such rapidity that it passed through the same gap in the gear teeth through which it was origi-

Fizeau (1819-1896)

nally projected. By increasing the speed of rotation of the gear, the light beam would eventually be blocked by the subsequent gear tooth. By knowing the distance and rate of rotation of the gear wheel, Fizeau computed the time required for the light beam's travel out and back. His determination of 312,000 km/sec (kilometers per second) was within about 5% of the accepted modern determinations of 299,792 km/sec.

In 1850, Léon Foucault made independent experiments using an improved design employing rapidly rotating mirrors to accomplish the same task, with greater accuracy. The rotating mirror would send out and then receive back a light beam reflected off a distant mirror. Over the time of its outward and return transmission, the rotating mirror would turn a slight bit. The returning beam was then reflected on a screen by which its angle of deflection allowed a calculation of the elapsed time over a known distance, and hence the light

Foucault (1819-1868)

speed. Foucault's measures were even more precise, registering at 299,796 km/sec, about 0.001% off from the modern determination, a truly remarkable feat with what today would still be considered an excellent apparatus.

This method was the subject of great interest for optical science, and was later taken up with refinements by Albert Michelson in the first two decades of the 1900s, for making even more precise light speed determinations. Michelson's measurements, discussed in the next chapters, remained the most accurate available through the 1930s – though with puzzling and rarely-mentioned *variations* in those light-speed measures. *All of these determinations of light's absolute velocity were averages computed from a wide range of variable readings.*

A further 1851 experiment by Fizeau was notable in its return to ether-theory measures, proving that light speed would vary according to the velocity of the medium through which it was transmitted. Using two pipes filled with water, and with glass end windows, water was forced to flow at a high speed, but in opposing directions within the two pipes. A unidirectional light beam was shone through them. In one case, water flowed in the same direction as the light waves, in the other case, water flowed against the light waves. *The result of this experiment proved that light speed was variable, depending upon the direction of motion of the medium through which it's waves were travelling.* The velocity of the water was additive to light waves moving in the same direction, and subtractive when moving against the water. When the water was still inside the pipes, the light-beam velocities were identical.

In 1853, Foucault made further proof of the changed velocity of light waves as they traversed through two different media, of water as compared to the open air. Both of these important experiments, by Fizeau and Foucault, using novel designs, added experimental proof not merely for the wave theory of light, but also that *light speed was variable according to the density and velocity of the medium through which it travelled.* From such experiments, ether was embraced by the world of science as a real thing, absolutely necessary to understand the transmission of light waves, and later electromagnetic waves.

In all these new experiments, however, the cosmic ether was variously described as a static and immobilized medium through which the Earth raced like a missile moving through a hard vacuum, or as a dragged phenomenon, implying the ether made intimate contact with planetary and stellar matter. Such touching contact with matter was necessary if entrainment of ether could occur, to form an ether layer of variable density around the planets, Earth and Sun. Newton's point about planets racing through a static ether with "...so small a resistance would scarce make any sensible alteration in the Motions of the Planets in ten thousand years" was for him an absolute theological necessity, to rid the ether of motional and material properties by which it could actually influence celestial motions, and thereby compete with his vision of deity as prime mover. And yet, as seen in the pre- and post-Newtonian era, ether was posited to touch and interact with matter, by which an entrained refractive layer might develop. Entrainment demanded a material, tangible ether, able to come into contact with and be moved by matter. Or, it required going back to the pre-Newtonian ideas of Descartes, of ether as prime mover, streaming and flowing in whirlpool vortex motions, carrying the planets and stars along with it as it moved. Either the ether had motional properties and pushed the planets and stars into motion, or it was contradictorily lacking in material properties, but nonetheless could be entrained into a layer of various density around the objects moving through it. And if the latter was true, then *how could planetary and stellar matter move through and drag such an ether without consequences to their own forward momentum, over billions of years?*

Figure 8.

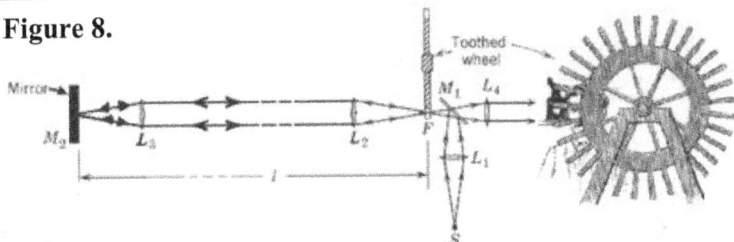

Fizeau's 1848 Speed of Light Experiment: A light beam was sent out to a distant hill near Paris, bouncing off a mirror, whereupon the light returned to cover an 8.6 kilometer distance. A toothed gear wheel was set to alternatively block or allow passage of the beam. When the wheel was unmoving, the beam of light made the trip out and back without interruption. When the wheel was rotated to a fast speed, however, at some point the light was blocked by the subsequent gear tooth, allowing calculation of light velocity over the distance.

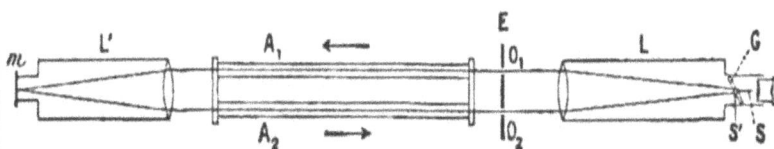

Fizeau's 1851 Moving Water Ether-Drag Experiment: Using two tubes of water with transparent end-glass, Fizeau proved that light speed was different when the water traveled with the direction of light than when moving against the direction of light.

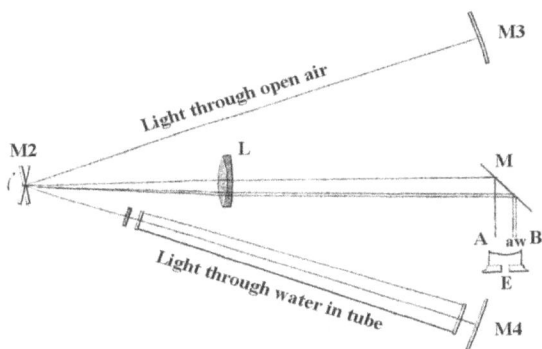

Foucault's 1853 Ether-Drag Experiment: Foucault proved that the speed of light was different in open air versus when directed through a glass-ended tube filled with water.

The Dynamic Ether of Cosmic Space

The *Positive* Results of the Michelson-Morley Experiment

The history of science records the July 8-12, 1887 ether-drift experiment of Albert Michelson and Edward Morley as a pivotal turning point, after which the energetic ether, filling all of cosmic space, was discarded by mainstream physics and astronomy. Thereafter, the postulate of "empty space" devoid of ether was embraced, along with related concepts which demanded constancy of light speed in all directions, in harmony with Albert Einstein's relativity theory. The now famous Michelson-Morley experiment continues to be widely cited today, in nearly every physics textbook, for its *claimed* "null", "zero", or "negative" results. These claims, however, are ***not true***, something easily determined by a careful reading of the original Michelson-Morley paper, which appeared in the *American Journal of Science* in November 1887. In fact, their experiment reported a *slight positive result,* later to be independently replicated by others, including by *both* Michelson and Morley, working separately from each other, with different research associates. Twentieth Century science nevertheless ignored all such positive evidence for the cosmic ether, as if psychologically compelled to make a wrong turn.

Albert Michelson
(1852-1931)

Edward Morley
(1838-1923)

Michelson's Initial 1881 Experiment

In 1873, the young Albert Michelson graduated from the US Naval Academy, pursuing further study of optics in 1880 with Hermann von Helmholtz at the University of Berlin in Germany. While in Germany, and with a grant from the American inventor of the telephone, Alexander Graham Bell, Michelson invented what came to be the most widely applied method for investigations of the properties of light, and for ether detection, a highly sensitive device known as the "interferential refractometer". This was later shortened to the *interferometer*. With it, one could measure light-speed variations down to the width of an individual wavelength. His first experimental use of it, aiming to detect the cosmic ether, was undertaken in late 1880 at the Astrophysikal-ische Observatorium at Potsdam.

Michelson's interferometer used two light beams directed at right angles to each other, forming an "X" pattern on a turnable platform, bouncing back and forth between mirrors along *two equal lengths of light path*. Figures 9 and 10 show the setup.

As the platform is turned, one of the light-beam paths will eventually be aimed *parallel* to the presumed ether wind, while the other light-beam path would simultaneously be directed *perpendicular* to that ether wind. By turning the whole interferometer, depending upon how directly the instrument was aimed into the ether wind, or perpendicular to it, the two light beams would develop different speeds of transit time out and back along the interferometer arms. This would be apparent after the two light beams were reunited to form *interference patterns*. The interference bands of light and dark stripes, or *fringes of light*, were visible through a magnifying optical eyepiece or telescope set on one of the interferometer arms, where a few or even a single light-fringe could be observed.

Any speed differential between the two perpendicular beams would be revealed when the apparatus was rotated, leading to a shifting of the interference fringes to the left or right of a central index pointer. This allowed the experimenter to count the amount of light-fringe shifting, which could then be computed to reveal a given speed of ether velocity. Any *changes in the velocity of light* impinging differentially upon the two light beams, one oriented directly into a presumed ether wind and another oriented at a 90° angle to that wind, could be determined. With enough turns of the instrument, over many days, the compass direc-

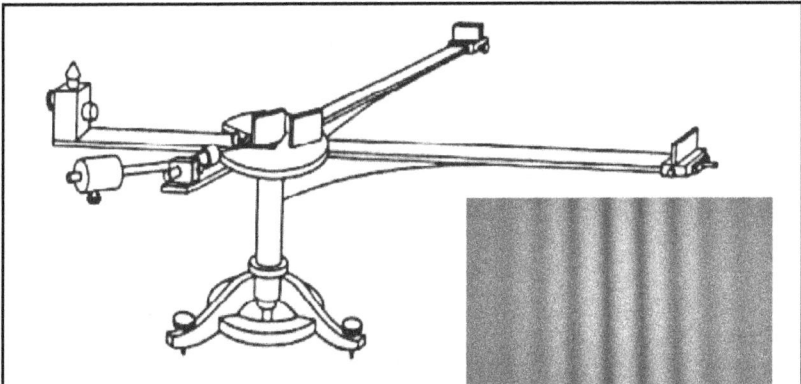

Figure 9. The Michelson 1881 Interferometer. A thin pencil-width light beam is split into two beams by a half-silvered glass mirror. The two beams then are directed along two perpendicular arms, each of one meter in length, at the end of which is fixed a full mirror, which reflects the beams back to the half-silvered mirror for recombining into a single beam. The two recombined light beams create interference bands or fringes, and a magnifying viewer then allows visual inspection of a few or merely one of the fringes. In a condition of no ether wind or drift, or no ether at all, as the interferometer is rotated on its base, the interference fringes would not move or shift their positions. With an ether wind, however, light speed becomes different along the two interferometer arms, depending upon the relative orientation of the arms to the ether wind, as the interferometer is turned. By directing the apparatus into or out of the ether wind, the optical fringes shift to the left or right, indicating the presence and magnitude of that wind, revealing a change in light velocity along specific directions. This early effort by Michelson proved to be a failure, primarily due to vibration problems and also a too-short light path to detect any significant ether-drift signal. (Michelson 1881)

tional *azimuth orientation* of the interferometer at the times of maximal and minimal fringe shifting could be determined. Of course, if the ether did not exist, or if the ether was fully stagnant at the surface of the Earth, without any motional wind, then the two light beams would not show any changes in light velocity nor shifting of the light fringes when the apparatus was rotated. It was an ingenious instrument and experiment.

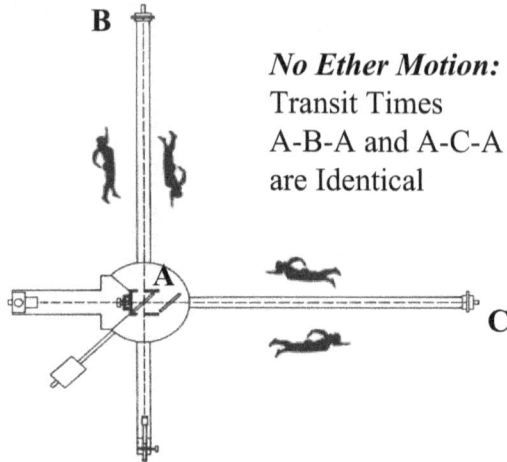

No Ether Motion:
Transit Times
A-B-A and A-C-A
are Identical

Figure 10. Michelson's Two-Swimmer Analogy for the Ether Drift Experiments. With *no ether flow* (above) two swimmers of equal strength or two light beams, going out and returning along the two arms A-B-A and A-C-A, take the same amount of time to make their respective trips. *With an ether flow* (below) moving from right to left, the swimmer on transit A-B-A takes less time than the swimmer of A-C-A. While swimmer A-B-A must compensate for a slight side-current in both directions, the swimmer on A-C-A must struggle against the full current for A-C, and does not fully regain that lost time on C-A. As described in the text, the overall transit time of A-B-A is therefore less than A-C-A This is true for light waves moving through or within any cosmic ether motion, be it a static ether, an ether wind, or a partially dragged ether. See the text for explanation of the triangle.

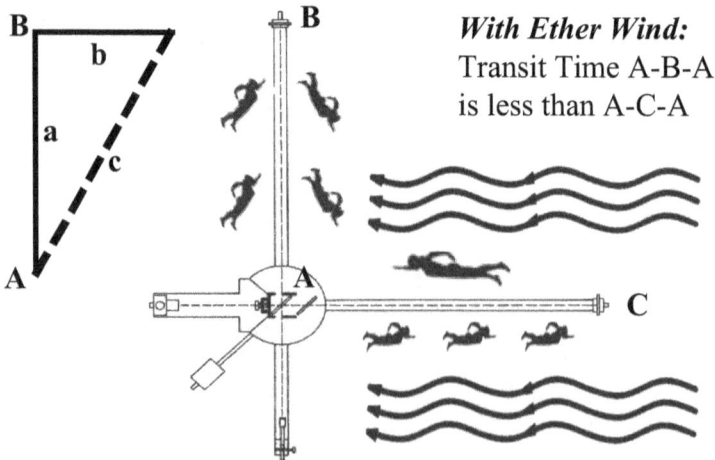

With Ether Wind:
Transit Time A-B-A
is less than A-C-A

As Michelson wrote in his published 1881 account, those first efforts in Germany were plagued by instrumental errors, and the original instrument lacked sufficient light path to achieve adequate sensitivity. It was nevertheless a sound methodology that, with improvements over the years, eventually led to detections of ether drift by Michelson-Morley, and by others such as Morley-Miller, and independently by Dayton Miller, Michelson, Galaev, Múnera and others, to be discussed.

The determinations of ether velocity are straightforward geometry, as shown in Figure 10, to which I can add the following simple maths. The top part of Figure 10 presumes no ether flow. If the two arms of the instrument represent identical distances that two identical-strength swimmers must travel, and with no water motion, they both take an equal amount of time to make their respective trips. If distances A-B and A-C are each 100 meters, and the swimmers can swim *very fast,* at 5 meters/sec, then they require 20 seconds to swim out, and 20 to swim back, for a total of 40 seconds each. With no ether-flow, or water flow as in this analogy, the times are identical.

With water (or ether) motion as in the lower part of Figure 10, swimmer A-B-A will take less time to make their trip out and back than swimmer A-C-A. Assuming a water flow of 3 meters per second from right to left in that diagram, swimmer A-C who swims at 5 meters per second but now facing into a 3 meters/sec current, will make headway towards point C only at 2 meters per second, slowed down considerably. It will now require 50 seconds to cover that 100 meter distance going out. On the return 100 meters, however, their net velocity increases when moving with the current, to 8 meters/sec. The return requires only 12.5 seconds. So their total transit time is 50 + 12.5 seconds, or 62.5 seconds.

Regarding swimmer A-B-A moving at 5 meters/sec, but with a cross-current of 3 meters/sec, we must apply the formula for a right-triangle to obtain the length of the hypotenuse, representing the actual distance A-B-A must travel, in both directions. Regarding the triangle inset in Figure 10, the formula is: $a^2 + b^2 = c^2$. We know the distance of triangle side **a** is 100 meters. Side **b** distance is also known, as follows. It originally took 20 seconds for swimmer A-B to cover that one-way distance. But now with a cross-current of 3 meters/sec, over that 20 seconds the swimmer would be pushed off course by 60 meters, each time going out and returning. We can then insert the values of length into the above equation:

$$100^2 + 60^2 = 10,000 + 3600 = 13600$$

By solving for **c**, the hypotenuse (deriving the square root of 13,600), we know the one-way distance the swimmer must travel, compensating for the cross-current, of 116.6 meters. While swimming at 5 meters/sec, it requires 23.3 seconds for going out the 116.6 meters, and another 23.3 seconds for returning, a total of 46.6 seconds for A-B-A. The swimmer A-C-A at 62.5 seconds therefore takes more time to make their voyage than A-B-A. This math is true for boats, swimmers, or light waves.

While the values for light speed are much faster, at ~300,000 km/sec, and a Michelson-Morley interferometer light path of only 22 meters in length, the ether velocity values calculated by them rested upon similar assumptions and methods of calculation. Other factors such as light frequency and the refractive index of glass and mirrors had to be considered, but overall the basics of the experiment were relatively simple and straightforward, even while requiring great skills in optics.

The Michelson-Morley 1887 Experiment

After his 1881 effort, Michelson returned to the USA and received an appointment at the Case School of Applied Science in Cleveland. There he continued with the ether-drift experiments, partnering in 1887 with Edward Morley from nearby Western Reserve University. As in the 1881 experiment, the basic principle was the same, to split a light beam into two parts using a half-silvered mirror, and send the two light beams down two different pathways, at right angles to each other. The relative velocities of the two light beams, after being recombined, could again be computed based upon the shifting of interference fringes.

The primary difference between the 1881 and 1887 interferometer was the length of the light path, the latter having a light path of 22 meters. The various optical apparatus of the 1887 Michelson-Morley interferometer – light source, mirrors, beam-splitters and focusing telescope – were mounted on a thick square slab of sandstone measuring ~1.4 meters on the diagonal, which was then floated in a shallow tank of dense liquid mercury. This allowed for a relatively smooth and frictionless, vibration-free rotation of the apparatus. During rotation, at specified compass markings on the base, the fringe shifts observed through a small magnifying telescope would be recorded. From those readings of fringe shifts, changes in light velocity associated with specific compass orientations of the apparatus were identified.

Other precautions were taken during the measures, which later turned out to be counterproductive to the goal of the experiment, actually blocking some percentage of ether flow and reducing the sensitivity of the instrument. For example, to keep out stray light, a heavy wooden cover was placed over the apparatus. To shield it from vibrations and thermal variations, the experiment was conducted *in the corner basement of the massive stone Pierce Hall building,* in which the old Case School Physics Department was located.

These physical obstructions – the brick-stone basement location and wood cover, as well as the low altitude of Case School in Cleveland Ohio (199 meters altitude) – would later prove to be critical inhibiting factors in the small but significant results of their 1887 experiment.

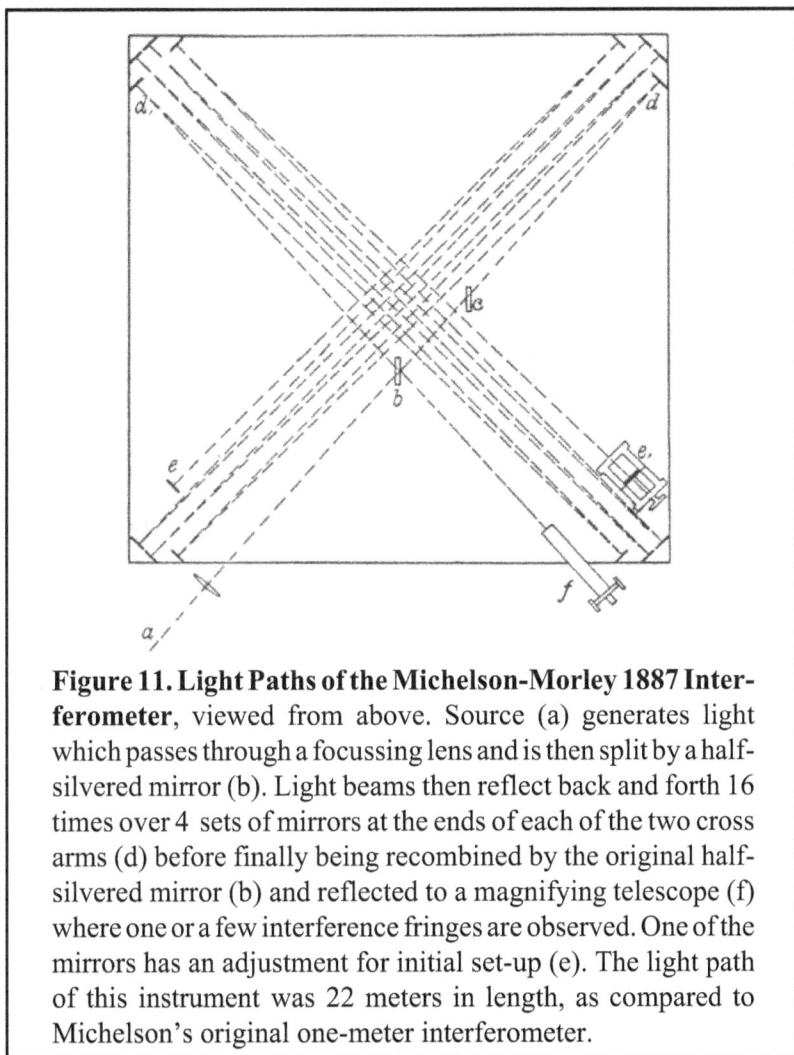

Figure 11. Light Paths of the Michelson-Morley 1887 Inter-ferometer, viewed from above. Source (a) generates light which passes through a focussing lens and is then split by a half-silvered mirror (b). Light beams then reflect back and forth 16 times over 4 sets of mirrors at the ends of each of the two cross arms (d) before finally being recombined by the original half-silvered mirror (b) and reflected to a magnifying telescope (f) where one or a few interference fringes are observed. One of the mirrors has an adjustment for initial set-up (e). The light path of this instrument was 22 meters in length, as compared to Michelson's original one-meter interferometer.

Figure 12. The Michelson-Morley 1887 Interferometer with an approximate 22-meter round-trip light-beam path, mounted on a sandstone platform in the basement of the old Case School Physics building (today, Case-Western Reserve University). This interferometer was 22 times more sensitive than the original one-meter Michelson interferometer used in 1881, as previously shown in Figure 9. The 1887 experiment was undertaken over four days only, on July 8, 9, 11 and 12, in a brick-stone basement with a protective wood cover placed over the light-beam paths, which was removed for this photograph and diagram. Such dense material shielding, as Miller showed, slowed down the movement of the ether. Even so, Michelson-Morley reported in their published results an ether velocity approaching 5 to 7.5 km/sec, and not a "null" or "zero"

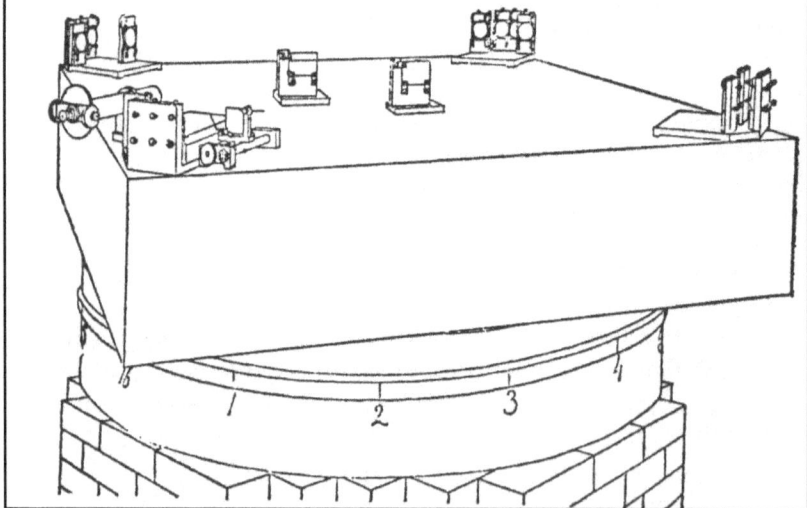

Their overall procedures were somewhat like trying to detect the existence of atmospheric wind by floating a small sailboat within an indoor swimming pool, located in a basement. Whatever ether wind might be present, it was reduced in velocity by the very precautions they took to control for presumably complicating factors.

While secondhand reports on the Michelson-Morley 1887 experiment repeatedly claimed it produced "no results", or a "null" or "zero" effect, *this was never the case!* In their final report on the results of their experiment, titled "On the Relative Motion of the Earth and the Luminiferous Ether" and published in the 1887 issue of the *American Journal of Science,* Michelson-Morley stated:

"...the displacement [of interference fringes] to be expected was 0.4 fringe. The actual displacement was certainly less than a twentieth part of this [0.02 fringe], and probably less than the fortieth part [0.01 fringe]. ...the relative velocity of the earth and the ether is probably less than one-sixth the earth's orbital velocity, and certainly less than one-fourth". (Michelson-Morley 1887 p.341. Brackets added)

With the Earth's orbital velocity at around 30 km/sec, that one-sixth or one-fourth fraction, of what was "to be expected", would place the measured ether wind maximum at *something approaching, or between 5 to 7.5 km/sec.* This was low by the standards of the static-ether expectations of 30 km/sec for Earth's orbital velocity, and even less than the several hundred km/sec anticipated from the Earth-Sun system's motion in a presumed static-ether galaxy. However, *it was not "zero", "null", or "no result", especially if considered as an expression of a partially Earth-entrained cosmic ether, or an active dynamic ether, which moves closer to the Earth's own solar orbital and interstellar velocity.* Michelson-Morley also considered that, for reasons of spatial geometry of Earth as it orbits the Sun, the measures made by them might have occurred at a seasonal period when the relative velocities between the Earth and ether were at a very low ebb. And so they stated:

"...only the orbital motion of the earth is considered. If this is combined with the motion of the solar system, concerning which but little is known with certainty, the result would have to be modified; and *it is just possible that the resultant velocity at the time of the observations was small though the chances*

*are much against it. The experiment will therefore be repeated
at intervals of three months, and thus all uncertainty will be
avoided." (Michelson-Morley 1887, p.341. Emphasis added)*

Being aware of the Fresnel and Stokes arguments about a partially
or fully entrained ether, or ether-drag effect, they also stated:

"It is obvious from what has gone before that it would be
hopeless to attempt to solve the question of the motion of the
solar system by observations of optical phenomena at the
surface of the earth. But it is not impossible that at *even
moderate distances above the level of the sea, at the top of an
isolated mountain peak, for instance, the relative motion might
be perceptible in an apparatus like that used in these experiments. Perhaps if the experiment should ever be tried in these
circumstances, the cover should be of glass, or should be
removed.*" (Michelson-Morley 1887 p.341. Emphasis added)

Unfortunately, the Michelson-Morley team never undertook any
further experiments, neither on mountain-tops nor at other seasonal
intervals. It is also important to realize that the amount of data they
collected in 1887 was quite small, involving *only six hours of data
collection over four days*, on July 8, 9, 11 and 12 of 1887, with a grand
total of *only 36 turns of their interferometer*.

They conducted the experiment in a dense stone basement location
with a wood cover over the apparatus, both of which would slow down
the velocity of any tangible ether of slight material composition. Their
notation about undertaking the experiment again at a higher altitude on
a mountain peak and with a glass cover was in fact an admission of this
possibility, that they might be dealing with *a material, matter-interac-
tive and Earth-entrained cosmic ether moving more slowly at lower
altitudes.*

These facts reveal how the 1887 experiment was *preliminary* in
nature, and hardly what one expects as the foundation for such a major
pivotal turning point in the history of science. Michelson-Morley knew
this, as otherwise, why write so clearly on the necessity of repeating the
experiment at other seasons, at higher altitudes, and with glass covers?

From our present perspective, these problems in the Michelson-
Morley 1887 experimental protocols, along with the relatively short
light path of the interferometer being used, guaranteed only a small (but

Figure 13. Miller's Review of the Michelson-Morley Data.

"In the original account of their experiment, Michelson and Morley give the actual readings for the position of the interference fringes in the six sets of observations. The upper one of the two long curves [at right, no arrows], shows the average of the three sets of readings taken at noon, and the lower long curve is the average for the three sets taken in the evening. These curves show the fringe displacements for a full turn of the interferometer, while the ether-drift effect being sought is periodic in each half turn. To find the latter effect, the second half of the long curve is superimposed on the first half by addition, which cancels the full-period effect and all odd harmonics, giving the shorter curve which is the desired half-period effect ... these curves are not of zero value, nor are the observed points scattered at random; there is a positive, systematic effect. These full-period curves have been analyzed by the mechanical harmonic analyzer, which determines the true value of the half-period effect; this, being converted into its corresponding value for the velocity of relative motion of the earth and ether, gives a velocity of 8.8 kilometers per second for the noon observations, and 8.0 kilometers per second for the evening observations." (Miller 1933a, p.206-207. Most importantly, the lower parts of the two above graphs, with added identifying arrows, are in close agreement.)

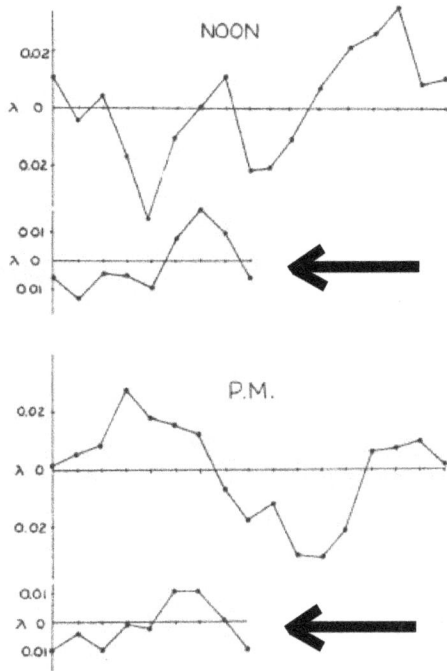

Fig. 3. Fringe displacements of the original Michelson-Morley experiments of 1887.

never "null") measured result. The result of ~5 to 7.5 km/sec was later reviewed and recalculated by Dayton Miller – another Case School professor and associate of Morley – and found to yield an average ether-drift velocity of around 8.4 km/sec. Miller's recalculated velocity was in close agreement with the ~10-11 km/sec maximum ether velocity he would systematically document some 30 years later, using a more robust experimental protocol and a more sensitive interferometer with an even longer light path. Miller also took his instrument high up on Mount Wilson, and ran the experiment over four seasonal epochs, in a small hut with open windows and glass covers at the level of the light path, just as Michelson-Morley had stated as necessary in their 1887 paper. Miller detected an ether-drift signal more clearly and definitively than anyone before or since.

As the history of science records, the original Michelson-Morley 1887 experiment was inaccurately heralded in most every scientific publication and newspaper of those days as a "null" or "zero" result. A host of speculations were thereafter stimulated, as to why the ether "could not be detected", *in spite of the fact that it **was** detected. Their velocity determinations approaching 5 to 7.5 km/sec (18,000 to 27,000 km per hour), is a considerable percentage of the general escape velocity of space rockets (~11 km/sec), as needed to reach full Earth orbit! That itself is quite a fantastic speed, an order of magnitude greater than the Earth's speed of axial rotation, and about 20% of the Earth's orbital velocity around the Sun.* Using Miller's 1933 revised analysis for the Michelson-Morley data, his average of 8.4 km/sec works out to be an even greater velocity.

The ether velocity detected by Michelson-Morley was never "null" nor inconsequential.

Michelson-Morley admitted their 1887 experiment to be only a first step of investigating the ether subject with the new method of light-beam interferometry. It was nevertheless greeted as a "defining negative result" in many quarters, with a chronic misrepresentation of "null-zero". How was it that only six hours of data collection on four days in 1887 was considered sufficiently robust for the majority of scientists – particularly the physics Mandarins in Europe – to push for and embrace the subsequent radical shifts in theory which dominated later 20th and early 21st Century conceptions of the universe?

The FitzGerald-Lorentz Theory and Morley-Miller Experiments

> "Strictly speaking, the condensation [of ether] must be still more considerable than the value we have found to be necessary. If the ether be attracted by the earth, it is natural to suppose that it is acted on likewise by the sun; thus the earth will describe its orbit in a space in which the ether is already condensed. In this dense ether, the earth must produce a new condensation." — Heinrik Lorentz 1899, p.446.

The years before the Michelson-Morley experiment of 1887 were characterized by a scientific discourse on the nature and properties of the ether, and its role in the properties of light and space. Nearly all had accepted the ether theory for most of their professional lives, and also accepted the wave theory of light, which demanded such a medium for light-wave transmission. Disagreements persisted on just what kind of ether might actually exist. Into that discussion came the 1887 result, variously described as "null" or "zero", but which as pointed out in the last chapter was a substantial quantity. A significant ether-wind velocity was recorded, of up to 5 to 7.5 km/sec by Michelson-Morley's own statements, or an average of ~8.4 km/sec as their data was later recalculated by Miller in 1933, using a new theory and understanding about Earth's net motion in space. The Michelson-Morley result was too small to accommodate the static ether of Newton, but it was significant and sufficient enough to warrant further investigation along the lines of a partially entrained ether-drag effect. Such an ether drag would by definition reduce the conventionally (at that time) "expected" velocity close to the surface of the Earth.

A trend was also set into motion following a new theory of "matter contraction", to dismiss the Michelson-Morley result as purely "null", and to *explain away* the cosmic ether itself, as if it were a nuisance. And if one could find a way to reject the ether, then the wave-theory of light could also be more easily rejected. The claimed but factually nonexistent "null result" led even those physicists and astronomers who embraced the wave-theory of light, and the luminiferous ether, towards self-doubt, and into rather ad-hoc mystical postulates.

FitzGerald and Lorentz Ignore the Positive Results of Michelson-Morley, and Postulate "Ether-Matter Compression"

In 1889, the Scottish physicist George FitzGerald published a letter in *Science* journal, asserting the "absence" of a positive result from the Michelson-Morley experiment, speculating that molecular forces within matter might be influenced by a current of moving ether. He postulated a tiny but significant shortening or compression of matter in the same direction as the ether wind, in proportion to its velocity. FitzGerald retained the ether concept, but argued for a theoretical "matter-compression", rendering the ether undetectable by Michelson's

FitzGerald
(1851-1901)

interferometric methods. This compression could, he imagined, contract the interferometer arm aiming into the ether wind, just enough to reduce and equalize the light velocity variance between it and the perpendicular arm of the instrument. His rejection of any small result from Michelson-Morley was partly founded upon Newtonian static-ether concepts, and also upon the proposition that the cosmic ether had certain electrical properties which could interact with matter at the atomic level. The electrical postulate was reasonable, but rejection of the actual results of Michelson-Morley was not.

In an 1889 article in *Science*, "The Ether and the Earth's Atmosphere", FitzGerald wrote:

> "I have read with much interest Messrs Michelson and Morley's wonderfully delicate experiment attempting to decide the important question as to how far the ether is carried along by the Earth. Their result seems opposed to other experiments showing that the ether in the air can be carried along only to an inappreciable extent. I would suggest that almost the only hypothesis that could reconcile this opposition is that the

lengths of material bodies changes, according as they are moving through the ether or across it, by an amount depending on the square of the ratio of their velocities to that of light." (FitzGerald, *Science*, 1889)

FitzGerald frequently lectured on this subject, which was also considered in a March 1892 lecture by Sir Oliver Lodge to the Royal Society ("Aberration Problems and New Ether Experiments"), later published in *Philosophical Magazine* in 1894. However, neither man provided or even suggested new experiments by which to test out the "ether-matter compression" postulate. No arguments or evidence was provided as to why the slight positive results actually measured by Michelson-Morley should be so casually ignored. FitzGerald's ideas nevertheless

Oliver Lodge
(1851-1940)

attracted attention in how a (wrongly) presumed negative result from the Michelson-Morley experiment could be understood within the context of static-ether theory. Somehow, motion through the ether would compress matter exactly (and conveniently) enough to render the ether undetectable. FitzGerald was rewarded for his efforts with an appointment to the Royal Society in the following year.

John William Strutt, 3rd Baron Rayleigh, better known as Lord Rayleigh, a decorated British member of the Royal Society since 1873 (and later President of it), weighed into this discussion in an 1892 article in *Nature*. He expressed reservations about the Michelson-Morley results, and concerns about how a fully stagnant entrained cosmic ether of the Stokes variety – which predicted a stagnant, unmoving ether at the Earth's surface – would affect current theories of stellar aberration. However, Rayleigh disagreed with FitzGerald, and argued for continued ether experimentation with the interferometer, much as Michelson-Morley stated in 1887 as necessary:

Rayleigh
(1842-1919)

"...Michelson's results can hardly be regarded as weighing heavily in the scale. It is much to be wished that the experiments should be repeated with such improvements as experience suggests. In observations spread over a year, the effects, if any, due to earth's motion in its orbit, and to that of the solar system through space, would be separated." (Rayleigh 1892)

In 1893, Stokes published a pamphlet, *The Luminiferous Aether*, reiterating his fully-dragged theory of a stagnant ether as the best explanation for the claimed poor results of Michelson-Morley. He also indicated the cosmic ether theory would retain validity if it could be considered as something more material and substantive in nature. However, in nearly every case, the small positive results of the Michelson-Morley experiment were being erased or misrepresented as "null" and insignificant.

Starting in 1895, the Dutch physicist Heinrik Antoon Lorentz also lectured and published papers addressing the claimed "negative result" of the Michelson-Morley experiment, and exploring the issue of ether-matter contraction. Nevertheless, in his 1899 publication on the subject, "Stoke's Theory of Aberration in the Supposition of a Variable Density of the Aether", Lorentz wrote in full support of a substantive cosmic ether, clearly contradicting the ether-contraction theory. He postulated a gaseous material ether that obeyed Boyle's law, and Newton's early ideas on ether condensation around matter. He contrasted the Stokes theory of a fully-entrained ether carried along at the same velocity as the rotating Earth's surface, to the Fresnel theory of a variable-density and partly entrained ether. Lorentz's thinking of this period were fully supporting of the cosmic ether, with open questions remaining only about its properties, and the extent to which it permeated and surrounded matter – as with the quote at the top of this chapter.

Lorentz pointed out how Stoke's version of an ether, which would become entrained as it moved down to the Earth's and the Sun's surface, must slowly condense and increase in density, eventually to be carried along at the same velocity as their surfaces. From Stokes, one could anticipate a zero result from any interferometer experiment

Lorentz
(1853-1928)

undertaken at the Earth's sea-level. In the end, Lorentz came down on the side of Fresnel, for a partially entrained ether, which would allow for its variable condensation and detection, with a higher ether velocity at higher altitudes. Lorentz' discussion suggested a dynamic ether in motion towards the planets and Sun, condensing around them into distinct layers, which suggests a vortex motion surrounding planets and the Sun, as already argued in Figure 4, given in the *Introduction.*

Nevertheless, Lorentz would neither accept nor even mention the actual result obtained in the Michelson-Morley experiment, the ~5 to 7.5 km/sec ether drift, as anything other than a fully "null" result. He could not let go of his static-ether bias, that their lesser positive result was meaningful. In the end, Lorentz also embraced the mystified "solution", of a never-demonstrated ether-matter contraction, similar to that of FitzGerald.

1898: Morley and Miller, Investigation of Matter Contraction and Ether Drift.

From the time of their original 1887 experiment to the turn of the new century, Michelson and Morley published nothing of significance on the ether-drift question. In 1888 Michelson published "A Plea for Light-Waves" in the *Proceedings of the AAAS*, wherein the "luminiferous ether" was mentioned, but nothing was said about his 1887 ether-drift experiment. By contrast, Morley teamed up with the younger Dayton Miller

Dayton Miller, c.1900

in a series of new optical and ether experiments. While these new experiments would eventually yield positive ether-drift detections, their own published papers of the period relied upon a-priori assumptions about the Earth's net motion in space which proved to be inaccurate, but which were corrected only some years later by Miller. This led them to initially report pessimistically on their own results. The delay in the more accurate reporting of their measured results was historically critical, given how the advocates of the ether-matter contraction theory continued to advance their formularies to the point of dogmatic entrenchment, leading everyone, including the young Einstein, to further wrongly assume the ether had not been detected.

Figure 14. The Morley-Miller 1898 Experiment for Magnetic Influence upon Light Speed, set in a basement room of Pierce Hall at Case School. This was the same location where the 1887 Michelson-Morley experiment had been conducted. Large batteries are seen on the right-side table, with connecting wires to an electromagnetic coil on the left-side table, through which one of two light beams passed, en route towards an optical interferometer. No differences were observed when the electromagnet around one light beam was activated, versus when it was off. It was a failed effort to detect variation in light speed due to strong magnetism.

The team of Morley-Miller would eventually build the largest, most sensitive light-beam interferometer ever constructed, and make significant detection of both ether wind and light-speed variations. In their early efforts, however, they mostly confined their investigation towards detection of the postulated FitzGerald-Lorentz contraction, of a slight compression of matter along its axis of motion.

In 1889, for their first joint endeavor, Morley-Miller constructed a stationary interferometer, using some of the original optical components from the 1887 Michelson-Morley experiment, to evaluate the effects of strong magnetism on light velocity. Their experiment was set up in the same basement-corner room where the Michelson-Morley experiment had been undertaken, in Pierce Hall at Case School, at 199 meters above sea level. This experiment indicated no significant changes due to a light beam's passing through the static field of a strong electromagnet. They published their results in two 1898 reports, in both *Physical Review* and the *Proceedings of the American Association for the Advancement of Science (AAAS)*, "On the Velocity of Light in the Magnetic Field".

Three years later in 1902, Lord Rayleigh tried but failed to detect ether-matter contraction in crystals, by changes in their refractive and polarization properties: "Does Motion through the Aether cause Double Refraction?" Another effort was undertaken by D. B. Brace in 1904, "On Double Refraction in Matter moving through the Aether". Like Rayleigh, Brace also did not detect any ether-matter contraction.

A related experiment was undertaken in 1903 by Fredrick Trouton and H.R. Noble, seeking to identify a preferred direction of ether flow in the rotational orientation of a 3000-volt charged parallel-plate capacitor, as suspended in a glass tube. Their paper "The Forces Acting on a Charged Condenser moving through Space" was published in the *Proceedings of the Royal Society of London* the following year, wherein they reported a negative result. By theory, the capacitor was anticipated to rotate in perpendicular alignment to the ether flow.

A few years later, Trouton and A.O. Rankine sought to detect a change in electrical resistance within a conductor oriented parallel versus perpendicular to the ether flow, as a function of its presumed change in length. To evaluate for such an influence, they constructed an ordinary Wheatstone bridge circuit, where four wire-coils were laid out in perpendicular pairs, forming a box-shape. By rotating the Wheatstone bridge circuit, two of the coils would theoretically be brought into parallel alignment with the ether flow, while the other two would automatically be perpendicular to that flow. Rotating the coil by another 90° would then reverse whatever small current might be detected. This experiment, published as "On the Electrical Resistance of Moving Matter", in the February 1908 issue of the *Proceedings, Royal Society of London*, also produced a negative result.

Both of the Trouton papers were brilliant in conception, but by modern standards lacked in proper electronics by which to detect or respond to the effects anticipated. And in both experiments, it would not have proven out an ether-flow contraction, but only that a moving ether had some dielectrical, electrical and/or magnetic properties. That was certainly a reasonable speculation. Two other problems afflicted their experiments. Firstly, in both cases of Trouton-Noble and Trouton-Rankine, their apparatus was enclosed in a metallic or wood container, and apparently placed within a structure in low-elevation London. So whatever the velocity of ether flow was at that location, it would have been predictably low, and additionally blocked by buildings and experimental enclosures.

All these and other experiment had a direct bearing upon the postulated FitzGerald-Lorentz contraction, as none had so far shown any kind of matter-contraction effect. *And none ever would.*

1900-1906: Morley-Miller Reproduce Michelson-Morley, with Positive Results

While all these failed experiments for detection of a FitzGerald-Lorentz "matter contraction" were underway, Morley and Miller again took up the subject of the basic ether-drift measurements. They had been persuaded to undertake a repetition of the Michelson-Morley experiment by Lord Kelvin, then leader of the influential British Royal Society, while attending the Paris International Exposition of 1900.

Their new experimental efforts began with construction of a rotating cross-arm interferometer similar to the original one used by Michelson-Morley, laid out in the manner of an "X", but in this case on a wooden foundation of white-pine planks. Flat iron plates were then bolted to the top center and ends of the wood planks for securing mirrors and other optical components. A round wooden float was added to the bottom center of the wood "X", by which the entire apparatus could float in a round tub filled with dense liquid mercury. This would allow for a smooth slow rotation. A light source and magnifying telescope to view the interference fringes were also mounted on the platform.

Figure 15 shows a top view of the setup, identical in function to the original 1887 experiment, but much larger, measuring 4.3 meters across. This would allow a bouncing of the light beams back and forth 16 times (8 times out, 8 times back) using clusters of four mirrors at the end of each interferometer arm. The two-way round-trip light beam path in the new instrument, of 64 meters, exceeded by nearly three times the original Michelson-Morley 1.5 meter platform with a 22-meter round-trip light path. The wood platform was chosen as a contrast to the original sandstone platform of the Michelson-Morley 1887 experiment, on the premise that if an ether-matter contraction existed, it might show up more clearly in soft pine wood than in sandstone.

With this new and more sensitive interferometer, they set out to investigate two major issues. Firstly they would try to improve upon the original ether-drift experiment of Michelson-Morley. Secondly they would try once more to detect a FitzGerald-Lorentz matter contraction. The experiments of Morley-Miller and their results were initially described in two publications from 1905, "Report of an Experiment to Detect the FitzGerald-Lorentz Effect", published in *Proceedings AAAS*, and "On the Theory of Experiments to detect Aberrations of the Second Degree", appearing in *Philosophical Magazine*. They wrote:

"Such a [FitzGerald-Lorentz] contraction can be imagined in two ways. It may be thought to be independent of the physical properties of the solid and governed only by geometrical conditions; so that sandstone and pine, if of the same form, should be affected in the same ratio. On the other hand, the contraction may depend upon the physical properties of the solid; so that pine-timber would doubtless suffer a greater compression than sandstone." (Morley-Miller, May 1905, p.66)

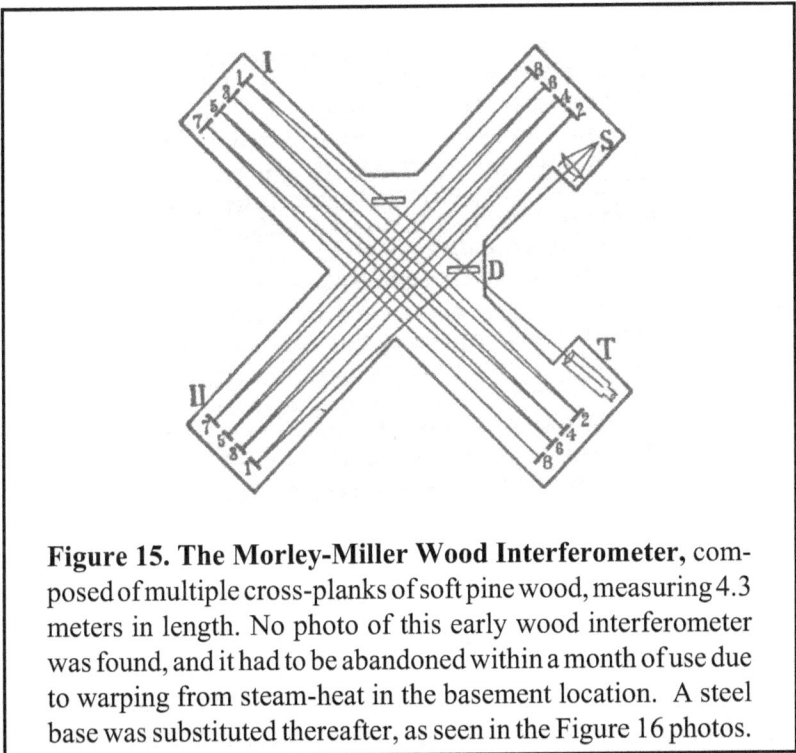

Figure 15. The Morley-Miller Wood Interferometer, composed of multiple cross-planks of soft pine wood, measuring 4.3 meters in length. No photo of this early wood interferometer was found, and it had to be abandoned within a month of use due to warping from steam-heat in the basement location. A steel base was substituted thereafter, as seen in the Figure 16 photos.

Proceeding from that theory, the new interferometer was set up in the same corner basement location where the original Michelson-Morley experiment, and their own prior magnetic-field investigation had been undertaken. Some of the same optics, such as mirrors, were recycled from the Michelson-Morley apparatus, as was the wood float and tank of mercury. Construction of the new interferometer and setting it up for stable measurements required more than a year, with experiments starting in August 1902.

While their initial results with the pine beam platform in August of that year provided for "good observations", the steam heat used in the basement location warped the wooden optical platform, requiring a rebuilding of the apparatus before further testing could proceed. For the rebuild, they used steel beam cross-arm supports. The steel beams, once constructed, were independently tested for any kind of length variation due to the Earth's magnetic field; none could be detected.

The rebuilt optical platform was of the same light-path length as the prior wood apparatus. Pre-dried pine-wood boards were laid down on top of the steel beams, with a brass tube/truss framework mounted on top of the wood boards. Four pine-wood rods were then placed inside the four long brass tubes forming each truss, whereupon the protruding ends of the pine rods were attached to the mirror supports. By this arrangement, any change in the length of the pine rods would determine the spacing between the groups of interferometer mirrors, of which there were four at each end of the cross-arm beams. Photographs accompanying their 1905 article in the *Proceedings AAAS* show the setup, including one where a pine-wood cover was temporarily placed around the light path, in agreement with the original Michelson-Morley experiment. By this method, the problem of wood warping was eliminated. Figure 16 presents these photos.

The total light path of the 64-meter interferometer yielded ~112 million independent light waves, which in turn produced interference fringes easily visible in the interferometer telescope, with a fine resolution down to tenths or hundredths of an individual fringe's motion against a fixed marker in the telescope field.

Their experiment resumed in June 1903, using the same basement location in Pierce Hall at Case School, close to the same monthly period as Michelson-Morley had been undertaken, with a new set of optical mirrors added to their instrument. Once again they could compare the results of their rebuilt pine-rod interferometer to the results from the 1887 sandstone-base interferometer of Michelson-Morley. They ran

Figure 16. The Morley-Miller 64-meter Steel Interferometer, in a basement room of Pierce Hall at Case School, c.1904. Top: A brass-tube and wood-rod truss framework connects the opposing mirrors, used for both ether-drift and FitzGerald-Lorentz contraction tests. Below: The light paths are covered in wood, as in the original 1887 Michelson-Morley experiment.

the experiment twice per day, around noontime and midnight, similar to the original Michelson-Morley experiment, but with many more turns of their new interferometer. Any persisting ether-drift signal, even at a similar low velocity as obtained by Michelson-Morley, would refute the Fitzgerald-Lorentz ether-matter contraction theory and confirm a real ether drift.

After completion of several years of experiments in the basement of Pierce Hall, Morley-Miller gave a lecture summarizing their work up through 1904 to a New York meeting of the National Academy of Sciences: "Report of an Experiment to Detect the FitzGerald-Lorentz Effect". This report was published a year later in the *Proceedings of the American Academy of Arts and Sciences*. They reported no evidence of ether-matter contraction, but a slight positive result for an ether-drift.

"If pine is affected at all, it is affected to the same amount as is sandstone. If the ether near the apparatus did not move with it, the difference in velocity was less than 3.5 kilometers a second, unless the [ether-matter contraction] effect on the materials annulled the effect sought. Some have thought that the former experiment only proved that the ether *in a certain basement-room* was carried along with it. We desire to place the apparatus on a hill, covered only with a transparent covering, to see if an effect could be there detected." (Morley-Miller *Proc.AAAS*, 1905, p.685. Emphasis in original)

Here, Morley-Miller reported similar but slightly lower results than the Michelson-Morley 1887 experiment, a positive ether-drift result which *refuted the FitzGerald-Lorentz contraction theory. That theory only made sense in the context of a truly null or zero ether-drift velocity, which was not the case in either the Michelson-Morley or Morley-Miller experiments.* In the last part of the above quote, Morley-Miller also repeated the basic problem known since 1887, that running the ether-drift experiments at low elevations or in the basement room of a stone building, was likely to block any significant flow of a cosmic ether with material properties, and thereby give very reduced results. Nevertheless, these early experiments by Morley-Miller produced *light-speed variations approaching ~3.5 km/sec.*

Euclid Heights: Significant Success but Computational Error

Morley-Miller resumed working through summer and fall of 1905, when the steel and wood-rod interferometer was moved into an octagonal hut on nearby Euclid Heights in Cleveland, at an altitude of 285 meters. This was about 100 meters higher than the prior efforts in the Pierce Hall basement, and away from all the stone buildings of the Case School campus. Glass panels were placed over the light-beam paths, which previously had been covered with opaque wood covers for temperature stability, much in the manner of the original Michelson-Morley experiment. Transparent eisenglass windows were also constructed in the hut, at the level of the interferometer light beams, so as to avoid any possible blockage of ether motion, on the assumption that ether would move more easily through transparent than opaque materials. *A total of 230 turns of the interferometer were made at Euclid Heights in 1905, yielding an 8.7 km/sec result.*

Miller's major paper of 1933 (p.215-217) gives the best account of the Morley-Miller experiments from August of 1902 through November of 1905. Over this period they conducted 995 turns of the interferometer, nearly 28 times as many as the 36 turns Michelson-Morley undertook. And like Michelson-Morley, a slight ether-drift signal was obtained, but of a higher velocity. In his 1933 review of the entire range of Morley-Miller experiments, Miller reported an average ether drift signal of ~9.2 km/sec, a *positive confirmation for ether drift effects upon light speed.* By contrast, the Morley-Miller efforts produced *no confirmation for an ether-matter contraction,* on the theory of FitzGerald-Lorentz. That was not surprising, given how the entire "contraction theory" was based upon the false assumption of a "null" result.

One of the problems exposed in the Morley-Miller experiments was the absence of a systematic method for data collection, often restricting their observations to a twice-daily routine, based upon theoretical expectations of a maximal ether-wind at those times. This led to a significant computational error. When data were correlated, relatively strong signals often occurred, but of opposing sign, plus or minus, which when averaged would cancel each other out to yield a much lower average value. (Miller 1933a, p.217) Figure 17, below, shows Miller's graph on this early method, about which he wrote in 1933 as "considered erroneous". For such reasons, including the early problems with the wood-base interferometer, the original 1905 reports published by Morley-Miller revealed primarily very small ether velocity values, or their results were reported in purely negative terms. Their 1905 publications were mostly devoted to the failure to identify the elusive "matter contraction" of FitzGerald-Lorentz. They also wrote many pages in 1905 in rebuttal to W.M. Hicks, who in 1902 had raised criticisms against the overall Michelson interferometer methodology, rejecting the Michelson-Morley results. After being rebutted, Hicks retracted his paper. However, as an advocate of an ether-vortex theory, he correctly pointed out how the *vertical components of a presumably gravitational ether* might not fully register on a horizontal interferometer. That issue remained important, nevertheless.

As Miller later explained it, the Morley-Miller experiments had been founded upon certain *a priori* assumptions which had not previously been questioned. In 1928 and 1933, he explained it thus:

"On the dates chosen for the observations there were two times of the day when the resultant of these motions would lie in the plane of the interferometer, about 11:30 A.M. and 9:00 P.M. The calculated azimuths of the motion would be different for these two times. The observations at these two times were,

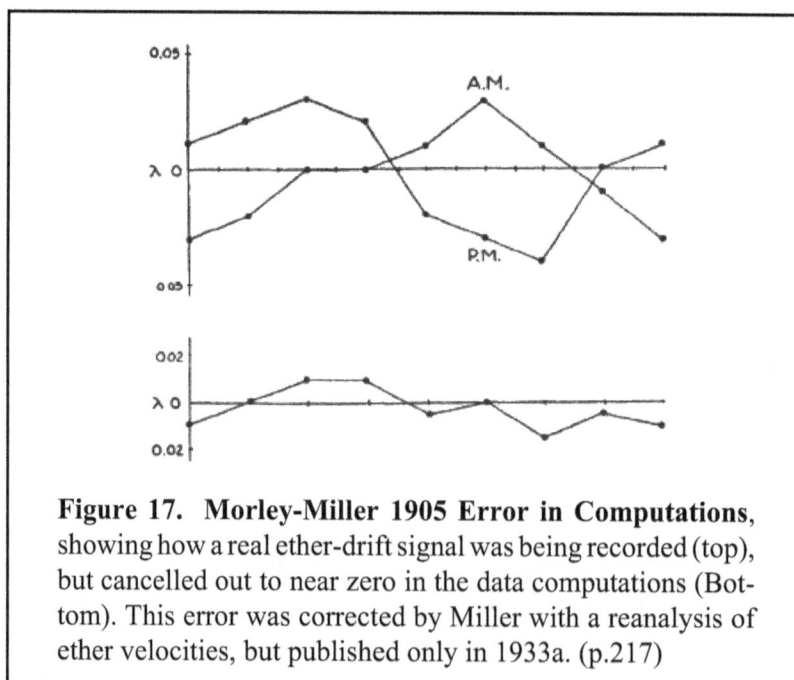

Figure 17. Morley-Miller 1905 Error in Computations, showing how a real ether-drift signal was being recorded (top), but cancelled out to near zero in the data computations (Bottom). This error was corrected by Miller with a reanalysis of ether velocities, but published only in 1933a. (p.217)

therefore, combined in such a way that the presumed azimuth for the morning observations coincided with that for the evening. The observations for the two times of day gave results having positive magnitudes but nearly opposite phases; when these were combined, the result was nearly zero. The result, therefore, was opposed to the theory then under consideration... it now seems that the superposition of the two sets of observations of different phases was based upon an erroneous hypothesis and that the positive results then obtained are in accordance with a new hypothesis as to the solar motion." (Miller 1928 p.353)

"Previous to 1925, the Michelson-Morley experiment had always been applied to test a specific hypothesis. The only theory of the ether which had been put to the test is that of the absolutely stationary ether through which the earth moves without in any way disturbing it. To this hypothesis the experiment gave a negative answer. The experiment was applied to test the question only in connection with specific assumed motions of the earth, namely, the axial and orbital motions combined with a constant motion of the solar system towards the constellation Hercules with the velocity of about nineteen kilometers per second. The results of the experiments did not agree with these presumed motions. The attention was given almost wholly to this velocity of the ether drift, and no attempt was ever made to determine the apex of any indicated motion. The experiment was applied to test the Lorentz-FitzGerald hypothesis that the dimensions of bodies are changed by their motions through the ether; it was applied to test the effects of magnetostriction, of radiant heat and of gravitational deformation of the frame of the interferometer. Throughout all these observations extending over a period of years, while the answers to the various questions have been "no," there has persisted a constant and consistent small effect which has not been explained." (Miller 1933a p.222)

The revisions reported by Miller in 1928 and 1933, included in Table 1, are in association with the new theoretical understanding. Unfortunately, Miller's recalculations didn't become public until those later dates. Miller seems to have been the more persistent member of the

Morley-Miller team. As the senior scientist of the pair, Morley might have been the one who wrote the more pessimistic reports of 1905, and applied the wrong theory of calculation in the first place. This remains unclear. However, it was certain that Miller was the one who caught and corrected the errors, and later went on over many years to investigate the ether-drift question in a more robust experimental effort, with a more systematic method of data collection.

Regarding their calculation error, we can visualize the problem they faced by comparing the horizontal interferometer to a wind-speed anemometer. It can detect atmospheric wind velocity best when exposed to a horizontal wind. When an *ether* wind blew horizontally across the cross-arms of the ether-wind detector (the interferometer), strong interference fringes would appear in line with its highest velocity. But unlike a person holding an anemometer, who can feel the atmospheric wind, one could not feel or otherwise determine the direction the ether wind was blowing. In another example, if the ether wind blew down on our heads from directly above, downwards and penetrating into the Earth, our horizontal interferometer would measure nothing, *no ether wind*, a big zero or null, even if that vertical ether-wind was "blowing" at 100,000 km/sec. It also follows, as a third example, that if a 10 km/sec ether-wind descended downwards at a 45° angle, obliquely from above, then the horizontal "anemometer interferometer" would measure a slower velocity, around 7 km/sec as com-

Table 1: Summary of the Morley-Miller Experiments		
(Data recomputed by Miller in 1933a)		
Experimenters and Dates	Number of Turns	Measured Ether Velocity
Michelson-Morley July 1887	36	~5-7.5 km/sec
Morley-Miller - Pierce Hall Aug.1902 & June-Sept.1903	505	~ 10 km/sec
July 1904	260	~ 7.5 km/sec
Morley-Miller - Euclid Heights Jul.Oct.Nov.1905	230	~ 8.7 km/sec
Totals and Averages for Morley-Miller:	995	~9.2 km/sec

pared to a fully horizontal 10 km/sec ether-wind.[3] That being the case, how could the net velocity and true axis of that ether wind be determined, if you did not already have a reasonable idea of its direction, to determine the best times for measuring?

A simple solution was found and applied by Miller in his later work on the ether-drift question, *by running experiments every hour over the full 24 hours of many sequential days, repeating this procedure at several other times of the year.* One could then know the orientation of the interferometer for getting both the strongest and the weakest ether-wind. As Miller stated, *all preconceptions had to be abandoned.* Only much later, after 1920, did Miller independently take such a systematic approach. Those later systematic observations allowed for a better determination and theory of the Earth's net motion in the cosmos. And from that, he was able to reconsider, analyze and recompute the ether velocities for the older Morley-Miller experiments. Clearly, a positive result for a real ether drift or wind was detected by Morley-Miller. Figures 17 and 18, and Table 1 give a summary of the various ether velocities and azimuths, as compared to Miller's independent post-1920 results, to be presented in the next chapter.

The Morley-Miller 1905 experiments ended when the Euclid Heights research location had to be abandoned. The property had been sold, and the new land owner asked for the interferometer and its house to be moved away. The large interferometer was then placed in storage, and the octagonal interferometer hut became a hot dog stand at football games, for the students at Case School.

Morley retired from Western Reserve University in 1906, moving to Connecticut, leaving Miller to pursue the question of cosmic ether independently. Nevertheless, had no further ether drift detections been made, the Morley-Miller experiments by themselves incorporated 995 turns of an increasingly sensitive interferometer design, with a corrected average ether velocity of around 9.2 km/sec. *This result clearly confirmed light-speed variations, and also in the process failed to show any indications for the never-demonstrated FitzGerald-Lorentz "contraction".* It also laid a foundation by which one of Einstein's central assumptions, of light-speed constancy, was proven incorrect.

Nevertheless, the computational error in the Morley-Miller data as published in 1905 left them vulnerable to criticism by their opposition, who also continued to misrepresent the slight positive results of both the

---

3. An ether wind of 10 km/sec divided by $\sqrt{2}$, based upon the formula for a right-triangle hypotenuse.

Michelson-Morley and Morley-Miller experiments. In a 1904 paper, Lorentz persisted with the untrue "null" interpretation of Michelson-Morley. He wrote, for example:

> "Michelson's well-known interference-experiment, the negative result of which has led Fitz Gerald and myself to the conclusion that the dimensions of solid bodies are slightly altered by their motion through the aether." (Lorentz 1904)

Lacking restraint of a real-world mechanism for transmission of light, and logical understandings of light-speed variation, Lorentz' 1904 publication went further into ad-hoc mysticism, conjuring up new properties for light and ether which also had no basis in experimental or empirical fact. He artificially separated light and ether into different "frames of reference", something possible only as a "thought experiment", happening inside his head, but not in the real world. He invented "time dilation" and other imaginings out of thin air, supported merely by mathematical formulations and theoretical necessities. Lorentz also split apart the once unified optical, gravitational and temporal functions as they occurred within the real natural world in ordinary Galilean/ Cartesian space and time. Gone were his prior references to the Stokes-Fresnel debate, about a fully or partially dragged ether. Gone were discussions about ether condensation and increased density around the planets and Sun, giving rise to aberration, refraction and gravitational effects. Lorentz offered no new experiments to confirm or test his post-1904 conclusions of an ether-compressed matter, nor for the other surreal add-ons to the original FitzGerald theory. Cosmic ether and ether-motion were forbidden entry into such a nether-world, as were light waves.

All those components of Lorentz' imaginary universe would later appear in Albert Einstein's 1905 equally imaginary special theory of relativity, to be discussed in a later chapter. Both Lorentz and Einstein continued to ignore what was actually stated in the 1887 Michelson-Morley paper, which clearly identified an ether-drift velocity and light-speed variance approaching 5 to 7.5 km/sec, as well as the initial Morley-Miller miscalculated low estimate of a 3.5 km/sec velocity.

The originally reported 1905 Morley-Miller small and miscalculated results were surely disappointing, but one must compare the utter and complete rejection of that result, along with the Michelson-Morley result, to the later quick and easy-happy embrace of *very tiny quantities*

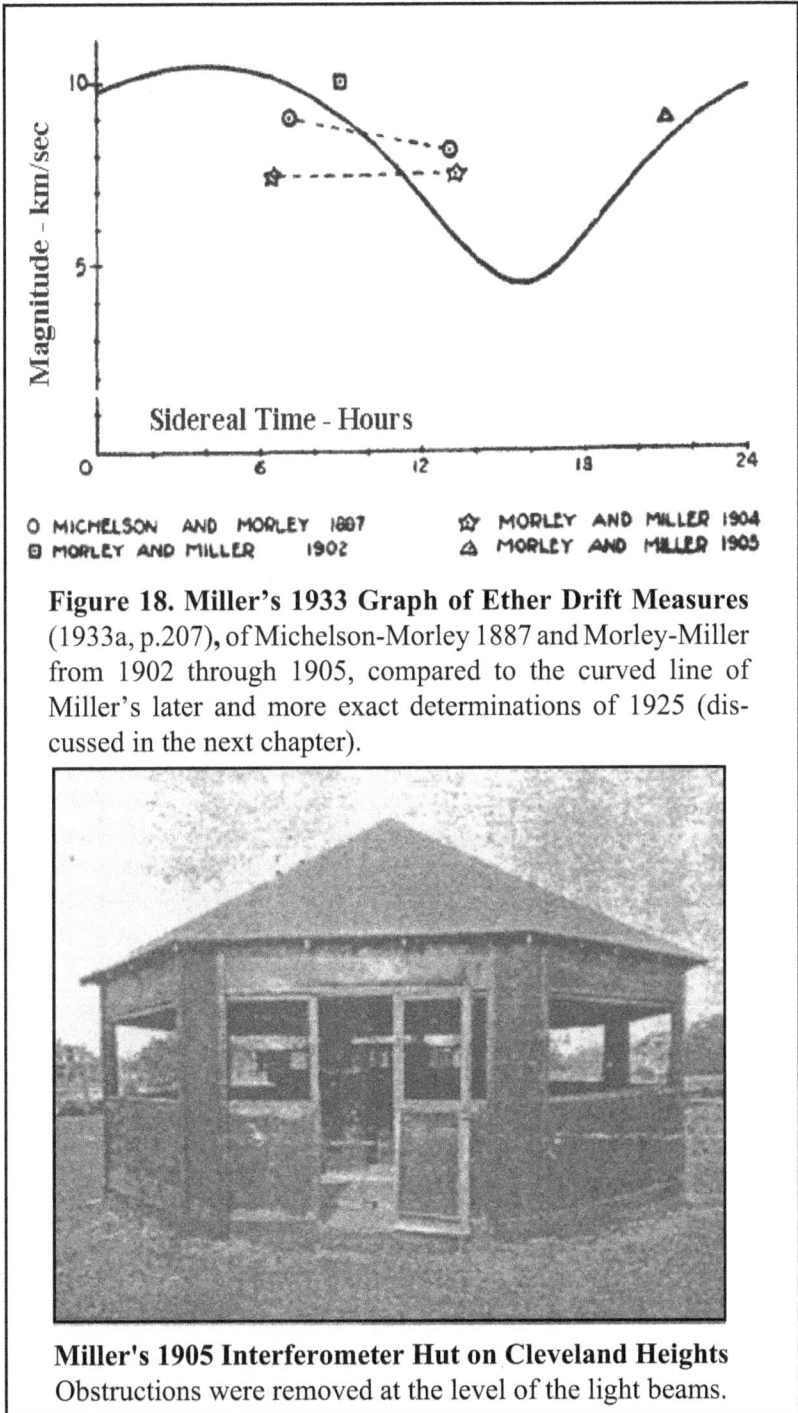

Figure 18. Miller's 1933 Graph of Ether Drift Measures
(1933a, p.207), of Michelson-Morley 1887 and Morley-Miller
from 1902 through 1905, compared to the curved line of
Miller's later and more exact determinations of 1925 (discussed in the next chapter).

Miller's 1905 Interferometer Hut on Cleveland Heights
Obstructions were removed at the level of the light beams.

of observed starlight bending during solar eclipses, and of shifts in the perihelion of Mercury, which were subjected to immense international media hype, in support of the Einstein theory. That will be discussed more thoroughly in the chapter on *Einstein Rising.*

Miller's recomputed determination of the Morley-Miller results was actually better than Michelson-Morley, with a 9.2 km/sec ether velocity, *more than 33,120 km per hour (20,580 mph). That is even closer to the general escape velocity of space rockets to achieve Earth orbit (~11 km/sec), and about a third of the Earth's orbital velocity around the Sun.* These understandings came too late for influence upon the scientific world of 1905, but certainly the obfuscation and discarding of Miller's results after the 1920s was not so easy to excuse or explain, except as a dogmatic insistence by Einstein and his followers, favoring mystical theory over empirical experimental determinations.

Miller's recomputations of 1928 and 1933 would also be accompanied by even more powerful direct and highly significant observational data from Mount Wilson. And yet, in those later years, the interest in those findings by Lorentz and Einstein continued to be "null". With the exceptions of a few worried statements from Einstein, the growing evidence for a real ether drift and variable light speed continued to be ignored and erased from mention in their published papers.

Another factor: As I reviewed their publications and biographies, after 1905 Michelson, Morley and Miller all appeared somewhat intimidated by how so much of the "Royal Society" of European highbrow physics ignored or down-put their findings. There was a strong emotional component to the growing scientific embrace of the never-proven mystic postulates of FitzGerald-Lorentz and Einstein. Within a few years after 1905, physics on both sides of the Atlantic would engage in an "emotional-drift" towards the speculative mystical Einstein theory of relativity, ultimately to become a stampede. Even Michelson occasionally began to use the "null" term to describe his own 1887 experimental results and to ignore some of his own newer data on light-speed variance, discussed in the following chapters. The 1905 publications of Morley-Miller also lapsed into such depressive "null" language, though I continue to wonder if the senior scientist of that team, Morley, had steered their published statements in that direction. Were they yielding to peer-pressure? It appears so.

Another fact: The Morley-Miller and later Miller experiments proceeded rather slowly and carefully, often understating their results in a cautious scientific attitude. Perhaps too cautiously. This contrasted sharply with the European quick and simplistic flights of fancy about unseen and never-demonstrated "matter contractions" and "space-time warps", which remained purely theoretical. Their mystic and entirely speculative theories were quickly published in top research journals controlled by their colleagues, where their quasi-Royal status appeared to outweigh any demands for experimental proof, or slow-going caution. Their ideas were widely discussed in serious tones, with "elegant maths" that always balanced out, even though *experimental proof or evidence was rarely offered to give substance to their postulates.* Meanwhile the ether wind or drift experiments were simply brushed aside and gradually subjected to erasure and silent treatment. Morley-Miller did make computational errors, which Miller corrected, but for which history discarded their work like road trash. By contrast, Einstein also made errors needing significant public correction, but afterwards gained even greater applause. *A battle between the American experimentalists and the European theoreticians was developing.*

The next chapter will present Miller's substantial and more definitive, independent work on the ether-drift question, notably as carried out atop Mount Wilson. Part II will provide details on Einstein's work, along with contrary evidence about the claimed "experimental proofs" offered in support of his theory of relativity. As I will show, those experiments are *not unequivocal*, and are just as easily, or *more easily* understood as the product of a partially entrained and variable density, motional-gravitational cosmic ether.

At this point in history, a centrally-important question originally raised by Michelson-Morley in their 1887 report, reiterated by Lord Kelvin in 1900, remained unanswered and untested. *What would the result be if such a sensitive interferometer was taken high up on a mountain, where ether-drag effects would be minimized, and ether wind maximized, and with data gathered over different seasons of the year?*

After the ending of the Morley-Miller investigations in 1905, Miller lacked funds and support to undertake more ambitious projects, as had been planned. He instead turned to other research, mostly in acoustical science. World War 1 also intervened and disrupted cross-Atlantic scientific debates. The ether-drift questions were put aside. A master of acoustical theory, and expert on the flute, with a growing collection of flutes from around the world, Miller investigated the subject of tone quality, and invented the *phonodeik*, the first apparatus to convert sound waves into visual images. He also developed a special *harmonic analyzer* to extract individual oscillating signals from apparently chaotic "noise". Additionally he contributed to development of the microphone and loud speaker, consulting with private manufacturers. During the war years, Miller worked with the military on the problem of shellshock. Some 16 years would elapse before he would resume work on the ether-drift question, in 1921.

During that period of Miller's other activities, Michelson and Georges Sagnac independently made new discoveries in optics and ether science. Einstein also attained celebrity status over this same period, with alleged experimental support in 1919 from the Eddington eclipse photographs. All these matters will be reviewed with open and objectively critical eyes.

Dayton Miller's Positive
Ether Drift Experiments, 1921-1926

> "I believe that I have really found the rela-
> tionship between gravitation and electricity,
> assuming that the Miller experiments are
> based on a fundamental error. Otherwise, the
> whole relativity theory collapses like a house
> of cards."
> — Albert Einstein, letter to Robert Millikan
> June 1921 (in Clark 1971, p.328)

In the decades following the Michelson-Morley experiment of 1887, the worlds of physics and astronomy were thrown into confusion, given how the cosmic ether had been a foundational theory for understanding the wave-theory of light, as well as a variety of astronomical and physical phenomena. While the Michelson-Morley experiment obtained a slight positive result, as already discussed, the phrase "null result" and similar misrepresentations came into widespread use when referencing their experiment. Conference lectures and published papers of that period, as by FitzGerald and

Dayton Miller
(1866-1941)

Lorentz, also previously described, carried forward with an increasingly mystified matter-contraction postulate, as a means to "explain" why the cosmic ether was not, or could never be detected – even though it had already been detected, repeatedly. Astrophysics thereby retreated away from real, tangible results on a critical experiment, in what psychologists might call *emotional denial*, substituting in its place a new metaphysics, which had its historical foundation in Newton's metaphysically-demanded static ether concepts.

The Dynamic Ether of Cosmic Space

In this chapter, Dayton Miller's exceptional work on the subject of ether detection will be detailed. After a hiatus which lasted from 1906 through the period of World War 1 until 1921, Miller returned to the ether-drift experiments with renewed vigor. Together with his work with Morley, his entire period of ether-drift investigations would eventually include a total of over 200,000 individual readings, from *over 12,000 turns of the new and highly-sensitive interferometer*, ending in 1926 with completion of his most important Mount Wilson experiments.

Revisiting the Morley-Miller Experiments, 1902-1906

Dayton Miller was the younger man, by nearly 30 years, of the Morley-Miller team. He obtained his physics doctorate at Princeton University in 1890, and by 1893 had been appointed as Chairman of the Physics Department at Case School of Applied Science in Cleveland, Ohio. Morley was then a professor of Chemistry at the adjacent Western Reserve University. Today, these institutions are unified and share the same campus, as the Case-Western Reserve University (CWRU). Miller also later served as President of the American Physical Society and the American Acoustical Society, and was inducted into the National Academy of Sciences in 1921. Like Michelson and Morley, he was no outsider to the mainstream of American science. He approached the new experimental tasks with enthusiasm and a history of solid experimental work in acoustics, optics, astronomy, mathematics and x-ray investigations.

Miller gained a small bit of early fame by making the first American x-ray photo, of his wife's hand. Mrs. Miller in turn made the first-ever full-body x-ray photo, of her husband. Miller also x-rayed broken bones of patients in cooperation with a local hospital. Within the Morley-Miller team, Miller attended primarily to the interferometer optics and measurements, while Morley focused upon the mathematical calculations. Together they produced a more extended and significant work than Michelson-Morley ever did, but not as important as the work Miller would later accomplish independently.

80

Miller's 64-Meter Steel Interferometer

Miller's interferometer was a refinement of the same 4.3 meter (14 foot) diameter instrument he first developed in cooperation with Morley. The round-trip light path was an overall 64 meters (208 feet), about three times the light path and sensitivity as the original 22-meter (72 foot) interferometer used by Michelson-Morley in 1887. These differences in total light path are just as important to the question of light-beam interferometry as are the differences in the size of optical lenses in large telescopes, where the diameter determines the light gathering capacity and sharpness of the images.

Four sets of mirrors were mounted on the end of each interferometer cross-arm, to reflect light beams – or narrow "pencils of light" as was the phrase often used in those days. The light beams, about the diameter of a pencil, were reflected back and forth 16 times horizontally to yield the round-trip light path of 64 meters. The basic operation of such a light-beam interferometer has already been explained over the last two chapters. While refinements were made to this large interferometer in the years after Morley's retirement, it remained true to the original basic concepts of the Michelson instrument.

Movements of a few fringes (in tenths to hundredths of a fringe, plus or minus in direction) were observed by one person who walked around with the apparatus while it was turned, starting and ending with cardinal compass points. The observer would speak out the readings at the ring of a new electric bell system, which automatically sounded when electrodes made contact at 16 equidistant intervals. An assistant then wrote down the readings on paper. That same walking-around observer also kept the interferometer turning by a gentle pull on an attached ribbon, though once it was set into motion, its mass and nearly frictionless rotation, floating in the tank of liquid mercury, would allow it to continue turning for an hour or more.

With an ether wind blowing steadily from one compass direction, the interferometer cross-arms would orient parallel and perpendicular with the ether wind two times each per full rotation, creating the *full-period effect*. The dual maxima and minima for each full rotation would then be divided in half, and overlapped, to create the more telling *half-period effect*. The interferometer could thus determine the maximum and minimum vectors of ether wind. With enough turns of the instrument, the *axis of net ether and Earth motion*, but not the absolute direction of motion along that axis, could be determined. Multiple sets of readings could then be taken at different times of day and year, organized to locate an ether-drift signal oriented to an identifiable set of sidereal-hour cosmic-galactic coordinates. Out of that procedure, with consideration of other astronomical findings, came more exacting determinations of cosmic ether wind and ether drift direction, or *azimuth*, in sidereal time.

The standard solar day, or civil clock day is almost exactly 24 hours, following the location of the Sun in the sky. The nighttime view of the heavens at any particular civil-clock hour changes over the course of the year, so what you see at midnight tonight is not what you would see at midnight 6 months previously, or hence. The sidereal day, by comparison, is fixed to the cosmic background of stars and constellations. A sidereal day lasts 23 hours, 56 minutes and 4 seconds, being 3 minutes and 56 seconds shorter than the civil clock or solar day. Each sidereal year is one day shorter than the solar year. This accounts for the slow changing progression of the star constellations and Milky Way Galaxy as seen overhead at night for the different months. At the start of any given month, a specific star or constellation will rise about 2 hours earlier than on the first day of the prior month. The sidereal cosmic clock is therefore a method of marking time by cosmic-celestial coordinates, and not by the "time of day" position of the Sun. Cosmic signals, such as ether drift, are anticipated to come from the background of cosmic-celestial space, and so the ether-drift data has to be organized in such a sidereal manner to be meaningful.

As noted above, the procedures of light-beam interferometry could identify the *axis of ether drift*, but *not the absolute direction of ether motion along that axis*. For that, one needs to logically compare the axis of ether-drift determinations against other astronomical observations related to Earth's seasonal position around the Sun, the Earth's movements relative to nearby stars, and other cosmic phenomena such as stellar aberration and parallax. So far, none of the ether-drift experiments had aimed at an independent "cosmic solution" without making *a priori* assumptions about the expected direction of ether wind or drift. By static ether expectations, for example, a 30 km/sec ether velocity was anticipated along Earth's orbital plane around the Sun, with a higher velocity of some unknown quantity, perhaps 200-300 km/sec from the solar system's motion through the Milky Way Galaxy. Miller would be the first to undertake extended ether measurements without significant reference to such assumptions.

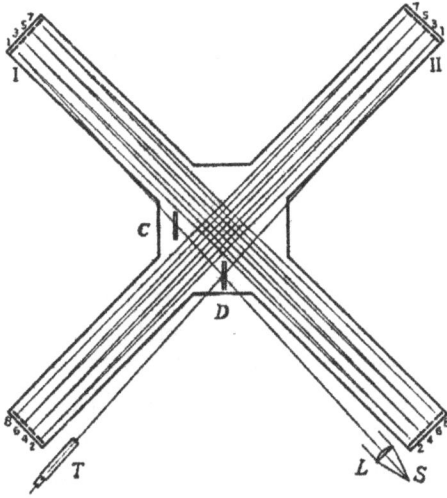

Figure 19. Light Paths of the Rebuilt 64-Meter Morley-Miller Steel Interferometer, with cross-arms of 4.3 meters, constructed in the same manner as Michelson-Morley, but of nearly three times the light path. Source (S) generated light which passed through lens (L) and was then split by half-silvered mirror (D). Light rays then reflected back and forth along two separate 4-meter paths (beams I and II), each with 8 mirrors (numbered 1-8) before finally being recombined by a half-silvered mirror (D) and reflected to small telescope eyepiece (T) where interference fringes were observed.

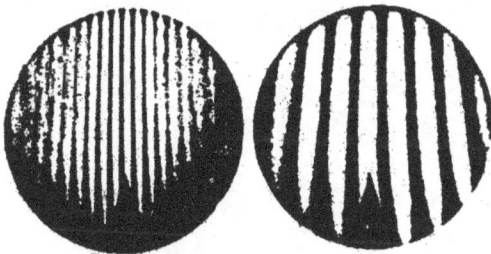

Figure 20. Light-Interference Fringes as seen in the steel interferometer telescope. Magnified by an eyepiece with precise graduated markings, one could observe the lateral movement or shifting of fringes as the instrument was rotated. Miller's larger apparatus used a 50x telescope, allowing individual fringes to be magnified more than what is shown above, sufficient for determinations down to tenths or hundredths of a single fringe.

Miller's 1921 Mount Wilson Results: *Breakthrough*

In 1919, the Einstein theory of relativity received its first strong scientific support in the solar eclipse measurements of Eddington and Cottingham. Their photographs of the eclipsed sun against the background of distant stars provided proof of the Sun's ability to bend starlight, changing the apparent positions of those stars. This was claimed to confirm Einstein's predictions, in yet another international media and scientific love affair. I cover this with full citations and analysis in the subsequent chapter, *Einstein Rising*.

The Eddington-Cottingham results and Einstein's claims brought Miller's attention back to the ether-drift question, if only because an entrained layer of cosmic ether around the Sun was an equally valid, but rarely mentioned explanation for the bent starlight identified in the eclipse photographs. With an invitation from astronomer George Hale, who founded the Mount Wilson Observatory in 1904, and with funding from the Carnegie Institute, Miller dusted off the 64-meter interferometer previously used in the Morley-Miller experiments. It was moved up the mountain and placed near to the domed observatory building at an altitude of 1750 meters (5740 feet). This was the highest altitude that anyone had so far operated a Michelson interferometer, and would ultimately be a major test for both ether drift and light speed variability, both of which severely challenged Einstein's theory of relativity.

Figure 21. The Mt. Wilson Observatory. Miller's "Ether Rocks" interferometer house was located ~100m from the telescope dome.

The exact location of Miller's new experimental building was close to Rock Crusher Knoll, so named for its role in the construction of the Mount Wilson Observatory, later dubbed "Ether Rocks". A concrete pad was poured at an unobstructed high spot, and a small house was erected in which the experiments would proceed. Four concrete piers were added to support the heavy steel interferometer and its tank of liquid mercury. The house itself had windows around the entire structure, from four to seven feet above the floor, at the level of the interferometer light beams. The house was also constructed with air-gaps in the rudimentary floor boards, wall panels, and eaves, to allow for natural air circulation and temperature stabilization. Removable canvas shields covered each window opening, to eliminate direct and diffuse sunlight, with black paper shields added over the canvas and open gaps to eliminate sun glare during daytime measuring. A canvas tent structure was later added to aid in this purpose. Precision thermometers were hung within the interior along with a barograph and thermograph, to record those variables and assist in thermal stabilization.

Figure 22. Miller's Interferometer House on Mt. Wilson (at arrow) perched optimally at nearly 6000 foot elevation, to catch the ether wind from all directions, and known in Miller's time as "Ether Rocks". Today, I am informed, there is no record at Mt. Wilson of Miller's extensive work, only a memorial plaque dedicated to Michelson and Einstein.

Miller's preliminary measurements in April 1921 included 350 turns of the interferometer, yielding once again a positive result for a real ether wind of around 10 km/sec. Before announcing these results, additional control procedures were implemented. A one inch thick layer of cork insulation was laid upon all the steel components of the interferometer cross-arms. An additional 273 turns of the interferometer were then made, also in April, described years later by Miller as "a periodic displacement of the fringes, as in the first observations, showing that radiant heat is not the cause of the observed effect". (Miller 1933a, 1933b)

Case W. R. U. Archive.

Figure 23. Miller's Rebuilt Light-Beam Interferometer, Mount Wilson, Ether Rocks 1921, measuring 4.3 meters across and 1.5 meters tall, was the largest and most sensitive of this type of apparatus ever constructed, with a mirror-reflected round-trip light-beam path of 64 meters. It was used in a definitive set of ether-drift experiments on Mt. Wilson from 1921 to 1926. Shown here fitted with 1 inch insulating cork panels covering the metal support structure, and glass coverings along the light-beam path. These insulation safeguards eliminated all measurable influences of ambient temperature differences upon the apparatus and the air within the light-beam path, but still allowed for detection of a real ether drift.

Miller returned back to Case School shortly after his April experiments, having recently been appointed to the National Academy of Sciences. Then on 25 May, he was visited by Albert Einstein. The two men apparently got along well, though Einstein could not then speak English. Miller was fluent in German, however, and so the two went on at some length discussing the various ether experiments *auf deutsch*.

According to Swenson, Miller felt "...that Einstein's visit was most pleasant and that the great theoretician was 'not at all insistent upon the theory of relativity'." (Swenson 1972, p.195) Given Einstein's prior ignoring of the positive results of Michelson-Morley and lack of curiosity regarding the Morley-Miller experiments, and his conduct as I shall describe later, I find Miller's statement that Einstein was "not insistent" about his theory of relativity to be rather naive, even though it appears the two men superficially got along well at the time.

After the very good preliminary results of April 1921, and Einstein's visit, and perhaps being stimulated by Einstein, Miller ventured to test materials with lower thermal and magnetic susceptibility than his steel interferometer. He removed the steel base platform and substituted another composed of concrete, reinforced with brass rods. All the connecting optical components were made of aluminum or brass, creating the first fully nonmagnetic interferometer, which also had a lower thermal expansion by comparison to the steel version. An

Figure 24. Miller's Concrete Interferometer, Mount Wilson, Ether Rocks 1921. Light beam paths are covered with glass.

additional 422 turns were made with this modified interferometer in December, with results "entirely consistent with the observations of April, 1921." [Dates of April 9-12 and Dec. 4-11, 1921.]

After his meeting and correspondence with Miller, and upon his return to Berlin, Einstein expressed alarm about Miller's April results, as recorded in a letter he wrote to Robert Millikan in June of 1921.

"I believe that I have really found the relationship between gravitation and electricity, assuming that the Miller experiments are based on a fundamental error. Otherwise, *the whole relativity theory collapses like a house of cards.*" (Albert Einstein, in a letter to Robert Millikan, June 1921. Reported in Clark 1971, p.328. Emphasis added.)

And Miller's early Mount Wilson results, summarized in the table below, were indeed nothing less than spectacular:

Table 2. Miller's 1921 Results at Mount Wilson

Dates	Number of Turns	Measured Ether Velocity
April 1921	350	~10 km/sec
April 1921 (insulated)	273	~10 km/sec
Dec. 1921	422	~10 km/sec
Totals & Averages	1045	~10 km/sec

It is interesting to note, just these three 1921 experimental runs by Miller, of 1045 turns of the interferometer in total, were nearly 30 times as many turns as the original 36 turns of the Michelson-Morley experiment. Miller also undertook additional improvements and control experiments during that time at Mount Wilson. As he wrote in 1933:

"Many variations of incidental conditions were tried at this epoch. Observations were made with the centering pin tight in its socket and then loose; with rotation of the interferometer clockwise and counterclockwise; with a rapid rotation of one turn in 40 seconds and a slow rotation of one turn in 85 seconds; with a heavy weight added first to the telescope arm of the main frame and then to the lamp arm; with the float extremely out of level because loaded first in one quadrant and then in the next quadrant; with the recording assistant walking round in different quadrants and standing in different portions of the house, near to and far from the apparatus. The results of the observations were not affected by any of these changes.

It was demonstrated that the use of the concrete base did not change the effect observed with the steel base either in magnitude or azimuth. The concrete base was less affected than the steel by change of dimensions due to changes of temperature; but this slight advantage was counterbalanced by the fact that it accommodated itself more slowly to a change of temperature. In spite of the fact that the concrete was considerably heavier than the steel parts which it displaced, it was much less rigid. Tests showed that a weight of 30 grams placed on the end of the arm of the [concrete] interferometer would produce a displacement of the fringes of one fringe width, while nearly ten times as much weight is required to produce the same effect with the steel base. The concrete base was abandoned and the original steel base has been used in all subsequent observations." (Miller 1933a, p.219)

Back to Cleveland, Miller's Additional Control Experiments at Case School 1922 - 1924

After completing his initial experiments at Mount Wilson, Miller had the large interferometer packed up and moved back to Case School in Cleveland for additional testing and improvement, in a new above-ground laboratory space in the Rockefeller Physics Laboratory building. As Chairman of Case Physics Department, Miller had personally planned that building, which was largely completed in 1904. He had travelled to Europe to purchase various new laboratory equipment used in research and lecture demonstrations.

Having results in hand from the first of what eventually became many Mount Wilson experiments, he gave his first lecture on the subject since the years working with Morley. He presented his "Ether-Drift Experiments at Mount Wilson Observatory" to a meeting of the American Physical Society in Toronto, in December of 1921. A published report also appeared in *Physical Review* in the following year, along with an April lecture to the National Academy of Sciences in Washington DC, plus a short note in *Science*.

For the next two years, Miller would occupy himself in a new program of testing and making refinements to the steel interferometer.

This included removal of the concrete, wood-rod and brass truss structure and restoring the original insulated steel beam system. He tried out new locations for the light source and viewing telescope, including removing them completely from the apparatus; the light beam was projected from a distance through a cleverly developed system of mirrors and prisms mounted above the centerpoint of its rotation. Those new methods were eventually abandoned, however, due to complications in keeping the system of complex optical components in proper alignment.

A still camera and motion picture method of recording the interference fringes was attempted, but the light of the interference fringes was too dim to register on available films of Miller's day. Electric arc, incandescent bulbs, mercury arc and acetylene lamps, and sunlight were all tried as light sources. In the end, Miller settled upon a small acetylene lamp, fixed to one of the interferometer cross-beams, close to its center of rotation. The fringe-viewing method evolved into use of separate objective and eyepiece lenses, without the telescope tube, mounted on another of the interferometer arms, allowing a magnification of 50 diameters.

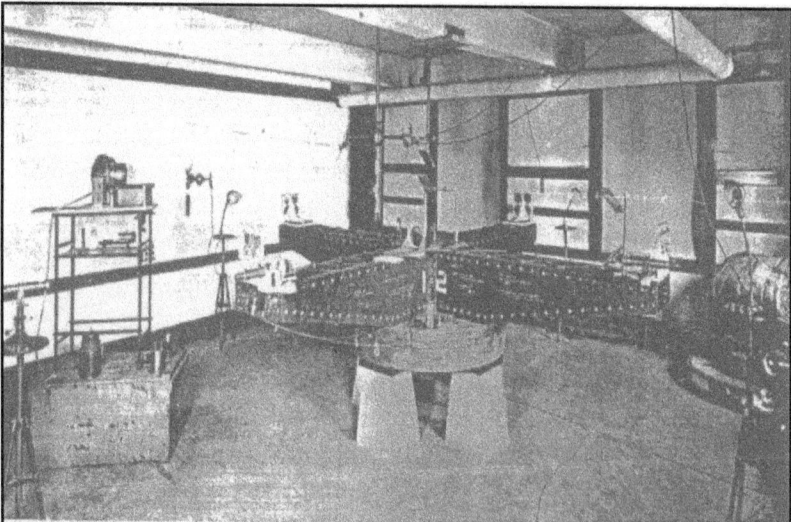

Figure 25. Miller's Steel Interferometer at Case School Physics Lab, 1922. The light source is mounted on a stand at the left side, with ceiling-suspended mirrors over central pivot, to reflect the light down into the interferometer.

The issue of possible thermal changes on the structure of the interferometer was always addressed with extensive control testing, from the very first days of the early Michelson interferometer, and Miller's work in this direction was intensive and precise. Parabolic electrical radiant heaters were used to heat up the room air, and also to focus heat upon one or another of the interferometer arms, whereby the exact effects upon the instrument and its fringe shifts, if any, could be identified. Such intense and focused air heating showed a slow but steady drifting of the interferometer fringes to one side, but no additional changes were observed from air heating during its rotation. Heating of the air in the light paths only resulted in fringe shifts when unequally distributed, as when one arm of the light path was covered with opaque cardboard while the other three arms were left uncovered. As Miller stated,

> "These experiments proved that under the conditions of actual observation, the periodic displacements could not possibly be produced by temperature effects." (Miller 1933a)

Miller nevertheless took exceptional precautions against temperature fluctuations when using the interferometer out in the field, such as covering the steel components with 1-inch cork insulation panels, and

Figure 26. Rockefeller Physics Building at Case School for Applied Science, completed in c.1904.

placing a glass enclosure over the light path. He also instituted a protocol for pre-turning the interferometer for about an hour before it was put into use, so as to equalize the temperature of all four of its arms to whatever small thermal variations might exist within the measuring room where it was stored and used. Other thermal control procedures were developed in the field, as described below.

These efforts by Miller to assess and remove thermal artifacts from his interferometer experiments were exceptional. Nevertheless, years after his death, Einstein's advocates would hammer away at Miller's positive results as being due to "thermal artifacts", ignoring everything he had done to eliminate such artifacts. Those attacks were initiated by one of Miller's former students, Robert Shankland, in cooperation with Einstein and several other of his followers, in an incompetent, biased and unethical affair detailed in Part II.

Miller's Return to Mount Wilson and Preliminary Experiment of 1924

With the control testing in Cleveland completed, the interferometer was transported back to Mount Wilson in July 1924. In August of that year, the interferometer house was moved to a new location a short distance from the cliff, to a nearby open grass-covered knoll. This was done to avoid possible vibrational disturbances and thermal effects from strong winds at the cliff side. New walls composed of insulating "beaverboard" (similar to modern mason or insulation board) were added to the house, as seen in Figures 27 and 28, replacing the older corrugated metal siding. A large canvas tent-cover was also added to provide additional thermal shielding of sunlight during daytime use of the interferometer, beyond the normal roofing material. Miller noted:

> "The interferometer... had the improved mirror mountings, protection from heat, improved light source, large viewing telescope and other refinements which had been developed in the laboratory tests at Cleveland in 1923 and 1924." (Miller 1933a, p.221)

With his revised and improved interferometer, and the modified new interferometer house, Miller began a series of fresh observations in September of 1924. Of these experiments, Miller wrote:

"[They were] undertaken in a wholly unprejudiced but very confident state of mind. The extended laboratory tests had involved every suggested source of instrumental and external disturbance and had proved that none of these was operative in the experiment. The method of observing was so developed that there was perfect confidence in the readings. It was felt that if any of the suspected disturbing causes had been responsible for the previously observed effects, now these were removed, the result would be a true null effect." (Miller 1933a, p.221)

Once again, the new results were positive, with no "null" results. Measurements were made over 4-6 September 1924, composed of eleven sets of readings with a total of 136 turns of the interferometer. Of his September 1924 results, Miller reported:

"...a positive periodic displacement of the interference fringes, as of an ether drift, of the same magnitude, about ten kilometers per second, as had been obtained in previous trials... The effects

Figure 27. Miller's Interferometer, Mount Wilson, 1924, Grass-Covered Knoll. Protective insulation was removed from the steel beams for this photograph. As before, windows were present all around the insulated shelter at the level of the interferometer light path, to allow for unimpeded flow of the ether-wind across the instrument at that level.

Figure 28. Miller's New Interferometer House, Mount Wilson, 1924, Grass-Covered Knoll. Above, the house with canvas-covered windows all around, and insulating beaverboard walls (wood fiber composite). Below, the same house is fitted with a tent cover over the roof and walls to further stabilize temperatures.

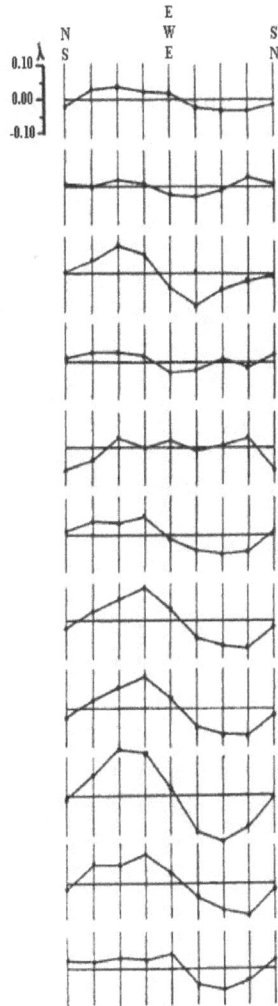

Figure 29. Miller's Calculations of Azimuth Versus Velocity of Ether-Drift, from 11 sets of observations made over 4-6 September 1924. Each line graph is composed from the overlap of two halves of the full 360° rotation of the interferometer, reducing the bimodal full period 360° effect to reveal the important half-period effect (see pages 55 and 101-103). Independent turns of the interferometer, when taken from different days and different times around the 24 hour clock, yielded these 11 different graphics, indicating a clear and repeating cosmic pattern, stronger at some times of the day than others. Because the measurements were taken over a few consecutive days, the ~4-minute daily sidereal variations did not significantly change. The data clearly show a maxima at RA ~6 hrs sidereal, more apparent in some of the graphs than others. The close alignments of maxima and minima in the curves proves the ether-drift measures are systematically induced, and not some kind of random artifact. These September 1924 efforts were considered preliminary so Miller did not elaborate upon them with greater detail, viewing his own later work over 1925-1926 as more centrally important. As shown later, a similar axis of ether drift was identified by Miller for the later Mount Wilson seasonal epochs. (Miller 1925, p.312)

were shown to be real and systematic, beyond any further question." (Miller 1933a, p.221)

Miller laid plans to make another series of interferometer tests during December 1924, but decided not to do so after considering how, in consultation with unnamed Mount Wilson astronomers, the "...tests during [that period] should give a resultant value for Earth motion near zero..." (Swenson 1970, p.65). Swenson's statement provided little detail about Miller's decision, which could have been motivated by a desire to not affirm a zero or low result, or simply because of expected heavy snow and bitter cold conditions on Mount Wilson in December. Whatever his reasons, from the current perspective of the solar system moving in spiral-form towards the center of the Milky Way Galaxy, a December-January ether-velocity minimum and June-July maximum are anticipated. This factor is discussed in Part III, in relation to how the Earth's spiral trajectory in space reveals variable orbital velocities quite different from Keplerian determinations.

Miller's Key Mount Wilson Experiments of 1925-1926

By the time Miller completed his preliminary interferometer tests in September 1924, he had accumulated data from over 3300 individual turns of the interferometer, *nearly 100 times the 36 interferometer turns of the original Michelson-Morley experiment.* His next task was to perform a full cycle of four series of readings with his improved interferometer, at intervals of about three months, as originally stated would be necessary by Michelson-Morley in 1887. In early 1925, using his newly refined interferometer and experimental house at Mount Wilson, Miller began that undertaking.

From 27 March through April 9th of 1925, Miller resumed measurements at Mount Wilson in the first of the four epochs of readings. At this time he also added sensitive thermometers in each corner of the interferometer house, to record differences in air temperature. From these thermometers, he noted:

> "...on numerous occasions the extreme variation of temperature was not more than 0.1°, and usually it was less than 0.4°; however, a variation of several degrees, while causing a constant drift of the fringe system, did not change the periodic displacements in azimuth or magnitude." (Miller 1925, p.312)

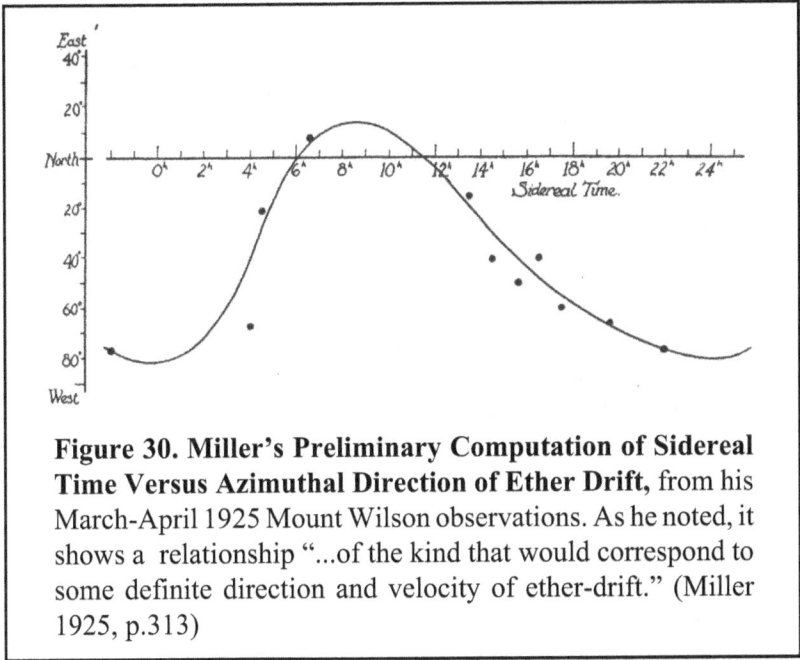

Figure 30. Miller's Preliminary Computation of Sidereal Time Versus Azimuthal Direction of Ether Drift, from his March-April 1925 Mount Wilson observations. As he noted, it shows a relationship "...of the kind that would correspond to some definite direction and velocity of ether-drift." (Miller 1925, p.313)

In other words, while the general slope of the center-line of the fringe-shifts sometimes slowly drifted, this did not produce the observed larger magnitudes and periodicity of fringe shifts, nor their azimuth, as the instrument was turned. In making the final computations for this particular April 1925 epoch, and all others, there was no complex mathematics involved in changing or reformulating the data:

> "There are no corrections of any kind to be applied to the observed values.... every reading of the drift made at Mount Wilson has been included at its full value. No observation has been omitted because it seemed to be poor, and no 'weights' have been applied to reduce the influence on the results..." (Miller 1925, p.313)

The results from this new round of March-April 1925 measurements, composed of 35 sets of around 100 turns of the interferometer, were also positive for a real ether drift. Shortly after completion of this round of measures, Miller was invited to give a lecture on his results to the National Academy of Sciences. In this lecture, entitled "Ether-Drift Experiments at Mount Wilson", he reviewed all the major ether-drift experiments since Michelson-Morley, including those of Morley-

The Dynamic Ether of Cosmic Space

Miller and his own experiments in Cleveland, as well as on Mount Wilson. His lecture was soon published in the *Proceedings of the National Academy of Sciences,* and contained his first preliminary evaluations of the sidereal day variations of ether drift. His presentation concluded with the following statement:

> "...there is a relative motion of the earth and the ether... of approximately nine kilometers per second, being about one-third of the orbital velocity of the earth. By comparison with the earlier Cleveland observations, this suggests *a partial drag of the ether by the earth, which decreases with altitude.* ... A complete calculation of the observations, now in progress, together with further experiments to be made in the immediate future, should give definite indications regarding the absolute motion of the solar system in space." (Miller 1925a, p.314. Emphasis added.)

Miller's 1925 lecture before the National Academy, with subsequent publication of his work in the *NAS Proceedings,* was followed quickly by an identically-titled paper in *Science* and *Nature* magazines, in June and July 1925. These papers summarized the whole range of his 1921-1925 ether experiments, including those of April and December 1921, September 1924 and March-April 1925, all at Mount Wilson.

With publication of his results in three top mainstream journals, news of Miller's work spread widely in scientific circles. He even received a $1000 prize from the Kansas City branch of the American Association for the Advancement of Science (AAAS). This brought him closer to the center stage of debates then raging over the Einstein theory of relativity, which was seriously threatened by the existence of light-speed variations due to an ether drift or wind. Local Cleveland and a few international newspapers also picked up on the story and reported it with the usual drama and errors, ever eager to cast the scientific debates as something of a boxing match. For example see the selected headlines in the adjacent text-box.

Word of Miller's March-April 1925 results reached Einstein, still in Berlin, who wrote worriedly to fellow physicist Edwin Slosson:

> "My opinion about Miller's experiments is the following. ... *Should the positive result be confirmed, then the special theory of relativity and with it the general theory of relativity, in its*

CLEVELANDER BOMBS EINSTEIN'S THEORY
Cleveland Plain Dealer, 29 April 1925

SCIENTISTS DEBATE
RECENT TESTS MADE OF EINSTEIN THEORY
One Indicates Light's Speed is Influence by Earth's Motion
Other Experiments Support Relativity
Speed of World Estimated at 125 Miles per Second
Washington Post, 29 April 1925

LOCAL MAN PROVES ETHER DRIFTS,
REFUTING EINSTEIN
Cleveland Times, 3 May 1925

current form, would be invalid. Experimentum summus judex.
Only the equivalence of inertia and gravitation would remain,
however, they would have to lead to a significantly different
theory." (Albert Einstein, letter to Edwin Slosson, 8 July 1925.
Hebrew University Archive, Jerusalem. Emphasis added.)

A similar statement was issued by Einstein in a letter to *Science*
magazine, on 17 July 1925, which was then circulated more widely. Six
months later, however, on 19 and 27 January 1926, Einstein retreated
from any sense of defeat, making off-the-cuff negative comments about
Miller's results. These appeared in the German and American press, the
latter of which prompted a terse response from Miller:

"If the results of Miller's experiments should indeed be confirmed, the relativity theory could not be upheld. Because in
that case, the experiments would question, that...the vacuum
speed of light was dependent upon direction. Thus the principle
of the constancy of the speed of light would have been proven
wrong, which constitutes one of the two cornerstones of the
theory. However, in my opinion there is hardly any probability
of Mr. Miller being right. ... Miller's results are actually hardly
credible because they claim the speed of light was strongly
dependent upon altitude above sea level. (Einstein, "My Theory
and Miller's Experiments", *Vossische Zeitung* 19 Jan. 1926)

The Dynamic Ether of Cosmic Space

"Speaking before scientists at the University of Berlin, Einstein said the ether drift experiments at Cleveland showed zero results... temperature differences have provided a source of error. 'The trouble with Prof. Einstein is that he knows nothing about my results.' Dr. Miller said. 'He has been saying for thirty years that the interferometer experiments in Cleveland showed negative results. We never said they gave negative results, and they did not in fact give negative results. He ought to give me credit for knowing that temperature differences would affect the results. He wrote to me in November suggesting this. I am not so simple as to make no allowance for temperature.'
("Goes to Disprove Einstein Theory", *Cleveland Plain Dealer*, 27 Jan. 1926)

While Einstein was to be appreciated for his honesty in admitting to the possible defeat of his relativity theory due to Miller's results, a troubling aspect of his German interview was Miller being referred to not as "Professor" or "Professor Dr." as was the usual usage in German society. Writing "Mr. Miller" without reference to his professional standing was a clear expression of disrespect, something which also appeared earlier in the attitudes of European highbrow "nobility" towards American science and culture in general, as mentioned in the last chapter. Perhaps this was the editorial license of the newspaper writer or publisher, but it wasn't the first nor the last time that Miller received such put-downs.

Unfazed, Miller continued his work, to eventually complete a total of four measuring epochs over 1925 and 1926, providing the confirmation Einstein was worried about. For him, even a 10 km/sec variation in light speed was significant enough to invalidate both his special and general theories of relativity. What were these new Miller results?

In the 30 April 1926 issue of *Science*, Miller gave a 10-page summary of his findings over 1925, writing:

"It is desirable to have observations equally distributed over the twenty-four hours of the day; since one set requires about fifteen minutes of time, ninety-six sets, properly distributed, will suffice. The making of such a series usually occupies a period of ten days. The observations are finally reduced to one group and the mean date is considered the date of the epoch. The observations made at Mount Wilson in 1925 correspond to

100

the three epochs, April 1, August 1, and September 15, and are more than twice as numerous as all the other ether-drift observations made since 1881. The total number of observations made at Cleveland represent about 1,000 turns of the interferometer, while all the observations made at Mount Wilson previous to 1925 correspond to 1,200 turns. The 1925 observations consist of 4,400 turns of the interferometer, in which over 100,000 readings were made." (Miller 1926, p.438)

This report in *Science* from April 1926 did not include the February 1926 measurements only recently completed, except as a notation at the end of the article indicating the February readings "are entirely consistent with the report here made." Miller also gave lectures to the National Academy of Sciences and to the American Physical Society, discussing his February 1926 epoch of experiments. A short note about it was also published in *Physical Review* in April.

Miller's published paper and lectures identified a general 10 km/sec ether-wind velocity, with some preliminary conclusions about the azimuthal direction of that velocity, close to the northern pole of the ecliptic. For reasons unknown, Miller was unable to carry out the experiments on Mount Wilson close to the December and June solstices, when for theoretical reasons the ether-drift velocities might be at a minimum and maximum – a consequence of which to his exact azimuthal determinations is explored in the next chapter.

One of Miller's Mount Wilson data sheets is reproduced in Figure 31, on the next page, presenting a continuous run of observations constituting 20 turns of the interferometer, on 23 September 1925, 3:09 to 3:17 AM at Mount Wilson. The sheet also reproduces the averages of these readings in the two inset graphic curves at the bottom. The upper of the two curves presents an average of each of the 16 data subdivisions for the *full-period effect,* which is bimodal. A full 360° rotation of the interferometer exposes its arms to two cycles of maximum and minimum ether-wind velocity. Miller cut that curve into two equal sections of 8 subdivisions, which were then overlapped and averaged again, yielding the lower graph, of a *half-period effect* of the ether drift. It is this half-period effect, in association with its sidereal time and orientations, from which the directional axis and velocity of ether drift can be determined. The upper left and right corners of the sheet record the start and end temperatures on the interior walls of the experimental house.

Table 3. Miller's 4-Epoch Ether Drift Determinations

Epoch Center Date	Veloc.Av.	RA South / North	Declination
1926 February 8th	9.3 km/sec	5h 14m / 17h14m	±69° 54′
1925 April 1st	10.1	4h 46m / 16h 46m	±70° 4′
1925 August 1st	11.2	4h 40m / 16h 40m	±72° 0′
1925 September 15	9.6	4h 54m / 16h 54m	±70° 11′
4-Epoch Average:	10.05 km/sec	4h 54m / 16h 54m	±70° 33′

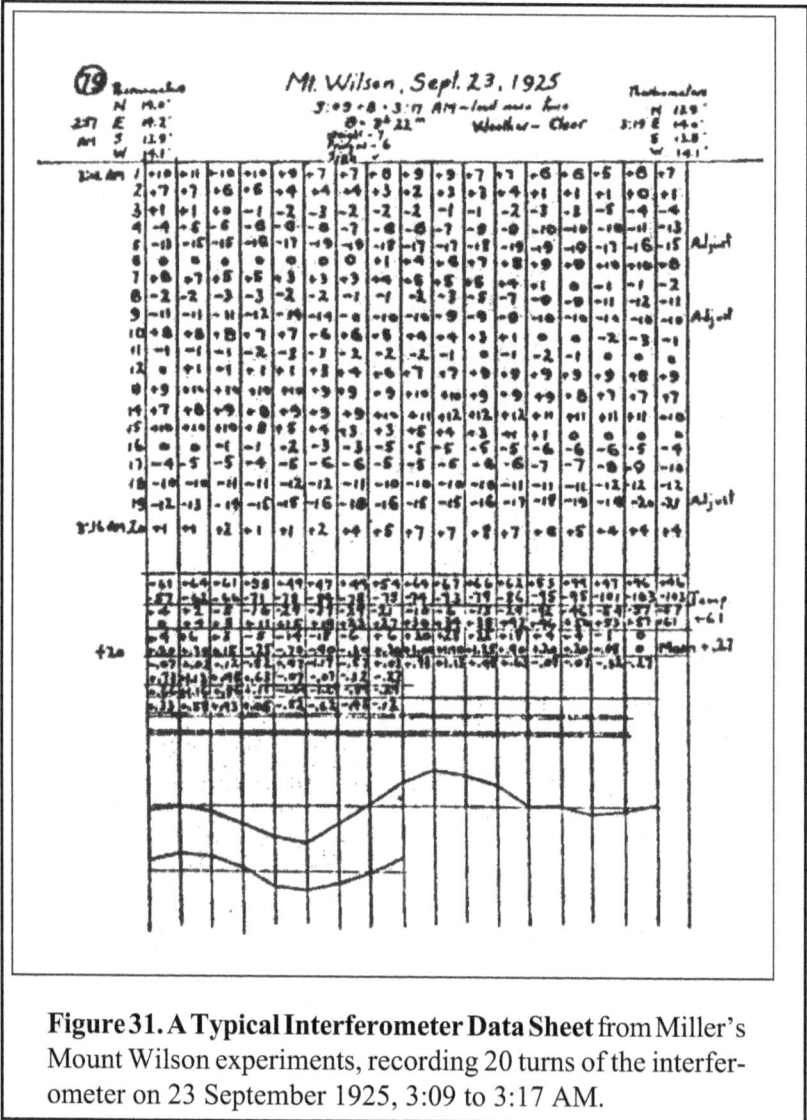

Figure 31. A Typical Interferometer Data Sheet from Miller's Mount Wilson experiments, recording 20 turns of the interferometer on 23 September 1925, 3:09 to 3:17 AM.

Miller's final calculations and most exacting results appeared in two additional publications, one being the Proceedings of a February 1927 Conference on the Michelson-Morley Experiments, held at the Mount Wilson Observatory, and published in *Astrophysical Journal* of 1928. A more elaborated 1933 paper by Miller also appeared in *Reviews of Modern Physics*, entitled "The Ether-Drift Experiment and the Determination of the Absolute Motion of the Earth." These two publications provide the best detail of Miller's findings and conclusions. From his 1928 conference paper in *Astrophysical Journal*, Miller summarized:

"A complete calculation has now been made, including the observations of both 1925 and 1926, which leads to the following conclusion: The ether-drift experiments at Mount Wilson show, first, that there is a systematic displacement of the interference fringes of the interferometer corresponding to a constant relative motion of the earth and the ether at this observatory of 10 km/sec, with a probable error of 0.5 km/ sec... toward an apex in the constellation Draco, near the pole of the ecliptic, which has a right ascension of 255° (17 hours) and a declination of +68°..." (Miller 1928, p.361)

In this report, Miller also mentioned his determinations were "just such as would be produced by a constant motion of the solar system in space, with a velocity of 200 km/sec..." This was an up-calculated estimate based upon the theory of a slowed ether velocity close to the Earth, due to entrainment effects. He never measured such a higher speed, it should be noted. This issue is addressed later on.

His 1933 paper presented slight revisions in his final calculations, for each of the four separate epochs of measuring. Each epoch covered a general 10-day period, identified by the center-date of that epoch. Table 3 gives a summary. (Miller 1933a, Tables III & IV, p.230 & 233). He also provided the azimuth in right ascension for both south and north poles of the ether-drift axis, also in Table 3. Declinations are identified as plus "+" for north, and minus "–" for south.

Miller's aggregated velocity and azimuth determinations from each of the four epochs of measuring at Mount Wilson, as calculated in his 1933 papers, are presented in Figure 32. That figure shows the average tallies from many interferometer turns of each separate 10-day epoch centered on the given date, along with an extracted curve of harmonic averages. Data is presented for each of the azimuthal sidereal-hours of measuring. The figure shows the variation of azimuth and velocity readings according to a given epoch. In both 1928 and 1933, Miller presented additional critical details on the periodicity of data for the four seasonal epochs organized by both cosmic sidereal hour and conventional civil clock hour. This is shown in Figure 33, on page 106.

Figure 33 is highly significant, exposing a hidden pattern in the data that is revealed *only when it is organized by sidereal hour cosmic coordinates*, as seen in the upper graph. The plot of his data organized by civil time, seen in the lower graph in Figure 33, is chaotic, without any discernible pattern to the ether-drift measures. The upper sidereal graph of Figure 33 shows a clear pattern, where the direction of ether drift swings back and forth by about 60°, as the Earth rotates, over the course of the day.

Figure 34, on page 107, presents additional detail incorporating the upper graphic of Figure 33, with both parts organized on the sidereal-day baseline. As Miller explained:

> "In accordance with the simple theory, the direction of the cosmic motion should swing back and forth across the north and south line once in each sidereal day, because of the rotation of the earth on its axis. When the observed azimuth of motion is charted, the resulting curve of directions crosses its own axis twice in each day."

Table 4 provides the actual *average daily azimuthal shift*, presented graphically in Figure 33, and on the page thereafter. The average epoch seasonal variation in the shifting azimuth in Miller's data, summarized in Table 4 and Figures 33 and 34, reveals a hidden aspect, of an *average annual vector* of 23.75°. This is very close to the Earth's axial tilt of 23.5°, suggesting the ether drift's specific motions play a role in creating the Earth's axial tilt, or is responding to it.

This reporting of an unanticipated Earth-axis correlation adds to the veracity of Miller's overall experimental approach and data, but may require a three-dimensional model and exercise for the general reader to comprehend the basics. A simple exercise can help, using an Earth globe set on a table, as given in Appendix 1.

Miller's actual data reveal both an annual overall ether drift with a 4-epoch average of ~10.05 km/sec velocity, and an interesting, significant 23.75° variation in the epoch-average daily swing of ether wind. He also identified a central axis of ether drift or ether wind, oriented

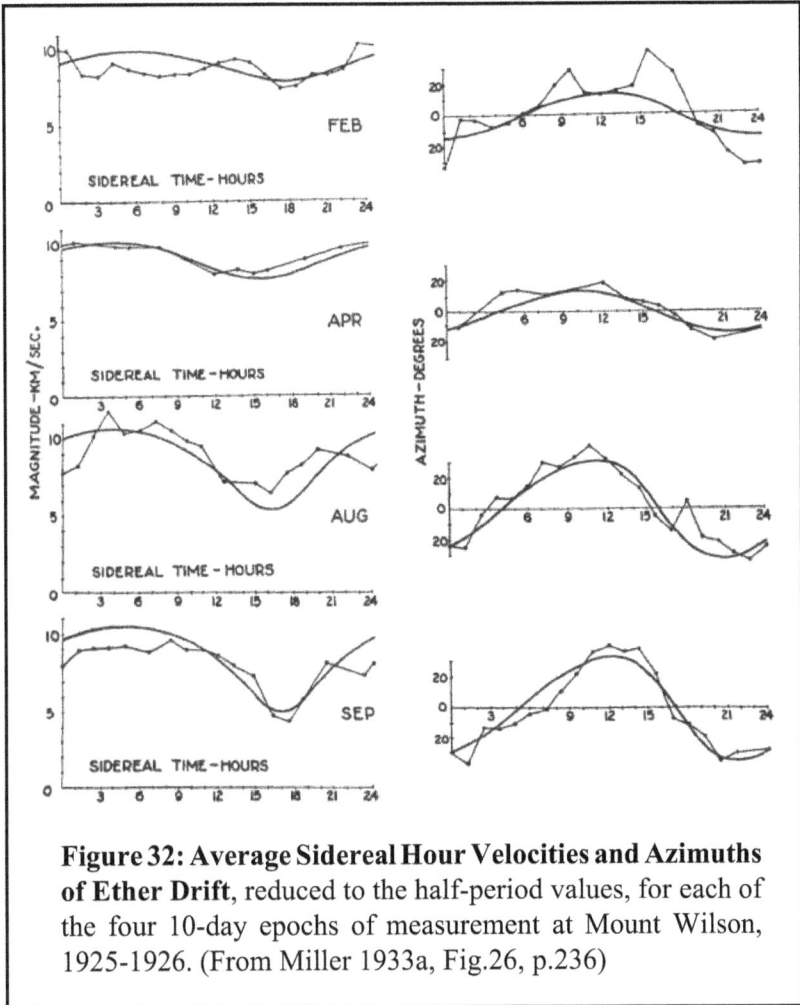

Figure 32: Average Sidereal Hour Velocities and Azimuths of Ether Drift, reduced to the half-period values, for each of the four 10-day epochs of measurement at Mount Wilson, 1925-1926. (From Miller 1933a, Fig.26, p.236)

Table 4. The Epoch-Average *Daily* Swing of Ether Wind

Epoch Center Date	Velocity Av.	Azimuth Shift, Av.
1926 February 8th	9.3 km/sec	−10° West of North
1925 April 1st	10.1 km/sec	+40° East of North
1925 August 1st	11.2 km/sec	+10° East of North
1925 September 15	9.6 km/sec	+55° East of North
Epoch Averages:	10.05 km/sec	23.75° East of North

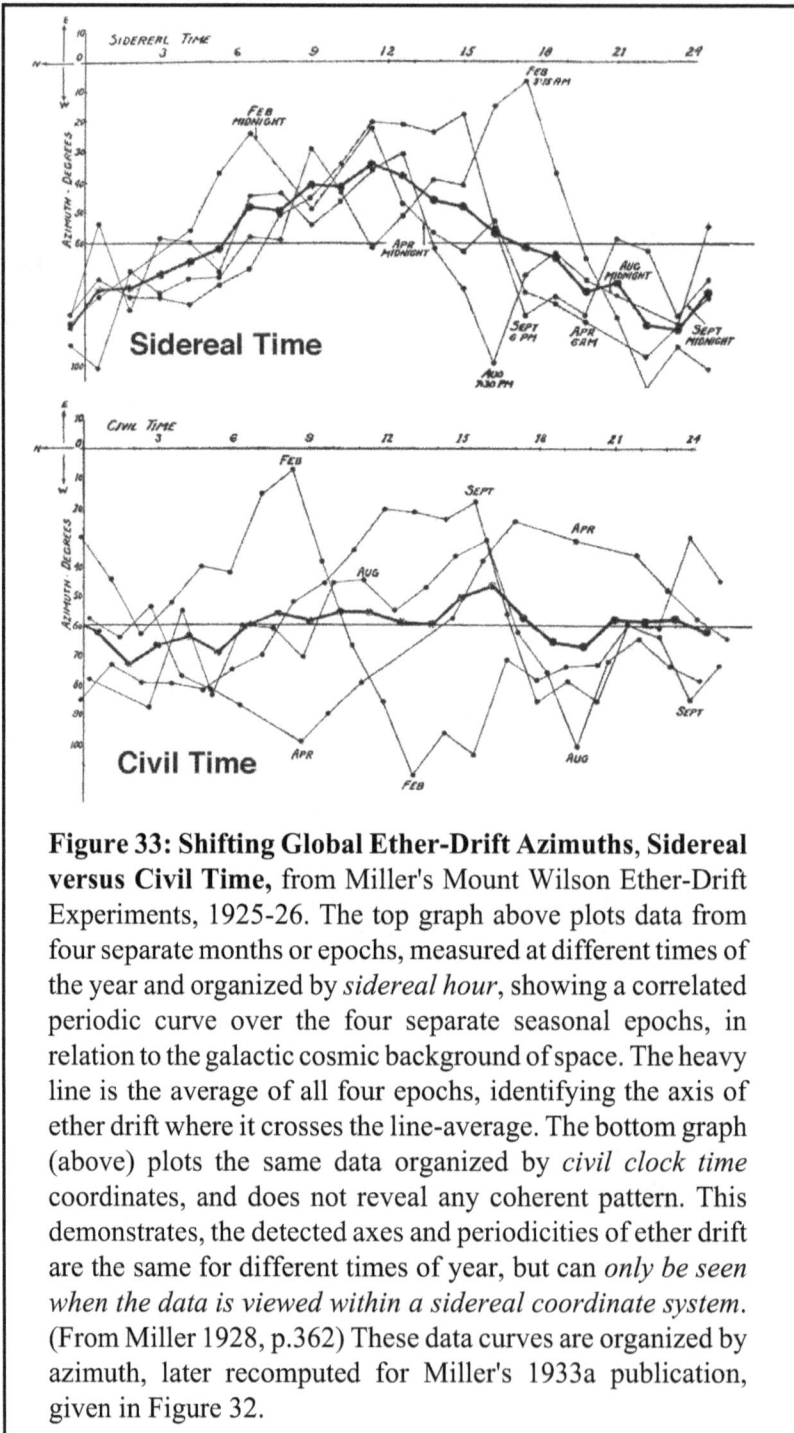

Figure 33: Shifting Global Ether-Drift Azimuths, Sidereal versus Civil Time, from Miller's Mount Wilson Ether-Drift Experiments, 1925-26. The top graph above plots data from four separate months or epochs, measured at different times of the year and organized by *sidereal hour*, showing a correlated periodic curve over the four separate seasonal epochs, in relation to the galactic cosmic background of space. The heavy line is the average of all four epochs, identifying the axis of ether drift where it crosses the line-average. The bottom graph (above) plots the same data organized by *civil clock time* coordinates, and does not reveal any coherent pattern. This demonstrates, the detected axes and periodicities of ether drift are the same for different times of year, but can *only be seen when the data is viewed within a sidereal coordinate system.* (From Miller 1928, p.362) These data curves are organized by azimuth, later recomputed for Miller's 1933a publication, given in Figure 32.

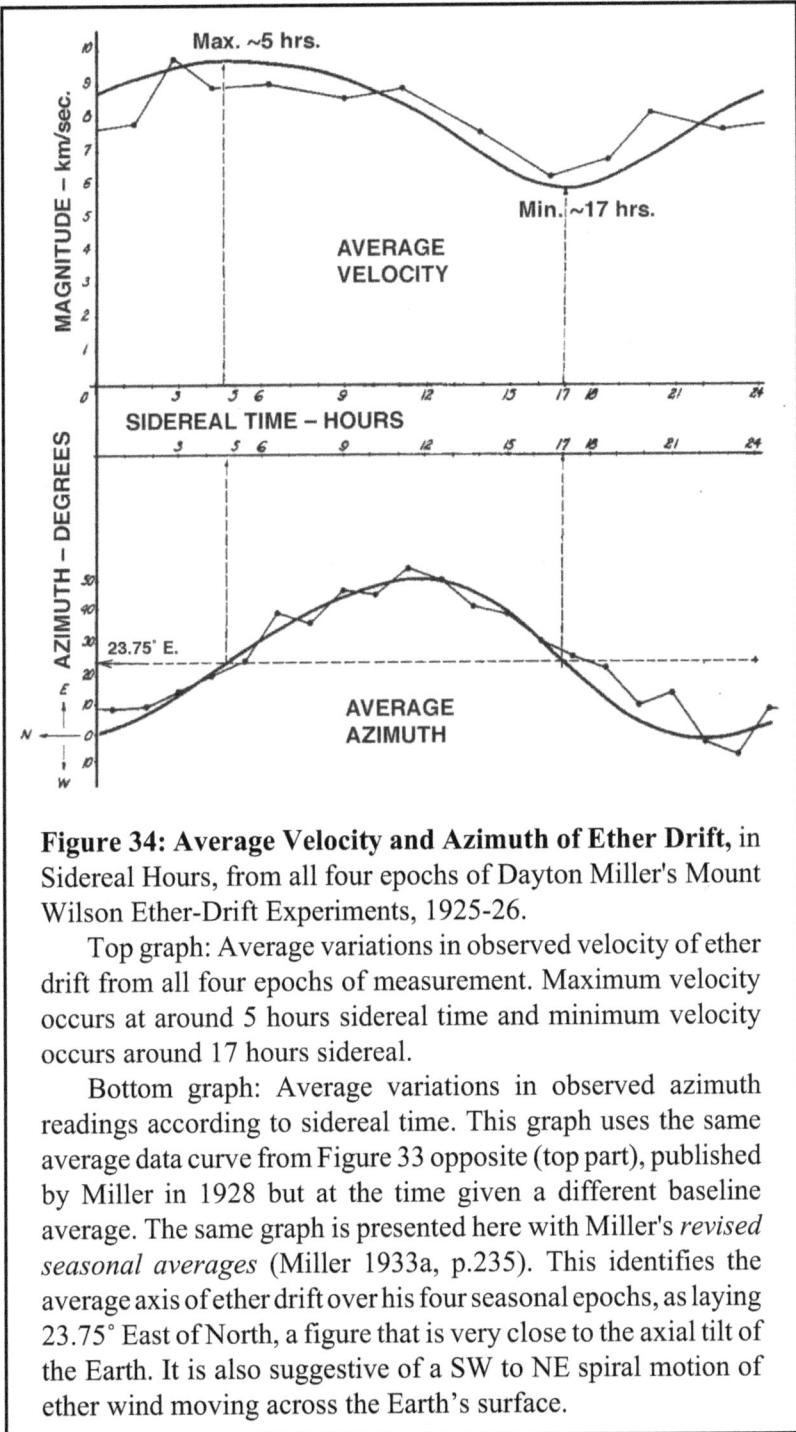

Figure 34: Average Velocity and Azimuth of Ether Drift, in Sidereal Hours, from all four epochs of Dayton Miller's Mount Wilson Ether-Drift Experiments, 1925-26.

Top graph: Average variations in observed velocity of ether drift from all four epochs of measurement. Maximum velocity occurs at around 5 hours sidereal time and minimum velocity occurs around 17 hours sidereal.

Bottom graph: Average variations in observed azimuth readings according to sidereal time. This graph uses the same average data curve from Figure 33 opposite (top part), published by Miller in 1928 but at the time given a different baseline average. The same graph is presented here with Miller's *revised seasonal averages* (Miller 1933a, p.235). This identifies the average axis of ether drift over his four seasonal epochs, as laying 23.75° East of North, a figure that is very close to the axial tilt of the Earth. It is also suggestive of a SW to NE spiral motion of ether wind moving across the Earth's surface.

close to the central axis of the solar-system ecliptic. The northern end of Miller's axis of ether drift is located only about 6° of angular distance from the northern ecliptic pole.

Miller also presented the margins of error in his data, in his 1933 paper, as follows: Ether velocity error margin of ±0.33 km/sec; right ascension (RA) error margin ±2.5° (each hour of RA is 15°); Declination error margin ±0.5°. Those error margins are small fractions of Miller's reported data.

However, a new problem came up in Miller's theoretical arguments. Five years earlier, by the time of the February 1927 Conference, Miller's determinations of the axis of drift were reasonably complete, based upon his full set of data for all four experimental epochs made at Mount Wilson. From this he calculated the net direction of Earth's motion, towards a northerly vector. Starting around 1932, however, he changed his mind about the direction of Earth's motion along his identified axis. As presented in his 1933 papers, *he shifted the azimuth of Earth's motion, from a northerly to a southerly direction.* While his velocity of ether-drift remained the same, at ~10 km/sec, and the axis of ether-drift remained nearly the same, the newly determined direction of drift *along that axis* represented a major change in his overall theory. Here are the changing coordinates, summarized:

Table 5. Miller's 1928 versus 1933 Determinations

Year	RA	Dec
1928 north axis of motion	17hrs	+68°
1933 south axis of motion	4hrs 54m	−70° 33′
1933 north antipode	16hrs 54m	+70° 33′

As he stated several times:

> *"The interferometer observations determine the line in which the [solar system] motion takes place but do not distinguish between the plus and minus directions of the motion in this line;* the choice between the plus sign, northward, and the minus sign, southward, must be determined from the consistency of the result when this motion is combined with the known orbital motion of the earth." (Miller 1933a, p225. Emphasis added)

Miller's 1933 change of net direction along his computed axis of ether drift placed the Earth's direction of motion towards the southern sky constellation of Dorado, near to the *south pole of the ecliptic* and close to the Great Magellanic Cloud of stars. That point is exactly on the opposite side of the celestial sphere from Miller's original northerly determination. However, there are solid reasons based upon newer experiments and astronomical determinations to reject Miller's theoretical abandonment of the original northerly solution, as I will detail in the next chapter.

In Figure 35, I plot Miller's 1928 calculation of the northerly axis of ether drift, along with other related cosmic factors, such as the northern pole of the solar system ecliptic, the modern solar apex of motion towards the star Vega, the Sun's rotational axis northern pole, and the sidereal vector of the central meridian of the Milky Way Galaxy. When viewed against modern determinations for cosmic motions of the Sun and Earth, Miller's original northern axis of ether drift is quite logical and agreeable. Additional evidence will gradually be presented to support my view. If you hold Figure 35 up to the sky, centered over the North Star Polaris, and at the correct time when the Milky Way Galaxy is directly off to your right side, the Earth will then rotate under this Figure 35 diagram in a clockwise manner, as if you leaned your head from left to right while holding the figure steady. In that configuration, the entire solar system moves off to a location towards your right, within this cluster of cosmic coordinates.

I remain in agreement with Miller on nearly all other factors, save for his departure from his own original northerly direction of motion. He stated, correctly, that the direction of Earth motion along the determined axis of ether drift, "...must be determined from the consistency of the result when...combined with the known orbital motion of the earth." Whether one agrees with Miller's 1933 southerly direction, or my return to his published 1928 northerly solution on this matter, *he clearly and consistently measured a cosmic ether drift of around 11.3 km/sec of maximum velocity, and 9.2 km/sec at minimum.* And his determination was along an axis line which was very close to both the northern and southern ecliptic poles, which identifies the averaged axis of the orbital plane around which all the planets in the solar system are revolving.

Miller's work produced a systematic finding of a real ether drift and light velocity variation, far more ambitious in scope than any ether-drift researcher before or since. Over many years he perfected and fine-tuned his interferometer, using it at different locations and at different times of year, but most decidedly at Mount Wilson. And as Einstein confessed several times, both privately and publicly, if correct, it was a death blow to both his special and general theories of relativity.

Miller's findings further exposed several new considerations. As noted in Figure 34, by Miller's determination presented in the lower graphic, the *average daily* azimuth of ether drift is shifted off to the East of North, by 23.75°. This is an average deviation of ether drift for all four measuring epochs combined, according to Earth's geographical surface coordinates, and is a different feature of his data quite apart from the longer-term full epoch averages in velocity or azimuth. The 23.75° daily average shift is very close to the Earth's axial tilt of 23.5°, as previously mentioned. Is this purely coincidental?

As shown in the upper graphic of Figure 33, when Miller's data from the seasonal epochs of measuring were averaged and plotted out according to sidereal day coordinates, the maximum ether-drift velocity was determined to be approximately 10 km/sec at 5 hrs sidereal, with a minimum velocity of approximately 6 km/sec at 17 hrs sidereal. This is partly what led Miller to think the Earth was pushing south through a static but entrainable ether, creating an ether wind maximum from that direction, as the Earth plunged southwards. However, these experimental observations can also be the consequence of factors briefly covered in my *Introduction* and fleshed out more thoroughly in Part III. This would be the mechanism of an *active, dynamic and material ether wind of a fluid and viscous, vortexing or spiral-motion nature. By this theory, the ether itself moves northwards, with a slight west-to-east component, and carries the Earth, Sun and planets along with it in the same direction. The ether acts as a gravitational pushing or "floating" force from the ~5 hrs towards the 17 hrs sidereal direction. This motional force also affects the Sun, and all the different planets, and by subordinate spiral-vortices, the orbits of various planetary moons.*

This latter interpretation of a dynamic ether is hotly controversial, as it is not merely congruent with the existence of cosmic ether, and with a northerly axis of motional wind or drift, but *also presumes that the ether itself moves in a lawful manner, with sufficient substance to push planetary and stellar matter around, and therefore is the prime mover, with gravitational properties.*

The 5 hrs versus 17 hrs sidereal positions, for maximum and minimum ether velocity, also form a plane that bisects the center core of the Milky Way galaxy. *The ether wind moves in a spiraling motion from south to north, with a west to east component, pushing the Earth from the 5 hr sidereal southerly axis towards the 17 hr sidereal*

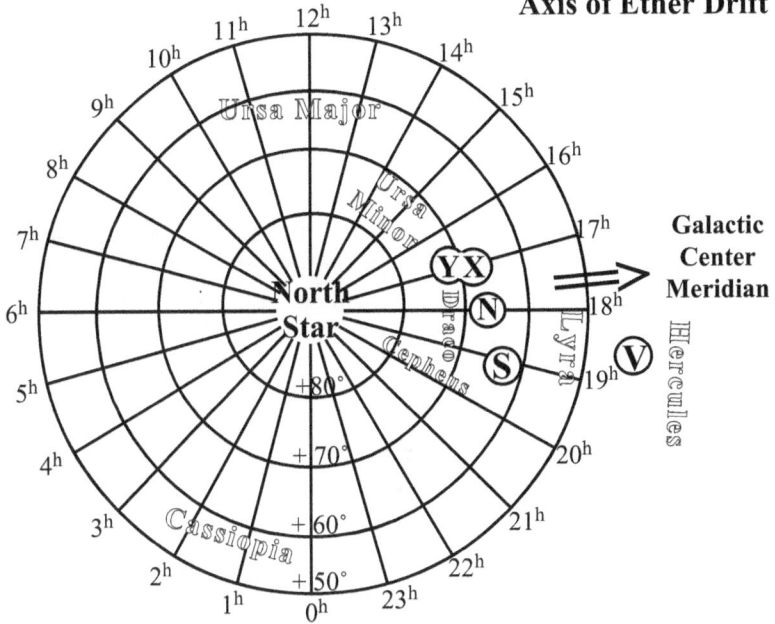

Figure 35. Miller's Original Determination for a Northerly Axis of Ether Drift

Looking North and Up to Polaris (Pole Star)
from Earth's Northern Hemisphere

Miller's Northern Axes of Ether Drift
1928: RA 17h Dec +68° North = (X)
1933: RA 16h 54m Dec +70.5° North = (Y)

Northern Pole of the Solar System Ecliptic = (N)
RA 18h 00m Dec +66.5° North

Modern Solar Apex Motion is Towards Vega = (V)
RA 18h 36m Dec +39° North

Sun's Rotational Axis Northern Pole = (S)
RA 19h 4m Dec +64° North

Arrow Indicates Meridian of the Center ⟹
of the Milky Way Galaxy RA ~17h 45m

northerly direction. This is so, even while the ether wind is strongest at the ~5-6 hr sidereal vector, and weakest at the 17hr vector, as generally depicted in my Figure 36. Additional figures will be given in the next chapters, drawing upon other lines of evidence, to document this motion more clearly.

Another factor to be clarified is the previously-mentioned issue of how Miller's ether-velocity determinations, when separated out into the four different seasonal measuring epochs at Mount Wilson, very much support the spiral-form motion of the Earth through the cosmos. Further, a good argument will be made in the next chapter, and also in Part III, that Miller's inability to make ether-drift measurements in the critical December-January and June-July periods – the times of postu-

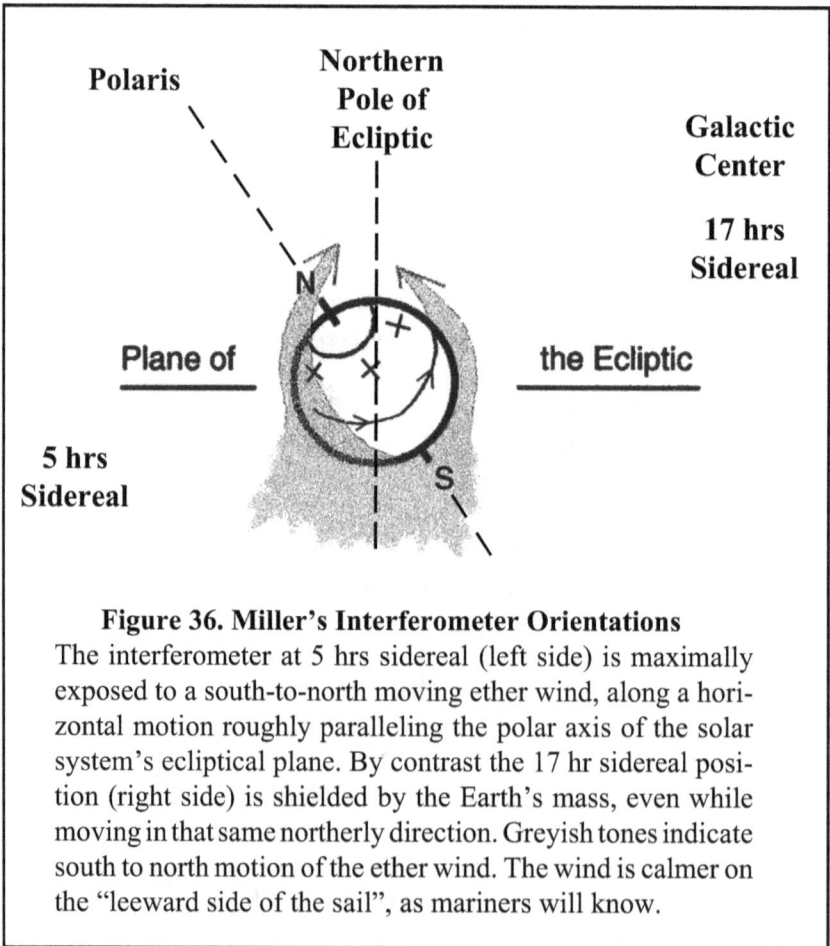

Figure 36. Miller's Interferometer Orientations
The interferometer at 5 hrs sidereal (left side) is maximally exposed to a south-to-north moving ether wind, along a horizontal motion roughly paralleling the polar axis of the solar system's ecliptical plane. By contrast the 17 hr sidereal position (right side) is shielded by the Earth's mass, even while moving in that same northerly direction. Greyish tones indicate south to north motion of the ether wind. The wind is calmer on the "leeward side of the sail", as mariners will know.

lated ether-wind minimum and maximum velocities, respectively – resulted in a final azimuthal determination that was pulled away from the conventional astronomical determination of the "Sun's Way" in the cosmos (towards Vega) than otherwise might have been the case.

My questions and personal views on these issues notwithstanding, Miller's findings overall are exceptional, and sufficiently profound for the modern astronomer and theorist to consider. More than anyone else, he proved that the ether drift and variations in light-speed not only exist, but can be objectively detected. That such motions through the cosmos could be identified by an experiment run on the surface of the Earth in a blind room, without reference to any celestial coordinates at the time of measurement, is quite remarkable by itself. The skeptic might claim these patterns are all just "one big coincidence". But read on, many more of such "coincidences" will be presented in the chapters to come, including from additional ether-drift experiments. Additional "coincidental" cosmological vectors will be added in Part III, from contemporary astronomy and other disciplines, including chemistry and biology.

Which Way Ether Drifting?
Miller's Mis-Step, and Last Years

Ether Confirmed,
Ether Velocity Confirmed,
Axis of Ether Drift Determined, but...
In Which Direction Does Earth Move Along That Axis?

The preceding chapter reviewed Dayton Miller's exceptional work on the ether-drift question, his confirmation of both ether and ether drift or ether wind, with a set of velocities and azimuths determined at four different seasonal epochs atop Mount Wilson. He also plotted the axis of ether drift, finding it close to the axis of the solar system's ecliptic plane. However, by the time of his comprehensive 1933 paper on the subject, he had reversed his long-standing view on the northerly direction of Earth's motion along that axis, and instead argued for a southerly direction. I object to his change in direction of motion along the ether-drift axis, but not to the axis itself. My claim requires a clear discussion of the evidence, both pro and contra.

As noted by Miller in the preceding chapter, the interferometer could determine the *axis of ether drift* using the Michelson interferometer, but *not the direction of ether motion along that axis.* For that determination, one needs to logically compare the axis of ether-drift findings against other astronomical observations related to the Earth's velocity and movements relative to nearby stars, and other cosmic coordinates and determinations.

After Mount Wilson

By 1926, after Miller concluded his four major seasonal epochs of ether experiments, he began to reveal his thinking as to the larger issues of the Earth's net velocity through the universe, as well as about an Earth pushing through a dragged ether. His writings on these matters reveal a level of comprehension and skill certainly equal to that of

115

Michelson, and far superior to that of Lorentz, Einstein, and other notable theorists of the time, who clung to popular falsehoods about the results of Miller's experiments.

According to astronomical knowledge of the mid-1920s, the Earth-Sun and solar system's net velocity and direction of motion through the cosmos were understood to be towards the northern apex constellation Hercules at a velocity of 19 km/sec. This determination was made by astronomers using telescopic methods with precise determinations of stellar aberration and subtle motions of nearby stars. Miller described these findings by others in a *Science* article of 30 April 1926, "Significance of the Ether-Drift Experiments of 1925 at Mount Wilson". Notable were the studies by Ralph Wilson of the Dudley Observatory on the proper motions of stars, and by Campbell and Moore of the Lick Observatory. Both studies determined the apex of the Sun's direction was towards the constellation Hercules, also termed the "Sun's Way" within the galaxy, at Right Ascension (RA) 18 hrs, Declination (Dec) +30° North, at 19 km/sec.

At the time of these studies, they constituted the best estimates for a net motion of the solar system, towards a northern apex that was agreeable with Miller's own northerly determinations. In his 1926 *Science* article (p.442), Miller expressed the view of Earth racing towards Draco near the northern ecliptic apex, at RA 17.5 hrs, Dec +65° north. While his right ascension and declination values were in keeping with his prior estimations, his reported velocity was significantly higher than previously, of ~300 km/sec. However, that was only a *theoretical value* (see Table 7). The *measured* ether-drift velocity was still only around 10 km/sec, which he presumed was the residual from a 20- to 30-fold reduction of a higher velocity outside of the Earth's atmosphere in open space. The measured velocity continued to strongly suggested a dragged ether with some kind of material properties, interacting with Earth's atmosphere and crustal material to slow down the ether drift. However, Miller was concerned that the different altitudes of actual ether-drift measures should have yielded a greater variance than the observed range of ~3 to 11 km/sec velocity.

Around late April or early May of 1926, Einstein wrote to Miller, following up on their earlier meeting in Cleveland. A friendly correspondence developed between the two, as Miller's findings continued to attract the attention of Einstein and his followers. In June of 1926, Miller travelled to London by ship and gave an invited lecture on his work to the Royal Society. He also may have visited Einstein in Berlin at that time, as expressed in his letter to Einstein of May 20th, but this meeting is not certain to have occurred, from any sources I have consulted.

In July of 1926, in an interview article for *Modern Science* magazine ("Measurement of Ether-Drift"), Miller expressed hesitations about the reality of an altitude-dependency for ether velocity, and hence, about the ether-drag theory itself. This puzzle would occupy Miller over the next several years, as did refinements in the precise calculation of ether drift azimuthal direction. Later chapters, on the more recent ether-drift experiments, will present evidence to document a strong altitude/ether-velocity correlation, further indicating that Miller's original ideas about ether entrainment were correct.

Miller's next major presentation of his work took place a year later, in early February 1927, at the previously mentioned Mount Wilson *Conference on the Michelson-Morley Experiment*. Other speakers at this historic event also presented lectures, such Michelson and Lorentz (summarizing their respective views), Roy Kennedy (who did his own ether experiments, to be discussed), E.R. Hedrick (presenting a math analysis), and Paul Epstein (more maths), with general discussion periods after each major paper presented. Miller's work was of primary concern to all. Nearly everyone present, except for Michelson, appeared to seek out reasons to dismiss Miller's work without ever addressing his actual experiments and results.

The *Proceedings* of this Conference, with all the major papers, were published in *Astrophysical Journal* of December 1928, almost two years later. In those Proceedings, Miller gave a brief historical overview of the various ether-drift experiments, including a discussion of which theories of ether drift could not be validated, and those that could. He summarized his tests for magnetostriction, radiant heat, gravitational deformations, and the various tests for the FitzGerald-Lorentz hypothesis of matter-contraction, indicating how none of those problems could produce the observed results, and how they were addressed and ruled out of his experimental results. Meanwhile, an ether-drift velocity of around 10 km/sec was rather constantly detected in those efforts. He stated:

"Throughout all these observations, extending over a period of years... there has persisted a constant and consistent small effect which has not been explained.." (Miller 1928, p.357)

The Dynamic Ether of Cosmic Space

In the decades prior to Miller's 1925 work at Mount Wilson, everyone including Miller was anticipating and searching for an interferometer signal indicating the 30 km/sec orbital velocity of Earth moving around the Sun, plus the velocity of the solar system of ~200 km/sec, headed towards the constellation Hercules. However, such velocities were never verified in his experimental work. In the *Proceedings*, Miller stated his results could not confirm the Earth's motion towards Hercules (close to Vega), but nevertheless at that time still considered the direction of ether drift to be towards a location closer to the northern pole of the ecliptic. These locations are reasonably close to each other. While the concept of an entrained ether layer close to the Earth gave Miller an understanding for the lower ~10 km/sec ether velocity, by comparison to the larger theoretically anticipated values, he remained somewhat unsettled about the precise azimuthal direction of ether drift, and continued to work on the question.

Miller Retires the Interferometer at Cleveland

By 1929, Miller moved the large interferometer from Mount Wilson back to his Cleveland laboratory at Case School, where further experimentation continued on a smaller scale. This subsequent experimental work could only be a shadow of his prior undertakings on Mount Wilson, but the additional Cleveland measurements agreed with his Mount Wilson findings. He gave a lecture on this to the National Academy of Sciences at Princeton University, the soon-to-be home of Einstein, with another lecture for the National Academy in Washington DC. A summary article was also published in *Science*, and republished the following year in the *Journal of the Royal Astronomical Society of Canada*. In that article he briefly restated his results from Mount Wilson, but primarily discussed how his observations and conclusions about the net motion of the Earth and solar system through space were in close agreement with other astronomical findings. These included additional findings beyond the studies cited by him in 1926, as previously noted. He wrote:

"Meridian circle observations of star places made by direct and reflected rays show peculiarities which are explained by assuming a motion of the solar system towards the sidereal time meridian of about seventeen hours. This effect has been found by the independent observations of Courviosier (Berlin) and by

118

Esclangon (director of the Paris Observatory). Esclangon finds evidence of similar motions in the observations of lunar occultations of stars, and still more convincingly in elaborate studies of earth tides (deformations of the earth's crust) and of ocean tides. In the latter work, he considered 166,500 observations, extending over a period of nineteen years. The well-known study of radial and proper motions of stars in our galaxy by Campbell (Lick Observatory) and by Wilson (Dudley Observatory) give a motion of the solar system towards the constellation of Hercules [close to Dorado], of eighteen hours right ascension. Stromberg (Mount Wilson Observatory) from an investigation of clusters and nebulae, finds evidence of a motion of the solar system with its apex at twenty-one hours right ascension and declination of 56 degrees north. By a study of the reflection of light, Esclangon finds strong evidence for what he calls an 'optical dissymmetry of space' with its axis of symmetry in the meridian of twenty hours sidereal time. This effect would be explained by an ether drift, and the results are in striking agreement with the ether-drift observations here reported. Many recent observations on cosmic rays show a very definite maximum of radiation coming from the direction indicated by the meridian of seventeen hours sidereal time. The very extensive observations of Kolhorster and Von Salis, and Weld and of Steinke, all show this effect. Observations made on the non-magnetic ship Carnegie show a maximum at seventeen hours sidereal time for the observations made between 30 degrees north and 30 degrees south latitude. There are several anomalies in astronomical observations of less definite character which, however, might be explained by the existence of an ether drift. Such anomalies occur in connection with the observed constant of aberration, standard star places and clock corrections determined at different times of day. There are at least twelve different experimental evidences of a cosmic motion of the solar system, all indicating the same general direction, and ten of them show a motion towards a right ascension lying between sixteen and one half hours and eighteen hours. Seven of these investigations give the declination as well as the right ascension, and thus determine the apex of the motion of the solar system. The various apexes all lie within a circle on the celestial sphere having a radius of 20 degrees.

This is a remarkable agreement considering the nature of the various observations involved." (Miller 1929; 1930)

Along similar lines, in 1931 Miller published a short note in a *Report to the Centenary Meeting of the British Association for the Advancement of Science*, in which he included a list of "Evidence on Solar Motion", that called attention "...to the results of several recent important experiments in diverse fields which seem to corroborate the indicated cosmic motion of the solar system." His 1931 list summarized the above lengthy paragraph from Miller's 1929 and 1930 lectures. Table 6 reproduces Miller's 1931 list, expanded by him to 13 different entries with azimuth and velocity data, including some material previously mentioned in his 1926 lectures and publications. They all provide independent supporting evidence for his interferometer determinations of variations in light speed, indicating the Earth's motion through the ether and through space. In all cases where a declination was provided, it was within the northern celestial hemisphere.

The *averages* for right ascension (RA) and declination (Dec) from these various determinations on Miller's 1931 are as follows:

Miller's 1931 Table Avg.:   RA 18h 02m   Dec +28°

These determinations compare favorably with Miller's slightly different 1928 and 1933 findings on the location of the northern polar apex of ether-drift axis, from his experiments at Mount Wilson.

Miller's 1933 Paper, and Theoretical Mis-Step

In July 1933 Miller published what would be his most complete and major paper on the ether-drift question, in *Reviews of Modern Physics*. It provided excellent information summarizing his entire body of ether research, as already covered in these pages. In that paper he also computed a 20-fold "k-factor" of *hypothetical* ether velocity increase at much higher altitudes (see Table 7). He postulated an approximate 200 km/sec ether wind at some distance away from Earth, out in open space. Miller felt this might explain why only a smaller ~10 km/sec velocity was actually measured, due to partial slowing or ether-dragging by the Earth's surface. This suggested to me a certain emotional kowtowing to conventional expectations at that time, which had been irrationally dismissive of the smaller ~10 km/sec actual velocity measurements from Mount Wilson.

Table 6. Miller's 1931 Table on the Absolute Motion of the Earth, a list of studies which independently agreed with his ether-drift measurements.

EVIDENCE OF SOLAR MOTION.

	α	δ	>		
Ether Drift	17h	68°	200	Miller	Science, 63, 433, 1926.
Meridian Circle Observations	16¼h	44°	—	Esclangon	Astro. Phys. Jour. 68, 341, 1928. Comptes Rendus 182, 922, 1926.
Meridian Circle Observations	18h	33°	—	Courvoisier	Comptes Rendus 182, 923, 1926.
Lunar Occultations of Stars	16½h	45°	700	Esclangon	Comptes Rendus 182, 923, 1926.
Angle of Reflection	20h	—	—	Esclangon	Comptes Rendus 185, 1593, 1927.
Earth Tides	16¾h	—	—	Esclangon	Comptes Rendus 182, 921, 1926.
Ocean Tides	16½h	—	—	Esclangon	Comptes Rendus 183, 116, 1926.
Motions of B-type Stars	20¼h	46°	—	Plaskett	Science 71, 152. 1930.
Interstellar Matter	20¾h	46°	—	Shapley	Nature 122, 482, 1928.
Star Clusters and Nebulæ	21h	65°	300	Stromberg	Astro. Phys. Jour. 61, 353, 1925.
Cosmic Radiation	17h	—	—	Kolhörster, Steinke, Büttner	Zeit. für Physik 50, 808, 1928.
Cosmic Radiation	17h	—	—	Ship Carnegie	Rept. Carnegie Inst. 1927–28, 255.
Radial Motions of Stars	18h	24°	20	Campbell and Moore	Lick Observatory, 1926.
Proper Motions of Stars	18h	27°	20	Ralph Wilson	Astron. Jour. 36, 1925.
Constant of Aberration	—	—	—	Doolittle.	
Clock Corrections	—	—	—	Boss.	
Clock Corrections	—	—	—	Lick Observatory.	
Star Places	—	—	—	General.	

Table Averages, without the top item from Miller: RA 18.2 Dec.+28 V. 248km/sec.

However, his 1933 paper also contained a major surprise, a serious error in my view, which probably caused his critics to react even more dismissively, and his dwindling number of supporters to groan in agony. Historian Swenson agrees, and wrote "This change was highly destructive to confidence in Miller's reports." (1970, p.67)

In his otherwise excellent 1933 paper, *Miller abandoned the well-established northerly apex of the Earth-Sun and solar system's net motion through the galaxy, in favor of a southerly apex*, at RA 4h 54m and Dec -70.5° south. This radical change was undertaken for weakly supported and illogical reasons, and lacked the clarity and independent scientific supports of his prior careful analyses for the northern apex. On page 224 of his 1933 paper, Miller revealed his reasons, namely the arguments of J.J. Nassau and P.M. Morse, as given in their 1927 paper in *Astrophysical Journal*.

Miller had ignored the Nassau-Morse arguments up into 1931, but no longer. The Nassau-Morse paper presented calculations, from harmonic analysis of the parallax determinations of 476 stars and their proper motions, claiming a southerly apex of motion for the Earth-Sun and solar system. In their determination, they used a new device, the *harmonic analyzer,* which Miller had invented, and to which Nassau had access, being a friend of Miller on the faculty of physics at Case School. This analyzer was a spin-off device similar to Miller's *phono-deik* apparatus for analyzing acoustical waves. Miller's perhaps too-enthusiastic adherence to the value of his invention may be the primary reason why he was biased to accept the Nassau-Morse results over the more logical northerly apex, independently confirmed by a variety of scientists, as presented in his own 1931 list (in Table 6). Nassau-Morse did not identify any new data on this question, merely a new method by which to analyze older sets of star-location data as gathered by many different astronomers. They also assumed, without evidence, that "the peculiar motions of stars are at random". This idea derives from older Newtonian concepts that had already erased any vestiges of a motional or dynamic ether, by which non-random movement might be organized within the cosmos. Such assumptions were later challenged by the very existence of the highly-organized spiral galaxies, such as Andromeda and our own Milky Way Galaxy. Many lines of evidence argued strongly against assumptions of purely random star-motions.

The Nassau-Morse selection of stars to be studied was additionally restricted to those with an estimated radial velocity (moving towards or away from the Earth) of less than 50 km/sec and with angular motions

Table 7. Miller's Observed Vs. Theoretical "*k*-Factor" Up-Calculated Velocity for Higher Altitudes

Epoch	Velocity-Obs.	Velocity-Calc.	*k*
Feb. 8	9.3 km/sec	195.2 km/sec	0.048
Apr. 1	10.1	198.2	0.051
Aug. 1	11.2	211.5	0.053
Sep. 15	9.6	207.5	0.046
Average	10.05 km/sec	203.1 km/sec	0.0514

Early ether-drift research anticipated an ether-wind velocity of around 200 km/sec, based upon Earth's orbital motion of ~30 km/sec and the Sun's motion through the galaxy of ~200 km/sec. Ether-drift experiments by Morley-Miller and Miller measured an ether velocity of 8 to 11 km/sec. It was theorized that there would be an even faster motion of the ether wind in space outside the Earth's atmosphere. From that concept, Miller developed a "factor of reduction *k*", setting the *average* value of *k*, for his 1933 Mount Wilson calculations, at *k* = 0.0514. By simple maths, Miller then recalculated upwards his measured values to a *theoretical* velocity that *might* exist at higher altitudes, and be more agreeable to conventional expectations. By this theoretical exercise, Miller could mathematically describe ether entrainment, which reduced ether velocity in his experiments by a factor of 20. Their presumed higher ether-drift values of ~200 km/sec in open cosmic space, were thereby contrasted to the measured values of ~10 km/sec. However, this theoretical exercise demanded that one accept the existence of an entrained or partially-dragged ether, by which such a reduction took place. This realization also demanded that all ether-drift experiments be undertaken, for optimal results, outside of heavy metal shields and stone buildings, and not within basement locations. The higher the altitude, and the more transparent was the interferometer along the plane of the light-beam paths, the better would be the measured results. (Miller 1933a, p.235) of less than 0.5 arc-seconds per year. In other words, by definition they selected the *least mobile* of stars available for study, which further colored their methods and conclusions. They calculated that this "cloud" of 476 relatively non-moving stars was, on average, *moving towards the same general northerly apex as per Miller's original determination.* However, as they were moving *faster* on average than our solar system was moving, they concluded, without justifications, that *the Earth must, then, be moving in the opposite direction to that cloud, towards a southerly apex.*

The Nassau-Morse conclusion had no logical basis, however, and rather beggars belief. They did not consider, Earth might also be moving towards the northerly apex, albeit at a slightly slower velocity. Given how our solar system carries two extremely large and massive planets, Jupiter and Saturn, as well as the Sun, that could be a reason for the slower motions of our solar system by comparison to other stars which might not have any planets at all. A comparison of our solar system's velocity to those of massive double-star systems in the nearby space might have been telling.

Also, no information was given by Nassau-Morse as to where or how fast the excluded faster moving stars might be going, by comparisons to the 476 slower stars. The velocity variance among those 476 stars was also not reported in their published account, which might otherwise have provided additional clarifications.

Why would Miller accept the very preliminary and questionable results of this *one study* by Nassau-Morse over all the other evidence he previously knew about and endorsed as recently as 1931, as provided in the above Table 6? Miller gave no clear answer. In his 1933 paper, he wrote the following vague statement:

> "Beginning in the autumn of 1932, a reanalysis of the ether-drift problem, and a recalculation of the observations made at Mount Wilson in 1925 and 1926 have now been completed. By adopting the alternative possibility that the motion of the solar system is in the cosmic line previously determined but is in the *opposite direction,* being directed to the apex near the south pole of the ecliptic, a wholly consistent solution has been obtained." (Miller 1933a, p.232)

The "opposing direction" Miller spoke of was in exactly the opposite direction of the universe as Miller's original northerly determination. Aside from the research and conclusions of Nassau-Morse, Miller's discussion revealed an important point not stated by him: *One must separate the "Earth's velocity through space" from the "velocity of the ether drift".* On average, they are moving along the same axis of motion, but not necessarily in opposing directions. Also, the seasonal-variations in sidereal ether velocities is quite important.

Critical to this discussion is the fact that modern astronomical observations have continued to confirm the "Sun's Way" or path through the Milky Way Galaxy towards a northerly apex. And these modern determinations have used far more stars than Nassau-Morse did back in 1927. For example, a 1946 study by O.R. Walkey, on "An Abstraction on the Solar Apex", discussed *eleven different astronomical indicators for the Sun's Way*, over the period 1933-1941, starting shortly after Miller's 1933 discussion. Those newer studies used star catalogs with *from 711 to over 32,000 different stars to make their determinations*. That's quite a bit more than the 476 stars used by Nassau-Morse. All of these eleven studies furthermore determined a *northerly* apex of solar motion through the galaxy, within a narrow band of right ascensions (~18 to 19 hrs sidereal), and 20° of declinations (+25° to +45°). These coordinates aim generally towards the star Vega in the constellations Lyra. Vega is very close to the modern calculations on the "Sun's Way" solar apex, which is at RA 18h 36m sidereal and Dec +39°. Both of these coordinates are nevertheless relatively close to Miller's computed axis of ether-drift.

Here are the coordinates for both the north and south ends of Miller's 1933 axis of ether drift, alongside a repetition of the similar values presented over the last few pages:

Milky Way Galaxy Center:	RA 17h 45m	
Modern "Sun's Way" Apex:	RA 18h 36m	Dec +39°
Location of Vega, in Lyra	RA 18h 37m	Dec +38.7°
Miller's 1926 North Apex:	RA 17h 30m	Dec +65°
Miller's 1928 North Apex:	RA 17h 00m	Dec +68°
Miller's 1931 Table Avg.:	RA 18h 02m	Dec +28°
Miller's 1933 North Apex:	RA 16h 54m	Dec +70.5°
Miller's 1933 South Apex:	RA 4h 54m	Dec −70.5°

The RA's of Miller's northern measured ether-drift results, lying between ~17 to 17.5 hrs sidereal, are close to the "Sun's Way" in the galaxy. The Dec values exhibit a wider range, but are nevertheless also

125

close, distributed from +28° to +70°, excepting for Miller's southerly axis of 1933. These coordinates are spread over a very narrow strip in the sky of 22.5° wide by 42° long in the northern celestial sphere, generally aimed at the center of the Milky Way Galaxy.

My investigations have turned up no evidence that anyone in astronomy or astrophysics today endorses a southerly polar apex of solar system motions. *For such reasons, I believe Miller's northerly apex, as he determined prior to 1933, is the correct one.* Let's review the basics in a few Figures.

In the preceding chapter on Miller's work, Figure 35 gave a preliminary indication of the overlapping evidence for a northerly apex of motion. Here, Figure 37 provides a simplified diagram of the Earth and Sun in motion, according to conventional astronomy, similar to what might be found in modern astronomy textbooks. However, if we study this graphic, we are forced to consider that the true nature of the Earth's motion is not a flat ellipse, but rather an *elongated off-center and open-ended spiral.* This important point about spiral motions forces a consideration that the velocity of the Earth around the Sun must be variable, *different from the usual Keplerian determinations of faster speeds at perihelion and slower velocity at aphelion.* Without appar-

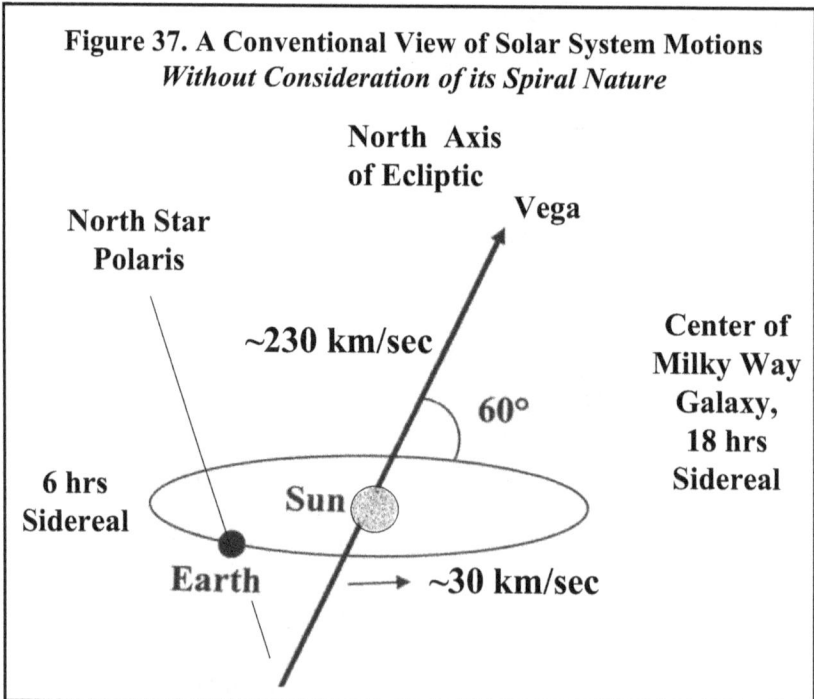

Figure 37. A Conventional View of Solar System Motions
Without Consideration of its Spiral Nature

North Axis
of Ecliptic

Vega

North Star
Polaris

~230 km/sec

60°

Center of
Milky Way
Galaxy,
18 hrs
Sidereal

6 hrs
Sidereal

Sun

Earth

→ ~30 km/sec

ently realizing it, Miller detected a variance in his ether-drift velocity determinations which were not congruent with Keplerian expectations, but which are in agreement with a spiral-form motion of the Earth in space. Miller may not have comprehended the implications. This spiral-form characteristic, which was most well-developed by Reich, will be presented analytically in Part III, but here we can explore it as a preliminary.

Today in 2019, the solar apex, or the "Sun's Way" through our Milky Way Galaxy, as it is called, is accepted to be in a northerly direction, close to the blue-white star Vega, the 5th brightest star as seen from Earth. While Miller's results on the ether drift is closer to the northern ecliptic pole than is the modern determination of Vega, both are near enough in proximity to accept Miller's findings on the northerly apex at face value. Also an entire cluster of northerly vectors remains close enough to sustain Miller's original theory of a northerly axis of drift, even while some mystery remains in the difference between his northerly ether-drift azimuth and the modern motional direction towards Vega.

Figure 38 on the next page presents a close-up star chart of the northerly region of interest, which can be readily compared to Figure 35 near the end of the last chapter. The star chart identifies Miller's northern axis of ether drift, alongside the northern polar axis of the solar system's plane of the ecliptic (which is defined by Earth's orbital plane). Also identified are the individual *northern orbital plane poles of the other planets*, and the Sun's northerly axis of rotation. By gravitational theory, the Sun ought to be a major force in determination of all the various planetary orbits. Instead, the Sun and its closest companion Mercury stand off in relative isolation from the axes of the majority of the planetary orbital planes.

The modern vector of the Sun's Way is also identified on this chart by arrows pointing to its right side, towards the star Vega, which is also close to Miller's 1931 list of factors documenting a northerly axis of Earth's motions, reproduced in Table 6. On Figure 38, objects appear far apart as the star chart shows them at a larger scale.

Beyond Miller's determination of the axis of ether drift, these and other factors, to be discussed in Part III, all indicate a powerful motional shift off towards Vega, which otherwise has often been interpreted as being merely an "apparent" motion within the nearby cluster of stars. For example, the tilt of the Sun's equatorial rotational axis pole diverges from the axis of the plane of the ecliptic by 6°, something known since the 1800s, but which has never been explained. The closest planet to the Sun, Mercury, has a similar 6° tilt of its *orbital plane*, suggesting the two are under a common influence. The orbital plane axes (not their rotational axes) of the other planets in the solar system are more closely clustered around the Earth's orbital plane axis.

Miller alluded to this curiosity, that the tip of the Sun's axis away from the axis of the ecliptic plane of the solar system might be related to his ether drift determinations. Again, Part III gives more discussion on this slight "core tilting" of the inner parts of the solar system.

Miller's axis of drift is also about 5 angular degrees removed from the northern ecliptic pole, and the ecliptic pole is about 6° removed from the Sun's orbital pole. All three, however, are about 26° from Vega, which also marks a direction towards the center of the Milky Way Galaxy. *All of these points in the cosmos are clustered together within a circle with diameter of ~26°, out of the full 360° range of possibilities, suggesting the cosmographical relationships are causally linked and meaningful.*

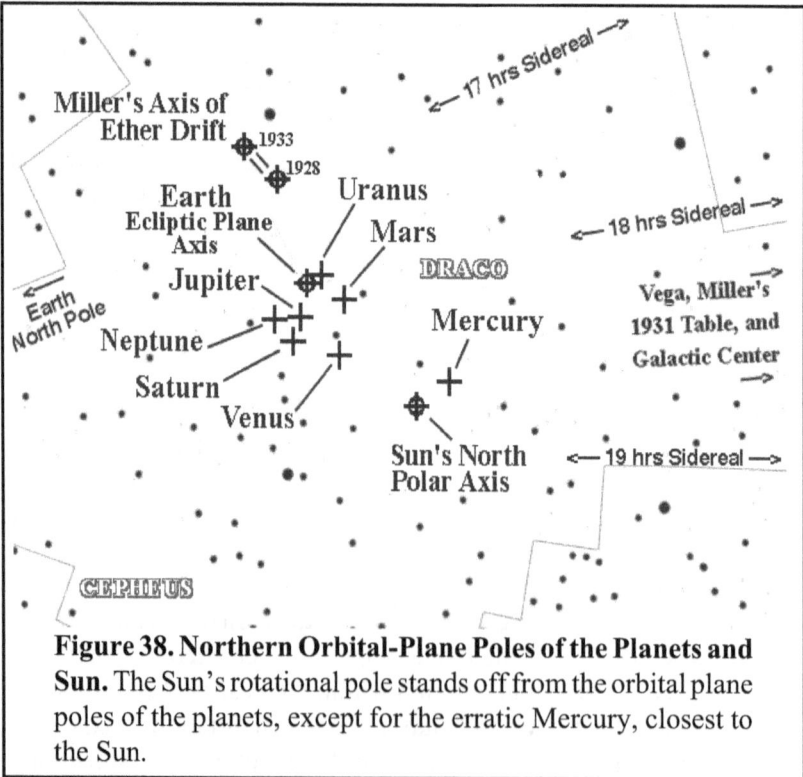

Figure 38. Northern Orbital-Plane Poles of the Planets and Sun. The Sun's rotational pole stands off from the orbital plane poles of the planets, except for the erratic Mercury, closest to the Sun.

As I will show in Part III, the *seasonal variations in Miller's ether velocity determinations* also go a long way in solving this puzzle, of Miller's final determination being somewhat off from the Sun's Way, even while providing yet another set of confirmations for his overall work.

After Miller's presentation of his major 1933 paper, his continued affirmations of the reversal to a southerly apex basically caused Miller's support to dry up and wither away. His critics became even more skeptical of his results. The conclusion cannot be avoided, that *Miller's shift to the southerly axis was a mis-step and error* that did damage to the acceptance of his overall findings. This was especially so, given how the supporters of Einstein were, at this point in time, busy erasing the concept of a tangible and dynamic, light-affecting cosmic ether in every way possible.

The sad part of this situation is that among all the different supporters and antagonists, Miller was the only one who had made extended ether-drift measurements at four different seasonal epochs, atop a high mountain, and with transparent covers and windows at the same level of the light-beam path. His critics were theoreticians, save for Michelson, who remained silent during the debates which swirled around Miller and the subject of ether drift. Miller stood basically alone, and carried the full weight of defending the concepts and evidence of cosmic ether and ether-drift motions.

In February 1934, Miller published a new paper in *Nature* magazine, following another talk the preceding year on the same subject to the *British Association for the Advancement of Science.* In that talk and publication, he again spoke about his reversal of the direction of ether drift towards the southerly apex, with a compromised and resigned-sounding statement that "...it seems necessary to accept the reality of a modified Lorentz-FitzGerald contraction, or to postulate a viscous or dragged ether as proposed by Stokes." What once was central to his theory, of a *partly* entrained and measurable ether drift, was now spoken about in resigned tones and second-guesses, also giving credibility to the mystical "contraction theory" which nobody had yet confirmed in an unequivocal manner, and which his own work with Morley had been unable to confirm some 30 years earlier. Also in 1934, Miller was appointed to the post of Honorary Professor of Physics. He clearly retained a high rank and standing in the American sciences, but his ether-drift work was rarely mentioned.

Miller's last years were marked by a few additional lectures on the subjects of ether drift and acoustics, along with various honors for his many decades of research and service to Case School in Cleveland. By 1936 he had been awarded five honorary degrees, from Miami University (1924 Doctor of Science), Dartmouth (1927 Doctor of Science), Western Reserve University (1927 Doctor of Laws), Baldwin-Wallace University (1933 Doctor of Laws), and from Case School of Applied Science (1936 Doctor of Engineering). These were aside from his earned physics doctorate from Princeton University in 1890.

In early March of 1940, a major Tribute was given to Miller at Case School, for his 50 years of contributions as physicist, with a front page dedication to him in the *Cleveland Plain Dealer* newspaper, which had run many articles supportive of his ether research and findings. A portrait painting of him was presented to Case School, praising his work and his friendly and kind demeanor. An undated document in the Case Western Archive included extracts from personal letters Miller received from scientists around the world, showing serious interest in his ether-drift findings. He was as much loved and appreciated for his dedication to science back then as he is today ignored or considered chasing illusions. Miller died in February of 1941, at the age of 74.

Sagnac and Michelson-Gale: Ether Detection by Rotation

Another variant of the ether-drift experiments employed a rotating platform which sent two light beams around an irregular "racetrack" path by use of mirrors, one moving clockwise and the other counter-clockwise. After completing the circuit, the two beams were recombined back into one beam, whereupon interference fringes would appear for observation. The experiment could evaluate for both the existence of an ether, and for changes in light speed dependent upon direction of rotation.

The success of these experiments, reported below but nearly forgotten or obfuscated by the Einstein followers today, provided even more direct proof that *light has variable velocities,* dependent upon the speed of the emitter and observer, but irrespective of whether the cosmic ether is static, is in motion along one or another preferred direction, or is even fully stagnant as per the Stokes concepts.

1913-1914: George Sagnac Proves Variable Light Speed and Ether

Enter Georges Sagnac, who undertook the original rotating interferometer experiment in 1913, only 8 years after Einstein's 1905 published papers on the subject of his new relativity theory. In this experiment, Sagnac created a rotating tabletop interferometer, turning at a speed of 2 revolutions per second. On the surface of the rotating table, two light beams were sent to bounce along different mirrors, so as to move either with the direction of the rotating disk, or in opposition to the rotation. The two light beams originated from the same light source, being split into two beams, much as in the Michelson interferometer. After moving around the rotat-

Georges Sagnac
1869-1928

131

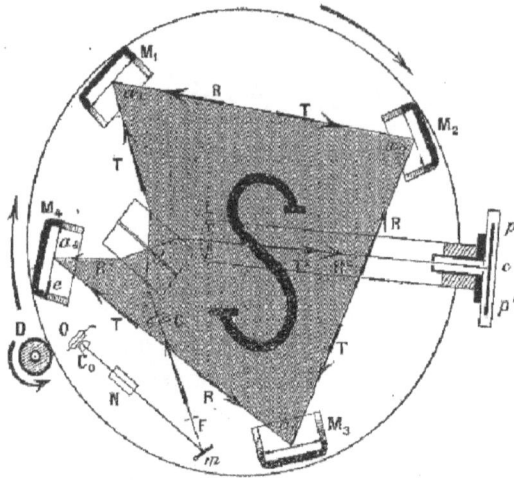

Figure 39. Sagnac's Rotating Interferometer. Sagnac's original diagram from 1913 is shown above, with a simplification of it presented below. The apparatus generates a light beam from A, which hits a half-silvered mirror C and is then split into two beams, one of which moves around the circuit with the rotation by mirrors C-D-E-F-C while the other goes against the rotation by C-F-E-D-C. The two beams are reunited finally at B, where interference fringes are displayed. When the platform is rotated the fringes shift, proving that light speed is variable, depending upon direction of rotation ($c + v > c - v$).

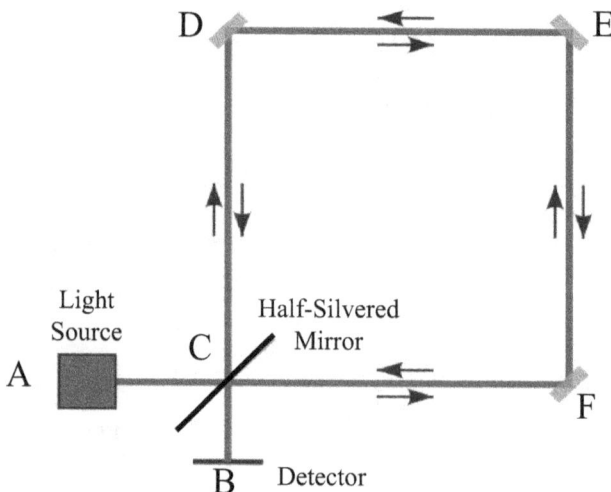

ing disk in their opposing directions, the light beams were recombined to reveal interference fringes as observed from a location above the center of the rotating apparatus.

Sagnac reported on his ether experiments in two papers (translated: "The Luminiferous Ether is Detected as a Wind Effect Relative to the Ether Using a Uniformly Rotating Interferometer" and "On the Proof of the Reality of the Luminiferous Ether") published in 1913 by the French Academie des Sciences, in *Comptes Rendus*. Even the relatively slow velocity of 1-2 revolutions per second was sufficient to show a measurable shifting of the light-beam interference fringes, distinguishable from rotational distortions such as due to centrifugal forces on the mirrors or other factors. As Sagnac described it:

"In a system moving as a whole relative to the ether, the propagation time between any two points of the system should change in a way similar to a stationary system subjected to an ether wind... and would contain light waves in a manner similar to atmospheric wind carrying sound waves. The observation of the optical effect of such an *ether wind relative to the* [stationary] *ether* will constitute a proof of the ether's existence, just as the observation of a wind relative to the atmosphere on the speed of sound in a moving system would constitute." (Sagnac 1913a. Brackets added)

"In Fresnel's ether hypothesis, the light waves ... are propagated in the ether of vacuum with a velocity... that is independent of the overall motion of the interferometer ...in the clockwise direction of propagation is altered along the closed circuit, as if the luminiferous ether were driven by a counterclockwise vortex when the circuit rotates in the [clockwise] direction." (Sagnac 1913b, See Fig.39. Brackets added.)

Optical fringes were photographed both before and after the turntable began rotating, demonstrating a clear fringe shifting which, as Sagnac put it, "...directly shows the existence of the ether."

Sagnac's interferometer could not detect a *cosmic sidereal ether drift*, and only an ether-wind velocity relative to the speed of its rotation, unrelated to the Earth's ether wind. However, it did prove that the velocity of light waves were variable in preferred directions, depending upon the velocity of their emitters and receivers – or, in this

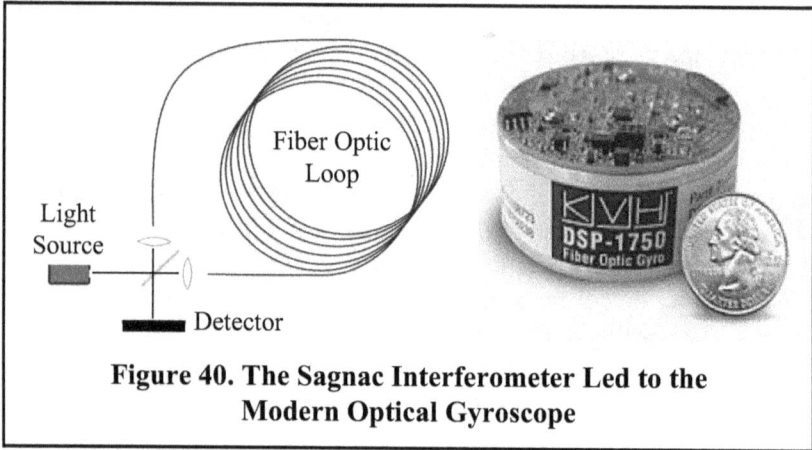

**Figure 40. The Sagnac Interferometer Led to the
Modern Optical Gyroscope**

case, the moving mirrors. This stood as direct evidence against Einstein's assumption of light-speed constancy, and for the existence of an ether medium. Sagnac's experiment was reproduced with successful results many times, notably by Dufour and Prunier in 1942.

Sagnac's work could not settle disputes between the advocates of an Earth-entrained ether, a dynamic motional ether, or the stagnant-static ether concepts, but his findings provided yet another serious challenge to claimed light-speed constancy. Those who argue in favor of Einstein also reject the positive results from Sagnac, as they do with all other positive evidence of ether drift. New technology developed from the Sagnac approach is today in widespread use, notably a class of fiber-optical gyroscopes as applied for navigation and other purposes.

1923-1925: Michelson-Gale Prove a Partially Dragged Ether, and Variable Light Speed

An experiment similar to that of Sagnac, but of much larger scale, was undertaken by the team of Albert Michelson and Henry Gale between 1923 and 1924, published in April 1925 in *Nature* magazine, and as a two-part article in *Astrophysical Journal*. The title in both cases was "The Effect of the Earth's Rotation on the Velocity of Light". These papers reported on what was, essentially, a very large "Sagnac" type of experiment, where the Earth turning on its axis was the rotating platform. The theory of this experiment had previously been detailed by Michelson in 1904, when he wrote:

"Suppose it were possible to transmit two pencils [thin beams] of light in opposite directions around the earth parallel to the equator, returning the pencils to the starting point. *If the rotation of the earth does **not** entrain the aether*, it is clear that one of the two pencils will be accelerated and the other retarded (relatively to the observing apparatus) by a quantity proportional to the velocity of the earth's surface, and to the length of the parallel of latitude at the place ... it is not necessary that the path should encircle the globe, for there would still be a difference in time for any position of the circuit. ... The system of interference-fringes produced by the superposition of the two pencils – one of which has traversed the circuit clockwise, and the other counterclockwise – would be shifted... in the direction corresponding to a retardation of the clockwise pencil, if the experiment were tried in the Northern hemisphere." (Michelson 1904. Emphasis added.)

In ordinary language, this was a test for different properties of ether, be it static and not dragged (Newton), or fully dragged and stagnant (Stokes), or something in between (Fresnel, Miller). The experiment was also based upon Michelson's remaining questions about the small result he had obtained in the original Michelson-Morley experiment of 1887, and also due to Miller's positive experimental results. Should the ether be *fully stagnant and entrained* at the Earth's surface, with a stagnant ether-layer being fully carried along with the Earth as it rotated, then there should be *no effect, no difference* in the velocity of the two counter-rotating light beams. However, if the ether was *unreactive to the material substance of the Earth* and was *not dragged* around by the Earth as it rotated, yet still served to act as the medium of light wave transmission, then a *maximal variance* would exist between the speed of the two counter-moving light beams, anticipated to be equal to the Earth's rotational velocity at a given latitude. Or, if the ether was *partially entrained,* then the value would range somewhere between zero (a true null) and the maximum value.

Michelson obviously could not organize a round-the-world experiment as imagined in his 1904 paper, but he finally was able to test the idea on a smaller scale, firstly in 1923 at Mount Wilson. There he set up a rectangular light-beam circuit of one mile in total length, with mirrors at the corners to allow for the two light beams to be projected in opposing directions.

Unfortunately, as Michelson described, this initial effort with light beams moving through the open air was plagued by disturbances in the readings: "...even under the best conditions, the interference fringes were so unsteady that it was found impossible to make any reliable measurements." Anyone familiar with viewing distant objects through binoculars will know this problem, of heat waves and air turbulence creating distortion effects. Michelson's efforts to resolve clear light-beam fringes in his interferometer optics were thereby thwarted. He subsequently planned a new experiment where the light beams and optical components would be enclosed inside protective pipes.

Michelson's second attempt was in 1925, in association with Henry Gale and assisted by Fred Pearson, in Clearing, Illinois. This was a flat region then occupied mostly by farmer's fields, and is today the location of Chicago's O'Hare International Airport, at latitude ~41.5°N and altitude of 198 meters (650 feet). A ~30 cm (12-inch) diameter steel pipe was laid out to form the light-beam circuit, sealed along the segments, and then pumped down to create a partial vacuum of around 12 to 25 Torr (0.5 to 1 inch of mercury pressure). The light beam and pipe circuit formed a rectangular path of ~613m by 339m (2010 feet by 1113 feet), for a total of 1.9 km (6246 feet) of light path distance in total. Three hours were required for the pipe system to be pumped down to the desired vacuum pressure, after which, as Michelson noted, "the fringes were perfectly steady, and as sharply defined as could be desired."

The longer sides of this rectangle were laid out in an East-West direction, the shorter sides in the north-south direction. A calibration circuit was also set up, too short along the east-west direction to meaningfully react to even a strong ether wind, and which by design produced a near-zero displacement of the light fringes. The calibration circuit results could then be compared to the light-fringe displacements observed within the larger rectangular circuit to determine the magnitude of fringe displacements, assuming they existed.

By this time, the stakes were high regarding the outcome of the experiment, as Einstein's new theory of relativity, like Stokes, also predicted no differences in light speed. Any zero result from this experiment could therefore be interpreted as either due to a Stoke's type of fully entrained ether, or an Einstein interpretation of light-speed constancy and no ether. If strong fringe shifting occurred, or even slightly so, it would be a death blow to the Einstein theory.

As described in their April 1925 publications, the theoretical fringe displacements for a fully stagnant ether, of 0.236 fringe, was calculated from the dimensions of the apparatus, the latitude, the wavelength of its sodium light source, the angular velocity of the Earth's rotation and the velocity of light. The measured experimental observations superficially appeared to be quite close, at 0.230 fringe, differing by an amount of only 0.006 fringe. (Michelson-Gale, Nature 1925) This made it difficult to distinguish absolutely between any of the different theories. However, their results could also be interpreted as an affirmation of the

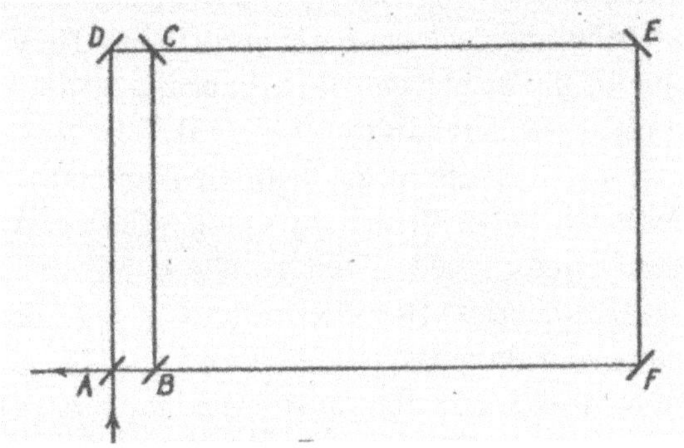

Figure 41. The Michelson-Gale Experiment. The above diagram shows the measuring paths, laid out on the ground inside metal pipes evacuated to a partial vacuum. Route A-D-E-F-A defined the clockwise path, with A-F-E-D-A for the counter-clockwise. The A-B-C-D-A pathway was a control loop designed to yield only zero fringe shifts, a true "null".

The Michelson-Gale Test Site in Clearing, Illinois, 1924

centrally-important objections made by Miller, that only a small result could be anticipated by surrounding the light beams with dense ether-blocking metal pipes. By Miller's criticisms alone, it was yet another example of experimenter bias self-destructing an otherwise important experimental design, by chronically refusing to take seriously the partial Earth entrainment of a material ether.

There were other problems in the MPP conclusions. In their longer paper to *Astrophysical Journal*, Part II where the assistance of Fred Pearson was acknowledged, additional details were provided. A total of 269 separate individual observations were aggregated together to extract the final fringe displacement average of 0.230. They stated "The calculated value of the displacement [assumes] a stationary ether *as well as* in accordance with relativity..." (p.143) Here Michelson-Gale affirmed that the close nature of the theoretical and measured values could be interpreted in either manner, by a Stokes fully-entrained ether, or by Einstein's empty-space. Their results could not distinguish between those two theories. However, a look at their data suggests Michelson-Gale had truly measured variations in light speed, significant enough to invalidate both the Einstein and the Stokes predictions and in favor of a variable density entrained ether.

Their paper in *Astrophysical Journal*, p.145, included a graphic showing the measured fringe displacement values. The final observed result of 0.230 fringe-shift was derived by a *data averaging procedure,*

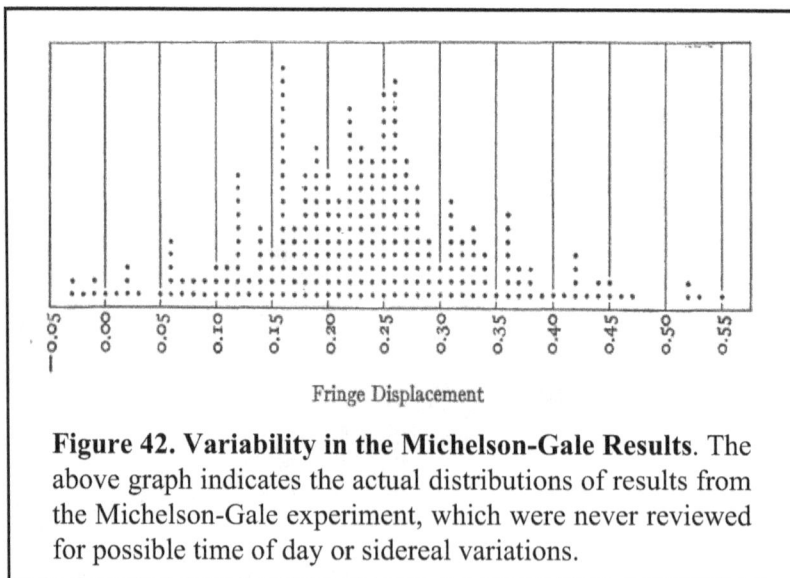

Fringe Displacement

Figure 42. Variability in the Michelson-Gale Results. The above graph indicates the actual distributions of results from the Michelson-Gale experiment, which were never reviewed for possible time of day or sidereal variations.

smoothing out its significant variation, ranging generally from -0.030 to +0.550 fringe, a variance of 0.580 fringe. This variation was more than twice the calculated and theoretically null 0.236 fringe value.

Given the Earth's surface rotational velocity at latitude 41.5° N is about 0.347 km/sec, the fringe-shift variation in Figure 42 might translate into a velocity variance between 0.035 km/sec, to 0.7 km/sec. The major unknown is, of course, the low altitude and use of the metal tube in which the light beams were enclosed, both of which would have significantly reduced ether velocity and added to Earth-entrainment effects. This great variation in light speed was never explained. Nor was it analyzed in any manner as to possible consequences of a real variation in ether wind velocity, moving in the same rotational plane as the Earth's equatorial rotation. Only a few years earlier, Miller had plotted the actual ether-drift direction along a general south-north axis, at ~10 km/sec, much faster than the Earth's rotational velocity. Was this stronger ether wind the source of the large transient variations observed in the Michelson-Gale experiment?

Michelson-Gale were content to report that the extracted average of 0.230 fringe was close enough to match the theoretical fully-entrained ether value of 0.236. Nevertheless, *their experimental result was prima facie evidence that light speed was significantly variable during the experiment, refuting Einstein's basic postulate and affirming the Fresnel-Miller expectation of a partially-dragged cosmic ether.*

We must ask, why didn't the Michelson-Gale team take seriously the cautionary points already known and emphasized by Dayton Miller by that time, as to the ether-blocking effects of dense metal pipe materials surrounding their light-beam interferometer? Certainly, possible technical difficulties at that period in history might have prevented the experiment from being undertaken with more optimal methods and materials. Simple glass panels could be used around the light paths, for example, sealed much like aquarium glass and be able to withstand a reasonable vacuum. This remains an important consideration for modern research, however, if they can shake loose from the ideas of Newtonian "absolute space" static ether, and the Einstein theory.

As things turned out, like the Sagnac experiment, Michelson-Gale could not tell anything about ether-drift directions, nor about static ether drag versus dynamic ether wind. However, the variations in their observations were significant, and should have been reviewed for possible diurnal or sidereal components. This was never done.

The Dynamic Ether of Cosmic Space

The followers of Einstein nevertheless latched on to both the Sagnac and Michelson-Gale experiments as their own, ignoring the very clear results in the case of Sagnac, and the seriously variable result of Michelson-Gale, which strongly suggested a partially entrained ether. The relativists claimed moving mirrors had changed the light paths, ignoring how this also would imply variable light speeds – accelerations and decelerations – as light waves bounced off the "moving mirrors". It has also been stated, with sophistry, that "special relativity does not apply to rotating systems", as if when presented with an experimental challenge, their theory is so flexible and malleable that it can be shaped and adjusted to meet any objection that might arise. Basic scientific method tells us that such flexibility and malleability are the *hallmarks of a bad theory, which when sufficiently tweaked claims to "explain everything" but can predict nothing,* at least, not in any exclusive unequivocal manner.

Michelson-Gale had step-by-step boxed themselves into a theoretical cul-de-sac. They failed to consider the possibility of a dynamic Earth-entrained cosmic ether wind, moving slower at lower altitudes, but not firmly fixed to the Earth's surface. This error was compounded by placing their light beams inside metal pipes, which further inhibited ether motion from affecting those light beams. And then, even when their own results showed significant variation in light wave velocity, their curiosity failed, and they didn't investigate further.

Within a few years after Sagnac and Michelson-Gale, Miller would undertake his work at Mount Wilson, and prove once more that the ether drift or ether wind was something of variable speed with definable maximum and minimum velocities and azimuths in cosmic sidereal space. His work showed once again how the interferometer results for ether-drift detection was affected by the density of materials surrounding the apparatus, including the density of the building in which it was located. That should have been a wakeup call for all the ether-drift researchers, and a blow to any experiment which had not heretofore taken those issues into account.

Michelson and Others Return to the Ether-Drift Question

1926-1928: Michelson, Pease and Pearson Confirm, but Nevertheless Deny an Ether-Drift

In apparent efforts to replicate Miller's results as obtained at Mount Wilson in 1925 and 1926, Albert Michelson, with assistance from F.G. Pease and F. Pearson (hereafter "MPP"), undertook a new set of ether experiments. Their results, with the title "Repetition of the Michelson-Morley Experiment", were published in the January 1929 issue of *Nature* magazine, followed by a nearly identical article a few months later in the *Journal of the Optical Society of America*. Unfortunately in both cases only a frustratingly vague and short report was given, exposing shortcomings well below the standards an optical expert such

Figure 43. Dayton Miller (left) **and Albert Michelson** (right) at a *Conference on the Michelson-Morley Experiment* held at Mount Wilson Observatory, February 1927.

as Michelson would normally have adhered to. The paper reported on three experimental attempts to detect ether drift using a device similar to the original Michelson-Morley interferometer, but with considerable self-defeating modifications. MPP declared a negative outcome on all three of their efforts, but a review of their actual data and discussions indicates seriously flawed methodology with significant positive results on the last of their experiments.

1926: The Michelson-Pease-Pearson First Experiment

This first experiment began in June 1926, after Miller had published and lectured on his positive ether-drift results from Mount Wilson. The new interferometer conceived by MPP was a top-heavy behemoth, constructed from *Invar* steel, an iron-nickel alloy with a low coefficient of thermal expansion. The light source was placed above the top-center of the instrument. Other optical components were constructed similar to the usual Michelson-type of cross-beam interferometer. However, the Invar machine allowed the observer to ride in a seat with the rotating platform, to read out the fringe shifts without having to walk around with it as it rotated. This method required a counterweight at the opposing end of the same beam upon which the observer was seated, equal to the weight of the observer. Unfortunately it proved to be unwieldy, needing constant adjustments. For example if the observer in the seat leaned sideways, forward or backwards, or significantly moved their legs or arms, it would immediately change the balance point of the apparatus, throwing everything out of kilter. So what might appear to be an "advantage" in such a "strong invariable steel" apparatus, avoiding to have the observer call out the measurements while walking, left the design susceptible to serious instrumental artifacts due to its own weight, and to human tendencies to move and shift around.

The unwieldy design of this instrument and a few other details were reported by historian Swenson:

> "By mid-1926, Michelson and his colleagues had completed construction of a massive Invar interferometer some 30 feet in diameter with a 55-foot lightpath, large enough to carry the observer in a bucket-seat during its rotation. However, difficulties with mechanical shear forces and strains during preliminary trials convinced Michelson that this device was simply too complicated and too massive." (Swenson 1970)

The MPP article did not mention these particular complications reported by Swenson, nor that the 17 meter (55 ft) light path was less sensitive than either Miller's 64 meter interferometer, or the original 22-meter Michelson-Morley interferometer of 1887. Neither did MPP provide details on the location, altitude, or structure of the building in which it was tested, nor any other information on these first experiments, except to publish them as if they were of significance, when they were not. This was also the first case I found where Michelson referenced his 1887 experiment with Morley as having "negative results", possibly due to the influence of his two assistants, who were known supporters of the Einstein theory of relativity. MPP wrote:

"Several hundred observations were made, all indicating the same negative result as originally obtained [by Michelson-Morley in 1887]. ...a displacement of 0.017 of the distance between fringes should have been observed at the proper sidereal times. No displacement of this order was observed." (MPP, *Nature* 1929a)

Figure 44. The Massive Michelson-Pease-Pearson Invar Interferometer, with a seat for the observer to ride while it was rotated. It was later revised and eventually abandoned due to unforeseen complications in maintaining its stability and calibration. This photo is from a model on display at the Michelson Museum, US Naval Academy, Annapolis, Maryland.

The Dynamic Ether of Cosmic Space

No data were presented by which we might know just what was the actual computed speed of observed ether drift. But we can make a comparison to the results from the original Michelson-Morley 1887 experiment and its 22-meter interferometer. In that case, Michelson-Morley anticipated a displacement of 0.4 fringe based upon expectations from the 30 km/sec orbital motion of the Earth around the Sun. The MPP interferometer of 1926 anticipated only a 0.017 fringe displacement from the same orbital velocity, indicating it was *20 times less sensitive* than the original 1887 device of Michelson-Morley. In 1887 Michelson-Morley measured 0.01 to 0.02 fringe displacement, which was something approaching 5 to 7.5 km/sec of ether drift velocity. With MPP using an instrument with such a lowered sensitivity, it suggests they were seeking to measure a phenomenon probably below the threshold of instrument noise – especially given how their cumbersome and top-heavy interferometer had to carry a human being with it as the instrument was rotated, increasing its instrumental error due to vibration and stress bending. Why would they bother to include such a vague report from a flawed instrument design in their 1929 published paper?

Given what details were provided for the second and third MPP experiments, described below, it suggests they fully ignored Miller's cautions about running such experiments inside heavy stone or metal containments. If their Invar interferometer had produced defendable scientific data, then why was this expensive apparatus, which might be considered a "luxury interferometer", seriously stripped down and eventually abandoned thereafter?

1927: The Michelson-Pease-Pearson Second Experiment

The second round of experiments by MPP began on an unspecified "autumn" date in 1927, once again without any detail as to its location, the actual dates, or data. The interferometer was identified as having a light path of 16 meters (~53 feet), using the optical components from the original Invar design, mounted upon a 7000 pound cast iron disk once used as the bedplate for polishing the 100-inch mirror for the Hooker telescope at Mount Wilson Observatory. With this new design MPP reported (1929):

> "In consequence of inadequate temperature provision (and probably unsymmetrical strains in the apparatus) the results, while not so consistent as could be desired, still show clearly that no displacement of the order anticipated was obtained."

Both these first two efforts reported "no displacement of the order anticipated". No further details were given by MPP for this second experiment in any of their published accounts, to my knowledge, and it beggars belief as to how any conclusions could be drawn from an apparatus which continued to suffer from admitted thermal and "unsymmetrical strains". Exactly what does that mean? One can only guess. It is strange that, while Miller spent *years* to test out and perfect his interferometer wherein such thermal or strain factors would be eliminated, he was nevertheless attacked by others (to be discussed later) who conjured-up nonexisting "artifacts" to dismiss his positive results. Meanwhile, the MPP team self-confessed to such artifacts and strains in their instrument and procedures, but nevertheless published their claimed negative result in major science journals, which later were widely cited as "further evidence of a null result".

1928: The Michelson-Pease-Pearson Third Experiment

The third experiment was undertaken with the same heavy disk interferometer used in the second experiment, but with a few modifications. It was conducted on a fully unspecified date (probably 1928), inside "...a well-sheltered basement room at the Mount Wilson Laboratory". The round-trip light path of this new instrument was slightly increased to 26 meters (85 feet), but still remained not even half as long as Miller's 64-meter (210 feet) interferometer. Worse, the problem of the Mount Wilson basement negated all other improvements in their design, given its heavy concrete construction, being something on the order of a bomb shelter. The base of this observatory was composed of massive concrete walls, to support the immense weight of the 100-inch telescope and its even heavier mount, as seen in the accompanying photo, in Figure 46. Nevertheless, having moved the apparatus to a higher altitude and using a slightly longer light path, a respectable quantity of ether wind was detected which may have approximated or even exceeded the result observed by Miller. *Incredibly, the results were unjustifiably reported by MPP in negative terms:*

"... precautions taken to eliminate effects of temperature and flexure disturbances were effective. The results gave no displacement as great as one-fifteenth of that to be expected on the supposition of an effect due to a motion of the solar system of three hundred kilometers per second. These results are differences between the displacements observed at maximum and

145

minimum at sidereal times, the directions corresponding to ... calculations of the supposed velocity of the solar system. A supplementary series of observations made in directions half-way between gave similar results." (MPP *Nature* 1929a)

In this case, MPP compared their results not only to Earth's orbital velocity of 30 km/sec, but to the presumed 300 km/sec velocity of the solar system through space. One fifteenth of a presumed ~300 km/sec "velocity of the solar system" is ~20 km/sec. Inexplicably, in their later article in the *Journal of the Optical Society of America* (1929b), they changed the "one-fifteenth" 1/15 fraction to "one-fiftieth" 1/50. Even so, one-fiftieth of ~300 km/sec is 6 km/sec, a not insignificant value.

While the skimpy and contradictory MPP account left out considerable detail, their last experiment on Mount Wilson indicated they actually did measure the ether wind, with a velocity of something under 6 to 20 km/sec. That velocity is within the same range of ether-drift as obtained by Morley-Miller and Miller independently.

In summary, all three of the MPP experiments published in 1929 were poorly reported, containing biased assumptions and unresolved experimental artifacts, especially in the first and second round of experiments where they admitted the interferometer was yielding spurious results due to problems of temperature and instability. But they included those results in their publication anyhow. They also expressed contempt towards the postulate of an Earth-entrained cosmic ether, which would clearly have slowed down any ether drift or wind inside the metal observatory dome and its concrete pier. That important possibility was dismissed by MPP *a priori*. Their published account failed to give important details about dates, times, locations and material surroundings of the interferometer. The MPP omissions were important, given how Miller had already reported that some times of day, and some days of the year, will yield results quite different from others, due to the variable speed of the ether-wind velocity over the course of a year, and how the rotation of the Earth brings the interferometer into a greater or lesser exposure to the ether's motion.

The MPP data interpretations and conclusions were cemented into static-ether assumptions, without reference to ether-entrainment due to altitude or stone/brick basement locations. Nevertheless, their results on the third trial confirmed Miller's results, albeit probably minimally (assuming the 1/50th correction was accurate), but again without the level of detail one would require for a full understanding.

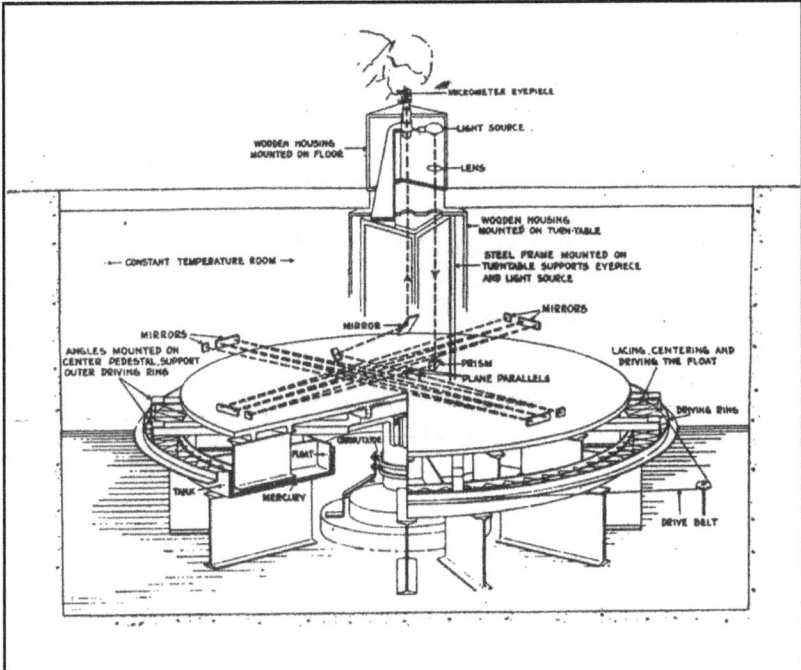

Figure 45. The Michelson-Pease-Pearson Mount Wilson Experiment. A successful detection of an ether-drift signal of an unspecified quantity approaching 6 to 20 km/sec was reported in their 3rd experiment, published in 1929. This positive result was inappropriately dismissed as a "negative" result because the experimenters had prematurely discarded the implications of an Earth-entrained and slowed ether wind. This experiment applied the largest light-beam interferometer ever constructed by Michelson, with a 26-meter round-trip light path. It was nevertheless far less sensitive than Miller's 64-meter interferometer. It is shown here, situated in a "constant temperature room" within the concrete base of the Mount Wilson Observatory, which by itself would also predictably block and reduce the measured result of ether velocity.

147

Figure 46. The Massive Concrete Base of the Mount Wilson Observatory, designed to support the new 100-inch telescope and its mount, under construction in the 1920s. This base has a flat circular top, the surface of which became the main floor inside the observatory dome. The dome was erected on the steel-frame structure surrounding the concrete base. The base structure was designed to support the heavy telescope and its even heavier, large mount. Many of the experiments at Mount Wilson claiming to "disprove" the ether were undertaken inside this massive concrete base structure, which would have blocked nearly all of the ether drift, even while providing some stabilization for temperature changes. Miller, by contrast, undertook his ether-drift experiments in a small house a short distance away, with windows all around at the level of the interferometer light beams. His approach for temperature stabilization was the use of insulation material in the house and on the interferometer itself, which was allowed to reach its own equilibrium with the ambient temperature inside his measuring house. The huge size of this observatory structure can be estimated by comparing the tall ladder near the square opening at the base in the above picture (in circle), with the completed observatory to the right.

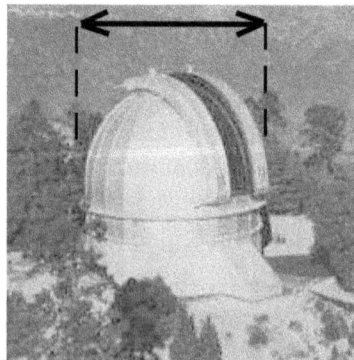

I also must question if Michelson took the lead in these experiments, or in writing up the published reports. Certainly he would not have forgotten his own original conclusions from 1887, and knew Morley and Miller had obtained better results in Cleveland and at Mount Wilson than he had when working with Morley. Michelson suffered several strokes late in his life, and died in March 1931. I have no information about when those strokes occurred, but they might have prevented him from evaluating and writing up the results of these experiments. Michelson remained friendly to the ether concept to the end of his life. By contrast, Pease and Pearson appeared to have been committed followers of Einstein's relativity, as exposed after Michelson's death. A possible disagreement between them might be inferred from a later 1930-1931 experiment, to be discussed below, in which the three men again cooperated, at Irvine Ranch in California.

1926: Roy Kennedy Blocks the Ether, Obtains Zero Result

Physicist Roy Kennedy made an effort to repeat the results of Michelson-Morley, and of Miller up to that time, using a small but conventional dual light-beam interferometer with an overall light path of only 4 meters. As revealed in his publication "A Refinement of the Michelson–Morley Experiment" in the *Proceedings of the National Academy of Sciences*, his experiment was no "refinement" at all, except in the sense of doing everything possible to kill off the cosmic ether motions.

The first phase of his experiment was undertaken inside a constant-temperature room within the stone-brick Norman Bridge Laboratory building, at the California Institute of Technology in Pasadena. Construction of that heavy stone-brick building had been completed in 1920. Kennedy's experiment proceeded in the early morning of two midweek periods of September, *inside the sub-basement of that building*. As he described his interferometer, "The whole optical system was enclosed in a sealed *metal case* containing helium at atmospheric pressure." No mention was made of the metal material, nor of the location where it was undertaken – that information was only years later reported by Swenson. From this effort, using an instrument of very short light path and consequent reduced sensitivity, cased inside a metal covering, placed in the sub-basement of a stone-brick building, filling the metal-enclosed interferometer with helium gas (which was never a part of the original Michelson-Morley experiment). Not surprisingly, no result was observed. Kennedy wrote:

> "There being no fluctuations in the field of view, it was unnecessary to take the average of a number of readings. As has been shown, a shift as small as one-fourth that corresponding to Miller's would be perceived. The result was perfectly definite. There was no sign of a shift depending on the orientation." (Kennedy 1926)

In the many pages of careful calculations presented in Kennedy's publication, not one of them addressed the issue of ether entrainment effects, much less the issue of using an interferometer of such a very short light path. Those factors had been "perfectly definitely" ignored.

Kennedy's second phase of this experiment proceeded on top of Mount Wilson inside the steel dome (or possibly, worse, inside the massive concrete pier structure as did MPP) protecting the main 100-inch telescope. "Because an ether drift might conceivably depend on altitude... Here again the effect was null." His report indicated absolutely no fringe shifts were obtained, which was hardly a surprise.

Shortly thereafter, a "Repetition of the Michelson–Morley Experiment Using Kennedy's Refinement" was published in *Physical Review* describing a new effort undertaken by K.K. Illingworth in 1927. Illingworth used the same metal-shielded interferometer with helium gas developed by Kennedy a year earlier. His experiment was also undertaken in the sub-basement of the Norman Bridge Laboratory in Pasadena. And he again announced, not surprisingly, "the same results obtained", though Swenson reports he did obtain a small result of one-tenth (one km/sec?) of what Miller obtained. (Swenson 1972 p.216)

Putting metal shields around an interferometer with minimal length light beams, and placing the instrument inside a typically massive brick, stone or concrete university building, then inside a sub-basement room designed for thermal stability, or setting it up inside a metal observatory dome or in the massive concrete space below it, was an excellent method for blocking an already entrained ether velocity down to a true zero result. My conclusion is, no significance can be drawn as to the existence or nonexistence of the cosmic ether, nor to ether-drift velocities, from either the Kennedy or the Illingworth experiments. Their use of helium gas also remains an unknown factor, but certainly these were *not* "Repetitions of the Michelson-Morley Experiment".

1926-27: Piccard and Stahel, Tiny Interferometer, Heavy Iron Shielding, Minimal Effort, No Result of Merit

Over 1926-1927, two Swiss scientists, August Piccard (later known for record-breaking high-altitude helium balloon flights) and Ernest Stahel, undertook experimental examinations of the ether-drift question. They attempted a Michelson-Morley type of interferometer experiment at high altitudes inside the small gondola of a large diameter helium balloon, of 2200 cubic meters volume. A second experiment took place within a laboratory in Brussels, and a third trial was made atop Mount Rigi in Switzerland. While their original short reports were published in 1926-27 in the journal of the French Academy of Sciences, *Comptus Rendus*, a more comprehensive report later appeared in the English language, "Realization of the Experiment of Michelson in Balloon and on Dry Land", in the February 1928 issue of *Le Journal de Physique.* In all these reports, laced with more than a touch of arrogance, they declared a zero or null result in every case. A review of their work and claims shows many serious problems, however.

The central problem was, Piccard and Stahel used a regrettably small interferometer shielded inside an iron and aluminum box enclosure. The total round-trip light path of their instrument was only 2.8 meters, with all the optical components fixed to an aluminum block measuring 43 cm square, and 4 cm thick. This compared poorly to Michelson-Morley's 22 meter and Miller's 64 meter interferometers. They used a movie-film recorder to capture images of the interference fringes as well as a set of markers indicating its azimuth compass orientation. From this, they claimed the ability to make extremely precise post-facto judgments of the actual results of their experiment than would be possible by direct viewing with the eye.

During the balloon experiments, unequal solar heating became a serious factor, in spite of having shielded their interferometer from temperature fluctuations by using a quantity of ice. This was in addition to *placing the entire interferometer inside the aforementioned iron box enclosure.* Originally they intended to pump the air down to a partial vacuum inside the iron box, but as they later stated, "...the container alone, by its great heat capacity (sic), homogenized the temperature so well that, even with atmospheric pressure, the thermal disturbances disappeared completely." Here I must note, iron has a low heat capacity, heating up to a higher temperature with very little thermal input than, for example, water, which has a very high capacity for absorbing heat with minimal thermal increase, making it an excellent material for heat storage. Did Piccard and Stahel make a *faux-pas* in their writing? Was it a translator's error? Or were they ignorant of thermodynamics?

When used in the gondola of a balloon, the turning of the interferometer was accomplished by rotating the entire balloon and gondola, to which the instrument was fixed. Two electrically-driven propellers produced a rotation once every 25 seconds. Both experimenters apparently rode in the gondola with the instrument, as they indicated the propeller-driven rotation could not be sustained for too long without making them airsick. The balloon experiments did not go well, by their own account, due to the lurching and swinging of the gondola, and also due to premature melting of the ice and other factors.

Their balloon experiment began with ascent on 20 June in 1926, with measurements taken from midnight to 4:00 hrs, and later at 10:00 hrs on the 21st, at altitudes of 2500 to 4500 meters (8200 to 14760 feet), respectively. The 4 AM measurements were plagued by thermal disturbances and so they were not confident in those results. Their 10 AM measurements were better.

> "At ten o'clock in the morning, to 4500 m, the visual observation only made it possible to note that the displacement of the fringes did not reach a tenth of the distance between two fringes, corresponding to approximately 30 km : sec."

Assuming their balloon experiment truly did detect a 3 km/sec result (one tenth of 30 km/sec) that is not "nothing" nor "null", especially when considering the limitations and inefficiencies of their self-defeating experimental design.

For their experiments on land, they firstly used their interferometer with undisclosed refinements in its "thermal protection" (more shielding?) in an unknown laboratory in low-elevation Brussels. This was undertaken at midnight on the 23rd, 25th and 29th of November 1926. Here, they suspended the interferometer with its aluminum base and iron cover on a "subtle rope" tied to an overhead iron beam in the laboratory ceiling, with a supporting pivot which came up from the floor. A small electric motor with propeller drove the instrument in a circle. The experimental data was recorded by direct observation on the 23rd, and by film-camera for the other two dates. On those two dates

"The first group of ten turns, measured by a collaborator not trained, had to be re-measured".

Their final effort was undertaken at the top of Mount Rigi, at an altitude of 1290 meters (4230 feet) on 16-17 September 1927, from 5 to 6 AM in the morning. Here, they undertook the experiment "under the roof of the highest hotel" without reference to possible ether-blocking by roof beams, tiles, and other materials for waterproofing and carrying of heavy snow-loads, as well as insulation and roof-interior construction materials. Presumably, the same metal covers were used in the Mount Rigi experiment, along with the same method of suspension from the ceiling by rope, as used in Brussels, although this was not clearly stated. Not unsurprisingly, with the same small and heavily-shielded interferometer placed inside an attic under-roof hotel room, they reported no results of merit.

One of the Piccard-Stahel experimental runs apparently did produce a velocity of 7 km/sec (cited by Miller in 1933,). However, this was averaged with the smaller results by Piccard-Stahel to yield a net ether velocity of 1.5 km/sec. The many problems aside, they made so very few observations with their inefficient apparatus it is remarkable they would attach any significance to them. Reading the details of this experiment, in the light of prior successful ether-detection experiments, the Piccard-Stahel effort appears comical. Were they really serious?

Where Miller undertook hundreds of turns of his interferometer over numerous seasonal periods, with exceptional care in construction and operation of his instruments, Piccard and Stahel revealed a few tens of turns in what was a rather lazy-bones effort, beyond all the other difficulties. In conclusion, they could not identify any periodicity of azimuths as anticipated, so they concluded a fully negative result.

Overall I have to say, the efforts of Piccard and Stahel were incompetently undertaken, with serious errors of basic assumption and often plain ignorance of what Miller had already published by the time when they planned and executed their experiments. They constructed a small interferometer and applied it in ways which were guaranteed to produce a near-zero result. And then, in sheer arrogance, they boasted about the superiority of their own brief experiments over the claimed sloppy methods and inadequate instrumentation of Miller, in a *put-down by spittle* which never should have been allowed in a scientific publication. After narcissistically extolling the virtuous excellence of their own efforts, they declared: "Our impression is that the apparatus of Miller does not have a precision sufficient for this kind of research."

(Piccard-Stahel 1928, p.59) Their own effort and written descriptions revealed not merely arrogance and incompetence, but comedy. I could only imagine Miller getting a good laugh out of their efforts, which I certainly did. Others have cited the Piccard-Stahel work in serious tones, however, which suggests they never read beyond their claimed "results". *Einstein applauded them.* (Einstein interview, 1927)

Their prejudice against Miller was also palpable in the tone of their writings, and how they referred to him as "Mr. Miller", rather than the generally-used and appropriately respectful manner common among high-level scientists and professors in Europe, such as "Professor Miller", or "Professor Dr. Miller". This was also a not-too-subtle disparagement, suggesting haughty attitudes aiming to dismiss and disregard the primary American scientist whose work seriously challenged their hero, Einstein. Piccard and Stahel went on to engage in lectures and pointed debates with further put-downs of Miller in a few science journals, but overall it held no meaning. They were dedicated to Einstein from the get-go, and conducted their experiments in such a manner as to guarantee the absence of significant results.

1930-1931: Michelson's Irvine Ranch "Speed of Light" Variability, Another Confirmation of Ether Drift?

The same team of Michelson, Pease and Pearson (MPP) previously discussed, went on to make basic speed-of-light measurements in 1930-1931, at the James Irvine Ranch, in what is today Irvine, California. This experiment was not organized to seek out any ether drift, and the "ether" word did not appear in their published paper. The experiment was further predicated upon the *a priori* assumption that there was no true variation in "the" speed of light. Their paper, entitled "Measurement of the Velocity of Light in a Partial Vacuum", was published in the 1935 *Astrophysical Journal.* Nevertheless, as they and others reported, *significant variations in light speed were observed.*

The experiment was undertaken with exceptional care and, unlike their prior 1929 papers on ether drift, was described in great detail in the above publication. The Abstract is reproduced here:

"The observations were made by the rotating-mirror method, the light passing through a steel tube 1 mile long, evacuated to pressures which ranged from 0.5 to 5.5 mm mercury. By multiple reflections the path length was increased to 8 or 10 miles. The distance was obtained by reference to a carefully measured base line adjoining the tube. The time was measured stroboscopically through successive steps by use of a tuning fork synchronized with the rotating mirror, a free swinging pendulum, a chronometer, and wireless signals from Arlington. There were made 2885.5 determinations of the velocity, the simple mean value of which is 299,774 km/sec, *with an average deviation of 11 km/sec from the mean.*" (MPP 1935. Emphasis added.)

The one mile long partially-evacuated steel tube, with an 8-10 mile light path, was constructed lying flat on the ground, oriented roughly southwest to northeast, at an elevation of around 20 meters above sea level. The method was simple, using a rotating mirror technique originally developed by Foucault in 1850, as previously described on p.41. While MPP detected an average light velocity as reported above, they also reported significant light-speed variations of around 11 km/sec in one standard deviation off the mean value of 299,774 km/sec. Their final calculations of the "absolute speed of light" were derived by *simple averaging* of a data set containing substantial variation. While MPP were careful to note the times and dates of every observation, no efforts were made to organize their results according to sidereal time, to see if a cosmic pattern existed.

In any case, significant variations in light speed were observed and reported in their published paper for those experiments, written up by Pease and Pearson after Michelson's death in 1931, and published in 1935. A newspaper account of these experiments, which appeared a year before the final publication of results, stated:

"Dr. Pease and Mr. Pearson say the entire series of measures, made mostly between the hours of 7 and 9 PM, show fluctuations which suggest a [variation]... of about 20 kilometers per second." (Dietz, *Cleveland Press* 30 Dec. 1933)

The published account of this new MPP experiment in 1934 confirmed the newspaper account, presenting data tables indicating significant variations in the "speed of light", of up to plus or minus 20 km/sec or 11 km/sec by standard deviation, with the data falling out into a typical bell-shaped curve. Their paper stated, "Attempts to explain these variations in velocity as a result of instrumental effects have not thus far been successful." In other words, *they were very confident in their equipment and procedures, and so could not understand why there was such variability in the observed velocities!*

Miller also commented on these results, suggesting they might have measured a stronger variation if they had not confined their light beams in dense steel pipes:

> "If the question of an entrained ether is involved in the investigation, it would seem that such massive and opaque shielding is not justifiable. The experiment is designed to detect a very minute effect on the velocity of light, to be impressed upon the light through the ether itself, and it would seem to be essential that there should be the least possible obstruction between the free ether and the light path in the interferometer." (Miller 1933, p.240)

Miller was the world expert on the ether-drift experiments by this point, but was almost systematically ignored in his advice on how to optimally detect the ether. However, even with metal-tube shielding, the extended light path of the MPP Irvine Ranch instrument probably

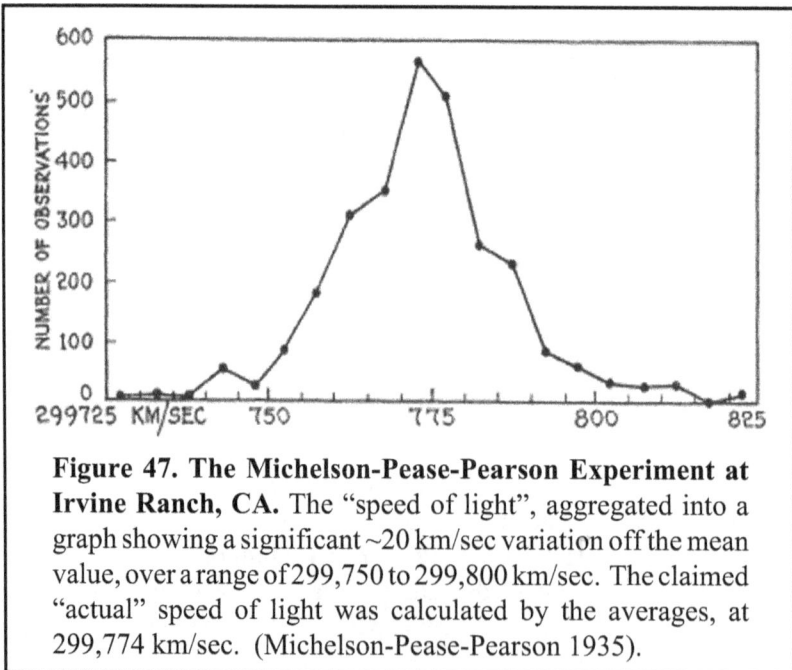

Figure 47. The Michelson-Pease-Pearson Experiment at Irvine Ranch, CA. The "speed of light", aggregated into a graph showing a significant ~20 km/sec variation off the mean value, over a range of 299,750 to 299,800 km/sec. The claimed "actual" speed of light was calculated by the averages, at 299,774 km/sec. (Michelson-Pease-Pearson 1935).

Figure 48. The Irvine Ranch Experiment. Above is a photo showing the one-mile vacuum tube stretching out into the distance, and the buildings holding the various apparatus in the foreground. Below is a diagram of the experimental apparatus. A single beam of light is reflected off a rapidly spinning (500+ revolutions per second), flat octagonal mirrored cylinder. The light beam is reflected down and back, several times, through a one-mile long, partially evacuated metal pipe. The beam then returns back to the same rotating set of mirrors, where slight angular deviations in the returning light beam can be observed, and the round-trip light speed calculated.

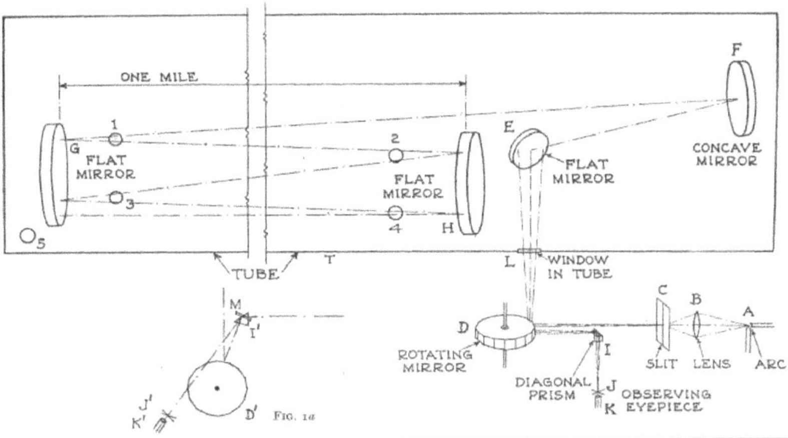

157

compensated a bit for whatever ether-blocking effects were created by the steel pipe through which the light beams travelled. And so, without planning to do so, they detected a portion of the cosmic ether wind.

Given that Michelson-Pease-Pearson made similar detection of variable light speed from their efforts at both Mount Wilson and Irvine, *their data inadvertently once again supported Miller's findings and challenged Einstein, but were never reported in such a manner.* It is also notable that Michelson's participation in both the Mount Wilson and Irvine experiments was the second and third time his work had detected light speed variations due to probable ether wind – the first time being his original work with Robert Morley in 1887, in Cleveland.

Michelson's health took a serious turn for the worse after 1929 (Livingston p.326-327) and he passed away in 1931. His declining health left it to Pease and Pearson to finish the experiments and write the 1935 *Astrophysical Journal* paper on the Irvine work. They subsequently gave up ether-drift and similar optical work in favor of preaching the Einstein Gospel.

1930: Georg Joos, Massive Basement Interferometer, No Results

Physicist Georg Joos teamed up with the Zeiss Optical factory in Jena, Germany, constructing a massive new Michelson-type interferometer and attempting to detect the ether drift by employing a list of high-quality but highly questionable, self-defeating refinements. His instrument stood about 3.6 meters in height (12 feet), having cross arms composed of plates of quartz-glass, of about 6 meters (20 feet) each in length. Such glass has a very low thermal expansion coefficient, and all optical components were mounted upon that quartz-glass platform, which in turn rested upon a metal framework. The total round-trip light path was 21 meters, slightly less than the original Michelson-Morley experiment, and significantly less than Miller's 64 meter instrument.

The light source of the Joos-Zeiss interferometer was placed above the center of the two cross arms with a recording camera located below that same center point. The instrument was suspended from an iron beam in the ceiling from which it could be rotated, with a pivot and motor for turning the interferometer set below the base near the camera. Vibrations during turning were dampened by use of fine-hair brushes which remained in contact with the exterior of the device as it was turned. Each rotation took about 10 minutes.

From this marvel of German engineering, one might think they had hit upon a design that included all the best ideas of the prior experimenters. They considered every detail, except for the most important point, *the ether itself.* Ignoring all that Miller had written about the issue of ether entrainment, specifically the necessity to expose the light beams in a transparent manner to the local above-ground environment, the Joos-Zeiss interferometer was placed inside a large cross-arm aluminum-alloy tank. The tank was then hermetically sealed, originally with the plan to pump the interior down to a partial vacuum; that idea was abandoned due to the inability to preserve a given vacuum pressure. The cross-arm tank was therefore sealed but kept at local atmospheric pressure. The instrument was also set up "in a cellar area of the Zeiss works..." in Jena, under one of their large brick and stone buildings at an altitude of around 150 meters (500 feet).

After operating the device for an unclear but apparently short period of time, Joos declared a negative result, of an "...ether wind smaller than 1.5 kilometers/second." Could Joos have ever detected an ether motion of higher velocity, given the placement of his interferometer inside a metal container of unstated thickness (photos suggest they were about 1 or 2 cm thick), and down in a cellar location in the low-elevation city of Jena? Again, not likely.

In 1933, Miller published his most ambitious paper to date, in *Reviews of Modern Physics*, summarizing his results as already described in a prior chapter. In spite of his own ignorance and failures, Joos wrote critically of Miller in a 1934 letter published in the 15 January issue of *Physical Review* (p.114), followed by a rebuttal from Miller. Their exchange is useful to review:

"Mr. Miller finds the cause of the discrepancy in the fact that I enclosed the optical arrangement in a metal case and worked in a massive building, as did the other experimenters cited by Mr. Miller. I did so of course, in order to eliminate disturbances caused by local and temporal variations of temperature. For if, assuming a length of the light path of 30 m, one calculates what difference in temperature of the two branches of the interferometer produces a displacement of 1/10 of a fringe... One gets the astonishing result that a difference of 1/500° is sufficient. The mere warmth of the body of the observer who in Mr. Miller's experiments, stands near the interferometer can produce such an effect. But the question whether the ether

159

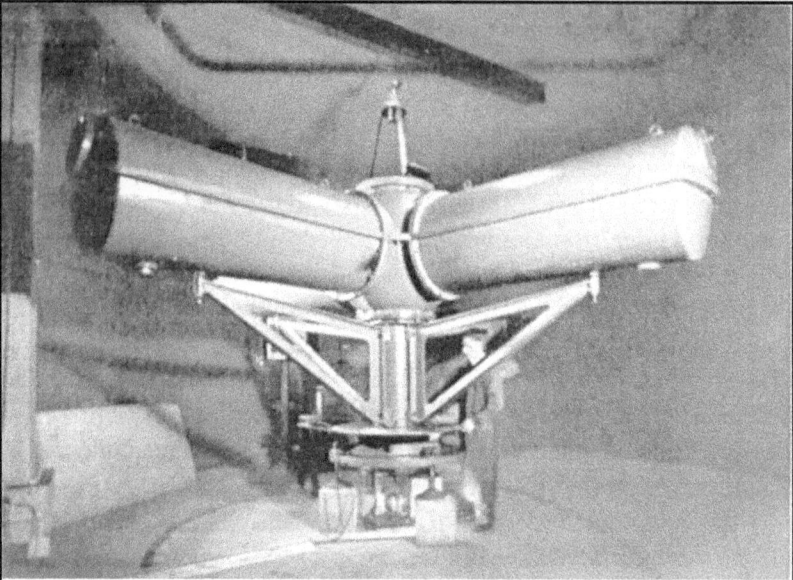

Figure 49. The Joos-Zeiss Interferometer, a massive, complex and fully-automated device, hermetically sealed inside a metal covering, placed in a basement at the low-elevation Zeiss Optical factory in Jena, Germany. No ether-drift velocity greater than 1.5 km/sec was observed, and nothing higher could rationally be expected given the multiple barriers to ether flow incorporated into the design. Below is the device with the metal cover removed.

penetrates the walls of a building, from the point of view of any ether theory, is decided by the fact that in the Sagnac and the Michelson-Gale experiments one gets the full displacement expected from the theory of a resting ether. To make use of this result in an experiment which, without the best protection against disturbances by temperature, is hardly performable... my experimental arrangement is apt to decide the question whether the ether drift exists or not and that it is not – as readers of the paper of Mr. Miller might be inclined to think..." (Joos/Miller 1934)

Note, once again, how the snobbish European physicist – this time writing from Hitler's Germany – showed no respect for "Mr." Miller's Princeton University doctorate and professional status as Chairman of the Case School Physics Department, in addition to his senior status in the ether-drift research. Ignoring the obvious put-down, Miller responded thus:

"A small change in the temperature of the air in the entire light path of the interferometer of the order of magnitude given by Professor Joos would produce a displacement of the fringe system of 0.1 of a fringe width, the *entire* light path being uniformly heated. When Morley and Miller designed their interferometer in 1904, they were fully cognizant of this fact, and it has never since been neglected. Elaborate tests have been made under natural conditions and especially with artificial heating, for the development of methods which would be free from this effect.

It should be borne in mind that the ether-drift observation does not depend upon any absolute reading, nor even upon a simple displacement of the fringes; it depends upon a *regularly periodic variation* in the position of the entire fringe system, and the period is *twenty-five seconds throughout.* [eg., the time required for one half turn of the Miller interferometer] The temperature would have to increase and decrease with periodic regularity in each twenty-five seconds! to produce the [Miller] results. Any irregular fluctuation will be eliminated in the long series of turns. The observer maintains a constant relation to the apparatus and if the warmth of the observer's body is effective, it would be a continual heating effect which produces

161

a *continuous* drift of the fringes, which is of no effect in the calculated results. The body cannot *cool and heat* the air, alternatively every twenty-five seconds, and by variable amounts which depend upon sidereal time.

The ether drift reported cannot be due to the heating of the house; elaborate analyses have been made to detect such effects. The effects are wholly independent of the sun's heat, of day and night, of summer and winter.

It seems quite sufficient that throughout the thousands of observations, the results are found to vary in both magnitude and azimuth in a systematic manner, depending upon sidereal time, and upon the varying combinations of cosmical and orbital motions, as is fully explained in the printed report." (Joos/Miller 1934, Emphasis in original. Brackets added.)

1932: Kennedy-Thorndike Confirm an Ether-Drift, but Deny It!

Roy Kennedy and Edward M. Thorndike pursued the question of Einstein's theory of relativity by using a novel interferometer design of two unequal light-path arms. They hoped to show a similar "null" while evaluating for Earth orbital effects. It appeared to be an effort to put additional nails into the coffin of the cosmic ether. Unfortunately for them, they detected a significant ether wind of ~10 to 24 km/sec.

The core of their apparatus applied the standards of light-beam interferometry, to produce interference fringes from the two unequal light paths. It was quite small in size, of around a half- meter in total, with a difference between the two light paths of about 30 cm (or ten inches). The optical platform was composed of fused quartz, which has an extremely low coefficient of thermal expansion. Unfortunately, the measured light path sections were enclosed in a spherical metal vacuum chamber, surrounded by a water-jacket with an additional spherical metal container, two forms of shielding which also could block the ether. This apparatus could, by Kennedy-Thorndike's estimation, maintain a stable temperature within the water jacket of no more than 0.001°C variation. It was placed in a "small dark room within a larger one". The smaller room kept the temperature stable to within 0.01°C, the larger room to within 0.1°C. A photographic system was set up to record the interference fringes.

With all the optical components set into an immobilized condition on the quartz platform, and temperatures so dramatically stabilized, the whole apparatus sat motionless inside a room within another room, using Earth rotation to turn their apparatus. Their published data was acquired over different months, from 1929 to 1931, January through October, at the California Institute of Technology, in Pasadena, at 213 meters elevation.

Kennedy-Thorndike sought a "null- zero" effect, to "prove" that light speed in their device was not influenced by Earth's motional velocity through an ether. And so they claimed a null result had been achieved. However, a reading of their paper reveals nothing of the sort. In fact *they detected and explicitly admitted to a short period effect of ~24 km/sec, and a long period effect of ~10 km/sec.* Nevertheless, their results – entirely in keeping with what is known about the cosmic ether drift as previously presented – were declared to be too small and irrelevant by Kennedy-Thorndike. In dismissing their own results, they made the following statement, revealing they had no conception of, or willingness to admit to an Earth-entrained or dragged ether:

> "In view of relative velocities amounting to thousands of kilometers per second known to exist among the nebulae, this [result of 10 to 24 km/sec] can scarcely be regarded as other than a clear null result; it is of the same order of precision as that of the Michelson-Morley experiment. It is perhaps best expressed at present in terms of a velocity, although of course the conclusion to be drawn is that the frequency of a spectral line varies in the way required by Einstein's relativity theory." (Kennedy-Thorndike 1932, p.416. Brackets added)

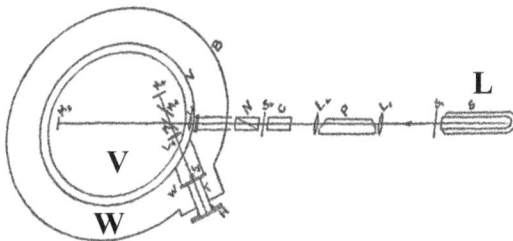

Figure 50. The Kennedy-Thorndike Experiment
Light source "L" sends beams of light into vacuum chamber "V" where beams interfere. Water jacket "W" and room enclosures kept temperatures extremely stable.

This incredible statement serves to illustrate how deeply ingrained was the concept of a static and immaterial ether, or the absence of an ether, which is an absolute prerequisite for the Einstein theory of relativity. In fact, they obtained a positive result, indicating that Earth velocity *does* have an influence upon light velocity, according to the favored direction of ether wind. But since it was around the same velocity as Michelson-Morley and Miller, who never claimed a "null", Kennedy-Thorndike erased all such results from existence in their minds, and used some magical words to make it all go away.

From the reports given in this chapter and summarized in Table 8 below, it is seen how one often finds the claims of "negative results" that were factually positive results. These papers are continually misrepresented by modern physics and astronomy as being negative in outcome, even though almost all showed positive indications for an entrained ether and light-speed velocity variances of significance. This is so, even though they were undertaken with far less precision and care than the original Michelson-Morley, and later Morley-Miller, and later still the independent experiments of Dayton Miller in Cleveland and on Mount Wilson.

Table 8: Velocities of Post-Miller Ether-Drift Experiments

Date	Researchers	Results
1926	Kennedy	true null
1927	Illingworth	true null
1927	Piccard-Stahl	1.5 to 7 km/sec
1928	Michelson-Pease-Pearson	6 km/sec
1930	Michelson-Pease-Pearson	11 to 21 km/sec
1930	Joos	1.5 km/sec
1932	Kennedy-Thorndike	10 to 24 km/sec

To these above results, additional similar positive ether-drift detections from more recent experimental efforts will be added in the next chapter.

Einstein once correctly wrote: "Experimentum summus judex" (Experiments provide our judgment). Unfortunately, it appears this wise consideration was never granted to the successful ether-drift experiments by his followers, only for the often deeply flawed and claimed "negative-null" efforts.

Michelson remained convinced of the reality of the ether throughout his life, never quite accepting the "new physics" and its mystical mathematics which arose in the wake of his supposed "null" results of 1887. Michelson's daughter Dorothy Livingston Michelson, author of *Master of Light*, a biography of her father, provides a moving account of Michelson's last years (Livingston 1973, chapter 17). His health had been deteriorating since 1929, firstly with a cancer requiring removal of his bladder, and several strokes thereafter. He died in May 1931, at the age of 79, shortly after a friendly and emotional visit from the 20-years younger Einstein.

Recent Confirmations
of an Ether Wind

Yuri Galaev, Kharkiv Ukraine Experiments, 1998-2003

Yuri Galaev is a radio engineer at the Institute for Radiophysics & Electronics, a part of the National Academy of Sciences of Ukraine, in Kharkiv[4]. His work, using both optical light and radiofrequencies to investigate the cosmic ether, produced a confirmation of Dayton Miller's work, "down to the details". His methods were unique, employing new designs, including both a radio wave analysis and a simplified Michelson-type apparatus. Like Miller, Galaev was

Yuri Galaev

one of the very few who embraced rather than ignored the material nature of the ether and the importance of *removing* shielding materials in the surroundings of the interferometer. This appears as a major reason for his success, and for the failure of so many others. He summarized the matter succinctly:

> "In 1933, Miller has marked the shielding property of metal covers in his work. However the scientific community did not react properly to such peculiarity, shown by him in this work... there was a lot of experiments with zero results obtained with the interferometers screened by metallic chambers by that time. ...proper significance [had not been given] to Miller's conclusions 1933 about the inapplicability of metal boxes in the experiments with an ethereal wind. Thus, proper checks of Miller's experiments weren't conducted yet until nowadays, in spite of numerous physicists' attempts to repeat his experiments! All his followers carefully screened the devices from an ethereal wind by metal chambers, and, according to A.A.

4. Kharkiv is the Ukranian city once called "Kharkov" during the Soviet era.

Atsukovsky's image expression, '...it's the same that to make the attempts to measure the wind, which blows outdoors, by looking at an anemometer put in a closed room'." (Galaev 2001, p.212)

When firstly learning of Galaev's work in 2003, I invited him to lecture on his findings at a meeting of interested scientists in California. Unfortunately, funding was not available to fly him over from the Ukraine. His work is hardly known in the USA. Here I will relate the essentials of his experiments, as taken from his two published English papers (2001, 2002), and from our correspondence. In some measure, my own theoretical understanding of his work with radio waves is incomplete, given his poor English descriptions of it, and my inability to speak Ukrainian or Russian. But it was clear that his radio wave experiment produced positive results, as did his optical-wave experiment, in a manner agreeable to Miller and other successful American ether-drift researchers.

Galaev's Radiometric Findings on Cosmic Ether, 1998-1999:

Galaev's work with radio waves ran over 5 months from September 1998 through January 1999. His findings first appeared in Russian and Ukrainian publications in 2000, followed in 2001 by an English summary "Ethereal Wind in Experience of Millimetric Radiowaves Propagation", in the journal *Spacetime and Substance*. The experiment was at run at Kharkiv, Ukraine, at 49.9° north latitude and 150 meters above sea level. Galaev's radiometric determinations were made by establishing an 8 millimeter radiowave link between two antennas erected on the rooftops of buildings on slightly higher hilltops, oriented generally East-West, with a distance between them of about 13 km. Figure 52 gives a cross-section, antenna "A" being located north of Kharkiv, antenna "B" near the village of Russia Tishky. Recording devices were placed at both antenna, for comparison of decibel signal strengths of the radiolinks, across the higher altitude direct line A-B, as compared to the indirect lower altitude, reflected A-D-E-C-B signals. As the Earth turned on its axis, the radio-link and the local terrain were exposed to changing cosmic factors. Once adjusted for changing weather conditions, the two signals should have been of constant comparative intensity, by the theory of no ether and "empty space". However, Galaev's experiment demonstrated a *sidereal hour variation in the signal strength of radiolink interference.*

Figure 51. Galaev's Radiowave Antenna used in his ether experiments of 1998-1999. Left is the rooftop antenna at Kharkiv (Point A, below), Right on a rooftop at Russia Tishki (Point B), separated by 13 km.

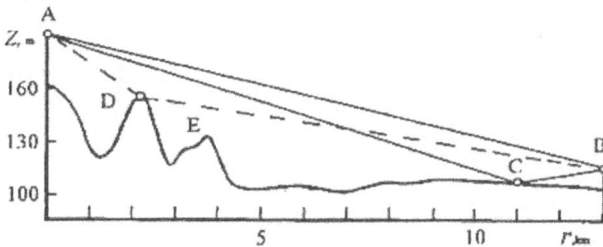

Figure 52. Galaev's Radiowave Experimental Diagram, showing vertical elevations of antenna versus their distances, with intervening topography. Signals were sent from points A and B, creating interference patterns off hills D and C. Variations in the interfering signal strengths were recorded.

Measurements were taken around the clock, for a total of 1288 hours of measurements over the 5-month period, from September through December in 1998, and January of 1999, with exceptions of weekends, holidays and irregular power-outages. Galaev's experiment revealed a clear non-random pattern in the sidereal hour radio-link interference, by which he calculated an average ether velocity of 1.4 km/sec, or 5040 km/hr. While this velocity is small by comparison to the work of Miller whose 10 km/sec computes to 36,000 km/hr, it is nevertheless a reasonable velocity, and certainly nothing trivial in the world of modern scientific measures. Figure 53 presents Galaev's sidereal-day radio-link determinations, in decibels, which appears to me as the *inverse of the actual ether velocity* at his location.

Galaev assumed the radio-link interference was in direct proportion to ether velocity. However, I would argue his radiolink interference values are bettter understood as being the *inverse* of ether velocity. A *faster ether velocity would suggest a lower ether density or viscosity,* allowing for a less distorted and cleaner, clearer radiowave transmission. A *slower ether velocity suggests a greater stagnation and entrainment, with an increased ether viscosity or density*, creating higher decibels of static distortion and interference in radiolink signals. An analogy would be how weaker winds aloft allow the lower atmosphere to become calm and stagnant. Higher winds aloft moves the atmosphere at all levels. From this, and by the dynamic ether theory, the Ukraine would be exposed to a south-north ether wind of higher velocity but lower viscosity and density at 5-7 hrs sidereal, thereby *reducing signal interference* at that time. Conversely, at 17-19 hrs sidereal, with the

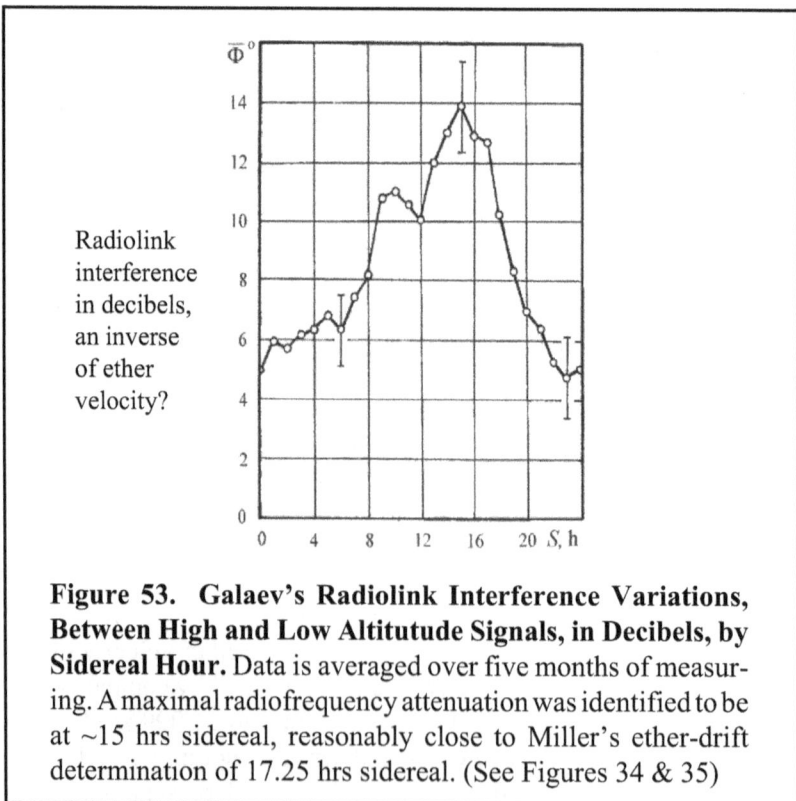

Figure 53. Galaev's Radiolink Interference Variations, Between High and Low Altitutude Signals, in Decibels, by Sidereal Hour. Data is averaged over five months of measuring. A maximal radiofrequency attenuation was identified to be at ~15 hrs sidereal, reasonably close to Miller's ether-drift determination of 17.25 hrs sidereal. (See Figures 34 & 35)

Earth shielding the radiolink from the same generally south-north ether velocity, the radiolink would experience a greatly slowed or stagnant ether motion, of greater density, thereby increasing the decibel amplitude of signal distortion and interference. Galaev's Figure 53, in any case, is composed from decibel data, of radiolink interference. My interpretation is also congruent with Galaev's optical ether-drift determinations, as presented below. A few years ago, I attempted to discuss this matter directly with Galaev, but lost contact with him following the Russian invasion of Ukraine. His email has gone silent and his whereabouts are presently unknown.

In any case, the overall lower estimates for Galaev's radiolink results appears due to his relatively low altitude of ~150 meters above sea level, by comparison to Miller's high-altitude 1850-meter location on Mount Wilson. Galaev also computed a theoretical "k-factor" (see p.123) altitude-adjusted Earth velocity for his location in Kharkiv, presuming his experiment was done at an altitude comparable to Miller's Mount Wilson location. That calculated theoretical velocity was ~8.49 km/sec, close to Miller's result.

Galaev's computation for the final components of net motion were similar to Miller's original 1925 northerly axis of ether drift. Miller computed a net apex of cosmic ether drift at 17.25 hours sidereal, while Galaev's determination was close, at ~15 hrs sidereal. The variance may be understood given how Miller's best determinations came from four seasonal epochs, of 10 days each, centered on April 1st, August 1st, and September 15th of 1925, plus February 8th of 1926, while Galaev's best determinations with radio waves transpired from September 1998 through January 1999. Galaev acquired more data from the critical wintertime period than others before him, *a time when by a dynamic ether theory, the Earth's net cosmic velocity through space becomes increasingly low,* as discussed in Part III. Galaev did not publish velocity determinations for the individual months, unfortunately. As previously discussed, standard astronomy sets the Sun's Way at 18.36 hours sidereal, with the Galactic Center meridian also being close, at 17.45 hours sidereal. Within these variations of measuring period and altitude differences, Galaev's computation for ether-drift direction agrees with Miller's. From Galaev's paper on optical interferometry, discussed below, further agreement was obtained on the sidereal hour maxima and minima of ether-wind velocity.

Galaev's second experimental paper, "The Measuring of Ether-Drift Velocity and Kinematic Ether Viscosity Within Optical Waves Band", was published in 2002, also in *Spacetime and Substance.* For this experiment, he constructed an optical interferometer similar but not identical to the Michelson method. Galaev's instrument in some ways had more similarities to the Fizeau and Foucault ether-drag instruments, comparing light velocity in air versus water, or in flowing versus stagnant water (see pages 40-43). Galaev's device split a single light beam into two *parallel* components, using various mirrors. This parallel motion of the beams evaluates for the difference in light speed **c** versus the speed of the Earth **E** through space, in a direct arithmetical manner (**c** + **E** or **c** – **E**), and is termed a *first-order* experiment. The

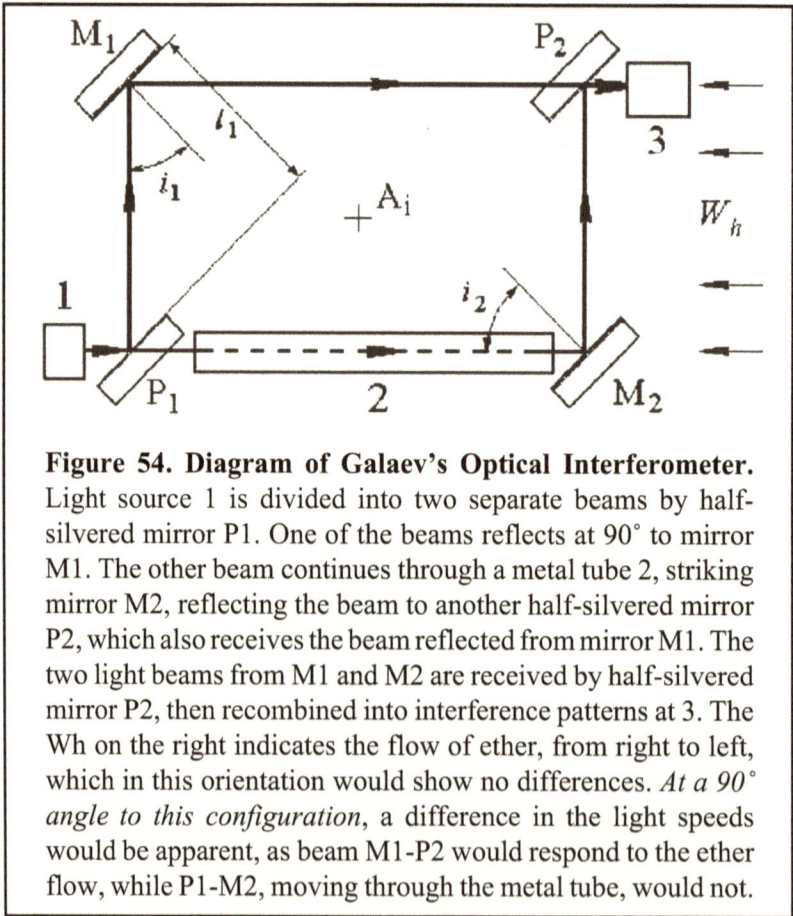

Figure 54. Diagram of Galaev's Optical Interferometer.
Light source 1 is divided into two separate beams by half-silvered mirror P1. One of the beams reflects at 90° to mirror M1. The other beam continues through a metal tube 2, striking mirror M2, reflecting the beam to another half-silvered mirror P2, which also receives the beam reflected from mirror M1. The two light beams from M1 and M2 are received by half-silvered mirror P2, then recombined into interference patterns at 3. The Wh on the right indicates the flow of ether, from right to left, which in this orientation would show no differences. *At a 90° angle to this configuration*, a difference in the light speeds would be apparent, as beam M1-P2 would respond to the ether flow, while P1-M2, moving through the metal tube, would not.

larger Michelson type of interferometer creates two light beams moving *perpendicular* to each other, in a *second order* experiment, contrasting the difference between the square of light speed versus the square of the Earth's velocity ($\sqrt{(c^2 - E^2)}$). (See Livingston 1973, p.74)

Like Miller, Galaev always emphasized the issue of the blocking of the ether by dense materials in its surroundings. Accordingly, his new interferometer *capitalized upon this blocking effect* in addition to using the older design of two parallel light beams. One of the light beams would constantly pass through a narrow diameter, open-ended thick-walled metal tube, shielding that beam from ether-wind influences when it was oriented perpendicular to the ether wind, but exposing the beam when oriented parallel to that wind, allowing the ether motion to flow through inside the pipe. The other beam remained exposed to the ether wind and open air at all times. As Galaev's interferometer was rotated, the metal tube would either expose (parallel) or shield (perpendicular) the light beam in the tube. The other light beam was constantly exposed to the atmosphere and ether wind, no matter how the apparatus was rotated. Rotation of the device could then allow determination of the velocity and axis of motion of the ether wind.

In Figure 54, a proposed ether flow is shown moving from right to left. In such a configuration, both light beams, in open air and in the metal tube, would be exposed to ether flow in an identical manner. The same would occur if the ether wind was moving from left to right, in the opposite direction. There would be no significant difference between

Figure 55. Galaev's Optical Interferometer, 1/2 meter in length.

the velocity of the two light beams in such a parallel-flowing ether wind. By turning the interferometer 90°, ether flow would be significantly blocked or reduced within the tube interior, subjecting that light beam to a near-zero ether velocity, while the open-air light beam would continue to be exposed to the ether wind. Galaev's instrument would thereby reveal different ether velocities dependent upon direction.

By rotating the apparatus through a full 360°, one could identify variations in ether velocity according to either civil clock time or sidereal hour. With enough data from multiple experiments, computations could then be accomplished to determine ether-drift velocity and the net motion of the Earth.

The Galaev interferometer had several advantages over prior apparatus, in that it was predicted to yield results over reasonably short light paths, without repetitive bouncing of light beams. It used a red laser for a light source, and was far less sensitive to mechanical disturbances than those of Michelson, Morley or Miller. From his unique approach, which *embraced rather than ignored the idea of a material entrained ether*, Galaev could identify the velocity and direction of cosmic ether wind at his location. He also identified specific properties of the ether, notably its "kinematic [kinetic motional] viscosity", indicating *the ether behaved similar to a hydrodynamic gas.*

Galaev's optical interferometer, fitted with a 2.5cm thick insulated cardboard cover and painted white on the exterior as seen in Figure 56, was moved to a rural countryside location north of Kharkiv, with an elevation of 190 meters above sea level. In the first location, the interferometer was set up on a tripod, above the ground surface by 1.6 meters, shaded by nearby trees and operated during part of August 2001. A second location, 4.75 meters above the ground on a rooftop, shaded by an umbrella, was used over the remainder of August, and through September, October and November of 2001. The two locations were no more than 15 meters apart horizontally, and both were set outdoors in the open air, away from any structure that might inhibit an ether flow at the level of the light paths. Those locations were selected in part to evaluate for variations in ether velocity due to small elevation differences. A third location in Kharkiv was employed during the winter months of December 2001 and January 2002, inside a room with windows on the upper floor of a brick building. The ground elevation was 130 meters and the building height added another 30 meters.

Measurements were taken episodically, with each measuring cycle lasting 25-26 hours, and with 2 to 4 such cycles per month. The instrument was first allowed to adjust to local temperature for an hour, with one separate measurement undertaken each hour. The measures were carried out by initially orienting the long axis of the instrument with the light source aiming towards the north. It was then turned slowly, stopping the rotation for a few seconds with a readout of fringe shifts every 15 degrees, for a total of 24 measurements per rotation, finally returning to the original north position. Approximately 5 to 7 turns were made for each of the hourly readings, which took about 10 minutes in total; an average of each hourly value was calculated for those 5-7 turns, after which the experiment paused until the next hourly readings. The pause at each 15 degree interval was to allow for natural restoration of ether wind within the metal tube, in keeping with the theory of metal shielding and increased viscosity of ether inside the tube when pointed in directions more perpendicular to ether wind.

Figure 56. The Galaev Interferometer on a Rooftop placed inside a thick cardboard cover box mounted on a tripod, with a rotating platform, near Kharkiv, Ukraine. The cover box walls have an interior layer of a soft insulating material, with a total 2.5 cm thickness. The cover box is painted white on the exterior for light reflection. It was shaded with an umbrella or placed under shading trees during periods of operation.

A total of 2322 measurements were made over the period from August 2001 to January 2002, with a total of 93 turns of his interferometer (2.5 times as many as Michelson-Morley) as follows:

Months 2001	Aug.	Sept.	Oct.	Nov.	Dec.	Jan. 2002
Readings:	792	462	288	312	240	228
Turns	33	19	9	13	10	9

The results of Galaev's optical experiment, as well as his radio wave experiment, were very much in keeping with Miller's earlier work. Figure 57 shows the variations of ether velocity by sidereal hour for *one day only* of Galaev's work in August of 1998 and 2001, compared to Miller's August 1925 seasonal epoch. The top two graphics are Galaev's determinations in optics (top) and radio waves (middle) for the one selected day. The bottom graph is from Miller's publication. The similarities are obvious, a maximum ether wind velocity was observed around 5 to 6 hours sidereal, with a minimum velocity around 17-20 hrs sidereal.

Recall that the ~5 hrs sidereal period, for the northern hemisphere mid-latitudes, places every laboratory more directly facing a strong generally southerly ether wind, than the ~17 hrs sidereal period, when the laboratory is higher up above the northern plane of the ecliptic, and is thereby shielded from the south-north motion of ether wind by the Earth itself. As such, a 5-hrs sidereal ether wind would push the Earth to "drift" in the 17 hrs sidereal direction, as described in the prior chapters. Figure 58 presents Galaev's optical and radio wave determinations for the absolute motion of the Earth, further confirming the Miller results.

From Figure 58, it is apparent that Galaev obtained much lower ether velocities than did Miller, with a peak velocity for his optical experiment of ~0.21 km/sec (~470 mph) and for radiowaves of ~1.5 km/sec (~3355 mph). These are nonetheless respectable velocities. Their lower nature is appears due to the *lower altitude and higher latitude* at Galaev's Central Asian location. Also his entire range of experiments took place over periods *with an expected lower ether wind velocity:* Optical from Aug. to Jan.; Radiowaves from Sept. to Jan.

In Figure 59, Galaev's graph indicates a dependence of ether-drift velocities upon altitude, in keeping with all what has been previously discussed. His optical experiments transpired in the low elevation region of Kharkiv, by comparison to Miller's and Michelson's (with

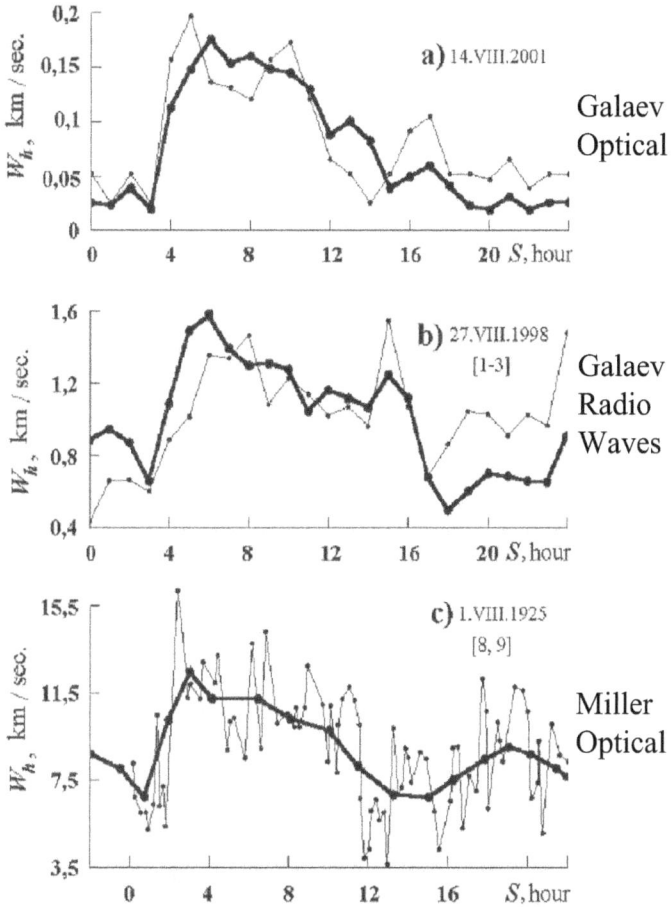

Figure 57. Galaev's Ether-Wind Velocity Determinations for *One Day in August*, Compared to Miller's Similar-Date Determination. The top graph is from Galaev's optical experiment on 14 August 2001, while the middle graph is Galaev's radio wave experiment from 27 August 1998. The bottom graph is from Miller's 1 August 1925 epoch at Mount Wilson. For this monthly period, from different years 1925, 1998 and 2001, all three data graphs share a common highest velocity of ether drift at ~5 hrs sidereal, with a minimum at ~17 hrs sidereal. This should not be confused with the absolute direction of Earth's motion in the heavens. *An ether wind blowing strong at 5 hrs sidereal will push the Earth in a 17 hr sidereal direction.*

Pease-Pearson) results at Mount Wilson. I present Galaev's Figure 59 graphic as he sent it to me privately, even though his own data points need a slight altitude correction of +150 meters above sea level; they are given on the graphic only as height above the ground surface. This altitude adjustment would change his graph only minimally. More detail on this altitude-velocity relationship will follow.

Galaev's new method not only confirmed Miller's prior work, but also provides additional evidence on the properties of cosmic ether:

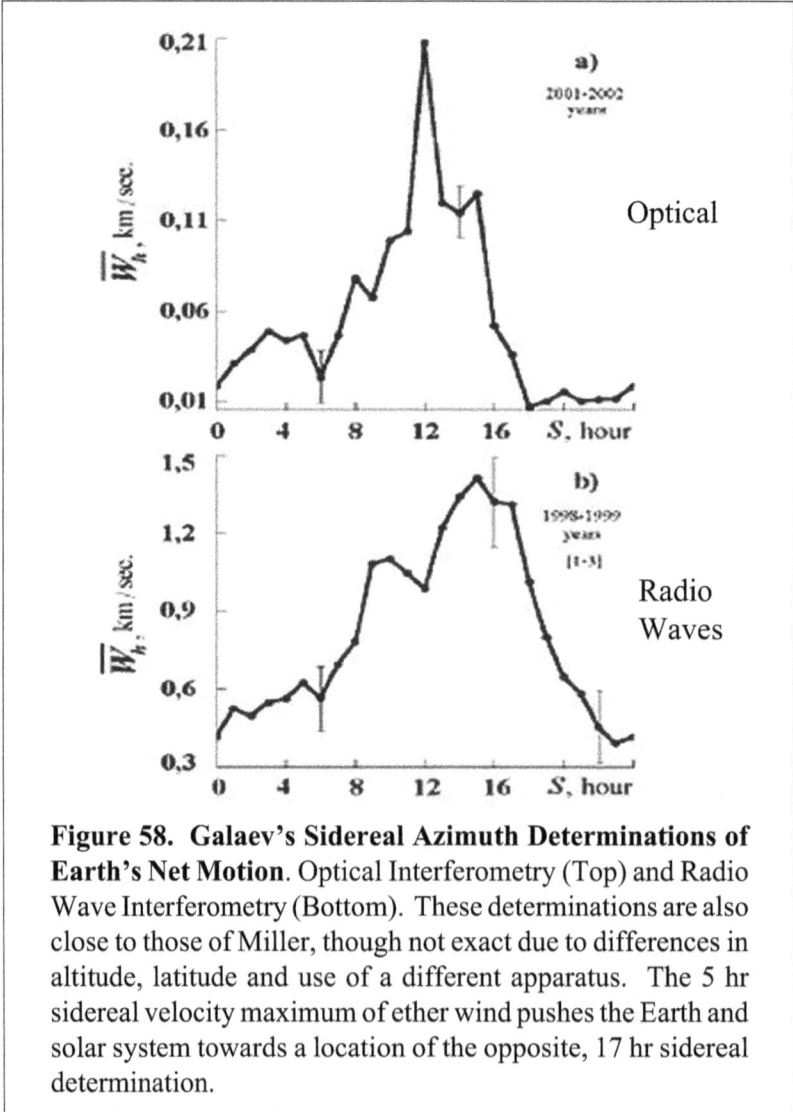

Figure 58. Galaev's Sidereal Azimuth Determinations of Earth's Net Motion. Optical Interferometry (Top) and Radio Wave Interferometry (Bottom). These determinations are also close to those of Miller, though not exact due to differences in altitude, latitude and use of a different apparatus. The 5 hr sidereal velocity maximum of ether wind pushes the Earth and solar system towards a location of the opposite, 17 hr sidereal determination.

1. The phenomenon of ether wind has relevance for both optical light waves and radio waves.

2. The ether has a slight mass, by which to interact with the Earth's hills and mountains, as well as with the vegetation and ground cover. It can be blocked or reflected by metals.

3. Ether velocity is faster at higher altitudes above sea level, and at higher elevations above the ground surface.

4. The ether has a *measurable viscosity* that is similar to a hydrodynamic gas. This not only gives the ether a capacity to flow at different velocities, but also to adhere to the surfaces of matter.

In later chapters, additional properties of the cosmic ether will be identified, including from meteorology, chemistry and biology.

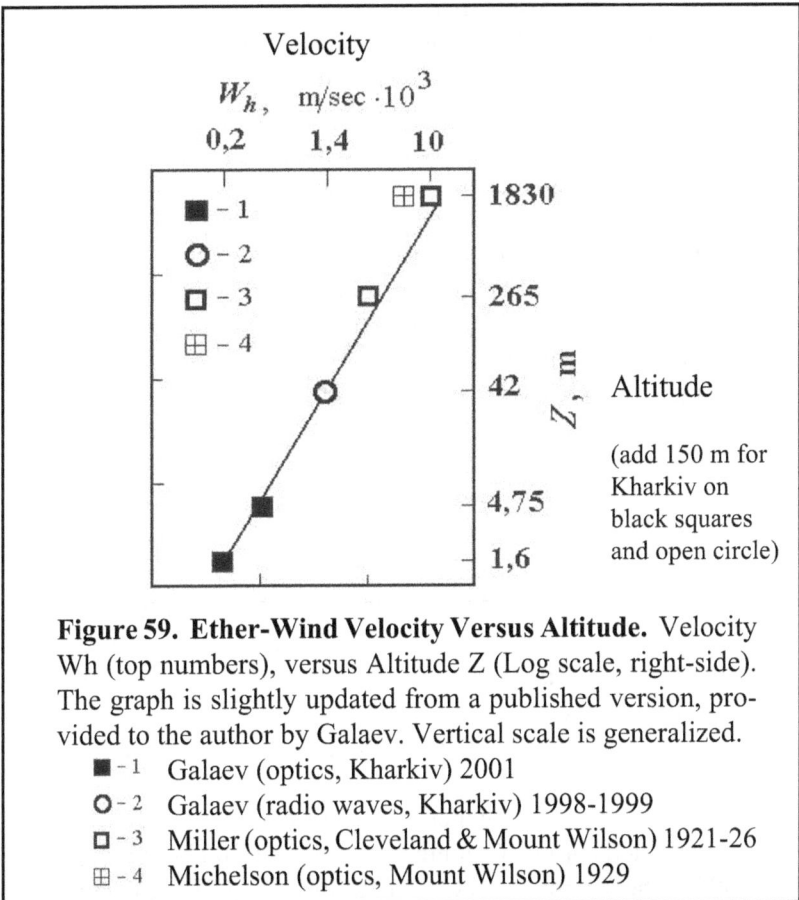

Figure 59. Ether-Wind Velocity Versus Altitude. Velocity Wh (top numbers), versus Altitude Z (Log scale, right-side). The graph is slightly updated from a published version, provided to the author by Galaev. Vertical scale is generalized.

■ - 1 Galaev (optics, Kharkiv) 2001
O - 2 Galaev (radio waves, Kharkiv) 1998-1999
□ - 3 Miller (optics, Cleveland & Mount Wilson) 1921-26
⊞ - 4 Michelson (optics, Mount Wilson) 1929

Héctor Múnera's Experiments, Bogotá Colombia, 1995-2009

Héctor Múnera is a chemical and nuclear engineer at the International Center for Physics in Bogotá, Colombia. Starting in 1995 and continuing over the following 14 years, Múnera led a team examining the historical ether-drift experiments, developing their own ether-drift apparatus and experimental protocol.

Múnera and his team constructed a stationary interferometer with a design similar, but not identical to the original Michelson-type of interferometer. It had two cross-arms, each being around 2 meters in length, one set

Héctor Múnera

north-south, the other set east-west, resting on a heavy concrete base in the lower floor of their facility. They used a polarizing filter and laser light sources. Light paths on their interferometer were enclosed in clear plastic tubing, with an insulated layer of polystyrene encasing the optical components and light paths. Two lasers, red and green, were initially used in parallel, eventually discontinuing the red laser. The interference patterns so developed were photographed once every minute, for 1440 photo-readings per 24 hours, allowing the turning of the Earth to expose the interferometer to different aspects of an anticipated cosmic signal.

The measuring sessions covered a few days each month, collecting data episodically over a full year. The photographed interference patterns were recorded on video, then converted by a computer program into brightness profiles, whereby fringe shifts were determined. Temperature, humidity and barometric pressure were also recorded, with a correction factor for these variables subtracted from the interferometer data. This correction procedure reduced the data to a residual fringe variation, which was larger than was observed by Michelson, Morley or Miller. A sidereal variability was recorded in the residuals, unrelated to known environmental factors. The abstract of Múnera et al 2009, described the experiments of 2003-2005:

> "A Michelson-Morley experiment with stationary interferometer operated during 26 months from January 2003 to February 2005 at the International Center for Physics (CIF) in Bogotá,

Colombia. ... After subtracting the fraction of [environmental] fringe-shift, we obtained a residual that ... represents, therefore, the fringe-shift variation with respect to the motion of earth relative to a preferred frame. The residual also exhibited a 24 h periodicity that was compared to a pre-relativistic model based on Galilean addition of velocity. We obtained the velocity of the sun that maximizes correlation between observations and predictions. Our value is $V = 365$ km/s, $RA: = 81° = 5$ h-24 m, Dec: $= 79°$." (Múnera et al. 2009)

The Múnera team accepted the Michelson-Morley and Miller experimental results, but interpreted their work, and their own, from the theory of a constant-velocity "absolute-space" static ether wind (Newton), abandoning the theory of an entrained ether drift (Miller).

While I acknowledge the experimental results of the Múnera team, I disagree with their theory. To make their theory work, they propose the historical ether-drift experiments of Michelson, Morley and Miller contained significant measuring errors, requiring them to have systematically overlooked large fringe shifts taking place in the plane of their interferometers. There is no evidence I am aware of to support this supposition, nor any other which supports their underestimating the actual number or amount of fringe shifts which occurred in those

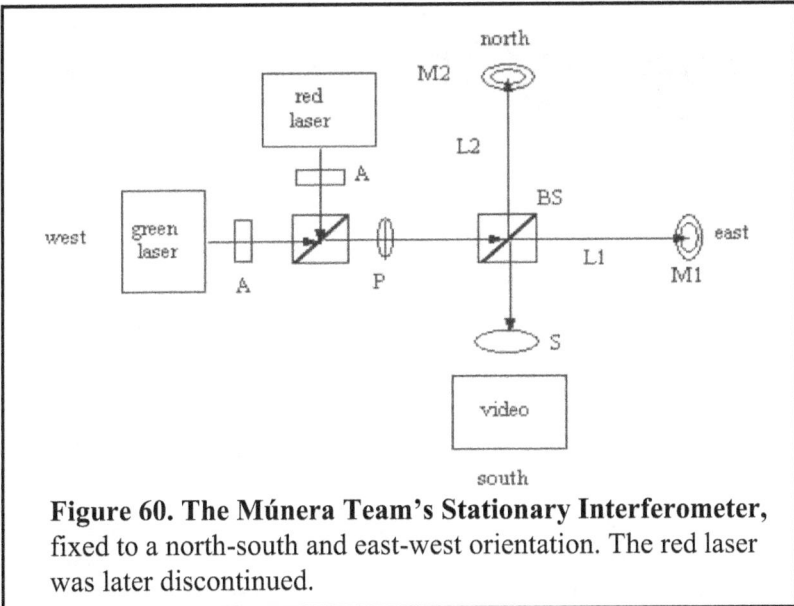

Figure 60. The Múnera Team's Stationary Interferometer, fixed to a north-south and east-west orientation. The red laser was later discontinued.

historical ether-drift experiments. Miller seemed very clear this could not be the case (1933a, p.211-212).

The Múnera team nevertheless adopted a theoretical procedure developed by Miller, to suppose what the ether-wind velocities might be higher up above the atmosphere, assuming Miller's own ~10 km/sec measured ether-drift values were the product of a material entrained ether close to the Earth's surface. As previously discussed (see Table 7 on page 123), Miller's theoretical approach developed a *k-factor* assigning an approximate 20-fold correction to amplify his actual measured ether velocity of ~10 km/sec, to a theoretical value of ~200 km/sec, presumed to exist at the edge of space. (Miller 1933a, p.234-236) However, *Miller was clear that his k-factor velocities were only a supposition, a theory, and not real measured velocities.* The Múnera team took Miller's theoretically higher value for outer-space altitudes as his actual final velocity result (Múnera et al 2009, p.80f), and made no mention of the effects of ether-drag or entrainment. I nevertheless believe it is possible to reconcile the Múnera team's experimental results with those of Miller.

The Múnera team reviewed Miller's 1933 "southerly apex" determination, but like myself, found the northern apex to be more accurate. Their determination of azimuth motions were reasonably close to that of Miller and others, but still a bit problematic. For example, their

Figure 61. The Múnera Interferometer, in Bogotá, Colombia

determination of RA at 5h 24m as the direction of Earth's motion in the cosmos, versus Miller's 17h 25m, requires explanation. I observe that the Múnera team identified that 5+ hr RA based upon it being the same direction as their laboratory was oriented at the time of highest ether-wind velocity. This sidereal hour vector is in agreement with Miller's determination of his own highest ether wind velocity, as previously presented in Figure 34, in the Miller chapter. However, the ~5 hr sidereal vector is *not the net motion of the Earth in space,* which is exactly opposite, at ~17hrs sidereal. By a material and dynamic entrained ether theory, *we view an ether wind blowing strong at 5 hrs sidereal as pushing or floating the Earth towards the 17 hr vector in cosmic space.* By this different theory, the Múnera measurements were exact, but misinterpreted on the sidereal hour vector due to adherence to Newtonian non-material static ether theory. Viewed in the dynamic manner, the Múnera team's results indicate a net motion of Earth towards RA 17h 25m sidereal, at Dec. +79° north, very close to Miller's determination of RA ~17 hrs and Dec. +68° north. By my interpretation, the Múnera results are therefore, like Miller's results, a very strong confirmation of the ether-wind and net motion of the Earth through space. Other interesting aspects of the Múnera team's work deserve further discussion, and a bit of theoretical adjustment, in my opinion.

Regarding their absolute ether velocity, it appears the Múnera team obtained their 365 km/sec velocity by up-calculations using Miller's k-factor. It is therefore reasonable to theoretically down-calculate the Múnera team's reported velocity by the same Miller k-factor of 20x. That would reduce their final determination down to a value of 18.25 km/sec. Given the high altitude of Múnera's Bogotá laboratory, this presumed lower velocity of 18.25 km/sec supports what both Miller and Galaev argued about an altitude-velocity dependence for the ether wind. For example, Bogotá Colombia stands at an altitude of 2,640 meters (8,660 feet), which is higher than Miller's Mount Wilson location of 1,740 meters (5,710 feet). Bogotá is also located at latitude 4.7° north of the Equator, by comparison to Mount Wilson's latitude of 34.2° north. Both factors, of higher altitude and lower latitude close to the Equator would by theory result in a higher ether-wind velocity for the Múnera experiment. Bogotá is also situated on a high mountain plateau, surrounded by tall mountains oriented generally North-South, suggesting any material gaseous-type of fluid ether wind, moving along the south-north axis, would be funneled into the Bogotá plateau, with an amplification of velocity. An ether-drift experiment conducted in

Bogotá would hence be even more clearly exposed to the full measure of ether wind at 5 hrs sidereal, as the ether pushed northward from a southerly apex.

In short, the Múnera team's experiments *provide a general confirmation of Miller*, within the original theoretical framework of an entrained material cosmic ether wind that increases in velocity with altitude above sea-level. From these data and the underlying theory, the Múnera, et al postulate of "absolute space" or static ether is unnecessary. From a recent correspondence, Múnera privately informs me he today has modified his views somewhat towards a more fluidic ether-gas, which would be more compatible with a Miller- or Galaev-type of entrainable ether.

Reginald Cahill's Ether Experiments, Adelaide, Australia, 2002+

Reginald Cahill is a professor of Physical Sciences at Flinders University, in Adelaide, Australia. His method of ether detection was unique, using stationary fiber-optic cables stretched across a distance within his university laboratory, sending a one-way light beam through those cables for determination of variations in velocity. The turning of the Earth provided exposure to the different cosmic directions. His work presented a "...*re-analysis of the old results from the Michelson-Morley interferometer experiments that were designed to detect absolute motion*".

In so doing, Cahill presented evidence indicating the ether wind could penetrate into glass fibers carrying light waves, influencing light speed. He reported a light speed velocity variance of ~400 km/sec, with a sidereal-hour vector at ~17 hrs. While the velocity of his ether-drift measures are very high by what is known from the work of Miller and others in prior chapters, his sidereal hour determination was in accordance with the results of those prior experiments.

As for his working theory, Cahill referenced a "quantum foam, neo-Lorentzian relativity theory" rather than the older ideas of a material and Earth-entrained cosmic ether. He nevertheless correctly criticized the claimed "null" result of Michelson-Morley, even while asserting Miller's result was plagued by thermal artifacts, a claim for which no defendable evidence exists. Einstein and his sycophant Shankland made similar assertions, notably in the Shankland, et al 1955 post-mortem, which as will be discussed in Part II, was deeply flawed and biased.

Cahill also advanced the Miller "*k*" correction value for his ether velocity calculations, but again without clarifying this was Miller's *hypothesis* of what a *theoretical* ether-drift velocity might be at higher altitudes in open space, outside of the Earth's atmosphere and above where Earth entrainment of ether wind might be occurring. Cahill thereby re-computed the lower velocities actually measured by Miller into a higher velocity, as if the velocities in the open space outside the Earth's atmosphere were occurring at Mount Wilson. Not so. By such mathematical amplifications and in accordance with his own theory, Cahill transformed Miller's measured ~10 km/sec velocity into a value of more than 400 km/sec, an increase of 40x! This was double the 20-fold *k*-value increase originally theorized by Miller, and which was also used in lesser measure by Múnera et al. (Cahill 2004, p.81)

As Cahill used a 40-fold multiplication factor to amplify Miller's ~10 km/sec up to a value of 400 km/sec, which was then agreeable to his own theory and fiber-optic determinations, we may legitimately revise Cahill's own measured result downwards, by a 40x *reduction* factor. Such a procedure yields the smaller velocity of ~10 km/sec, which compares favorably to Miller's own ~10 km/sec determinations. Whatever the case may be, Cahill's experiment was the first known attempt to investigate the cosmic ether in the southern hemisphere and to use fiber optics for the light-path.

Cahill also examined a number of similar cosmic velocity experiments, as I have in this work, which used methods *other than* the cross-armed Michelson-type of light-beam interferometer, most of which produced results very close to that of Miller. (Cahill 2003, 2004, 2014) His latter 2014 paper was quite comprehensive on this issue, emphasizing how all such findings violated the basic assumption of the isotropic "empty space" required for Einstein's relativity theory. Cahill was consequently a serious critic of the Einstein theory, but also had dropped much of the older terms and logical understandings of the early ether-drift experiments. While I fully agree with his criticisms of the modern "empty-space" Einstein-relativity view of the cosmos, I find his postulated theoretical alternatives unconvincing. "Quantum foam" is not something affirmed in other empirical contexts and so I consider it hardly better than Einstein's postulated space-time, or Newton's absolute space.

In 2016, Cahill's methods and maths were criticized by Jay Seaver, an Einstein supporter, in a slapjack experiment from which Seaver claimed to have obtained a null or zero result. While Seaver made a few constructive points, his own new experiment was without merit, and in my opinion, neither valid nor even serious. He asserted the false "null" claim about the Michelson-Morley experiment, and failed to cite Miller at all, primarily referencing an amateur, student-level dismissal of Miller, for which Seaver made his own exaggeration of its significance. Seaver also ignored the important ~17 hrs sidereal variation in Cahill's data, and how that was in agreement with other ether investigations, about which he apparently knew nothing.

Experimental Summary

Based upon what has so far been presented, I developed a new graphic in Figure 62 below, summarizing the dependence of ether-drift velocities upon altitude, for all of the major positive ether experimental results. Table 9 gives a summary of the different experimental determinations used to create Figure 62, starting with Michelson-Morley in 1887, and running into the 21st Century work of Galaev, Cahill and Múnera. This new figure shows a clear velocity dependence upon altitude, with a *very approximate* 1 km/sec increase in velocity for each ~150 meters of altitude. Added to this interesting correlation is the fact that, where the azimuth of ether-drift motion was determined in these same experiments, they all pointed towards a very similar direction in

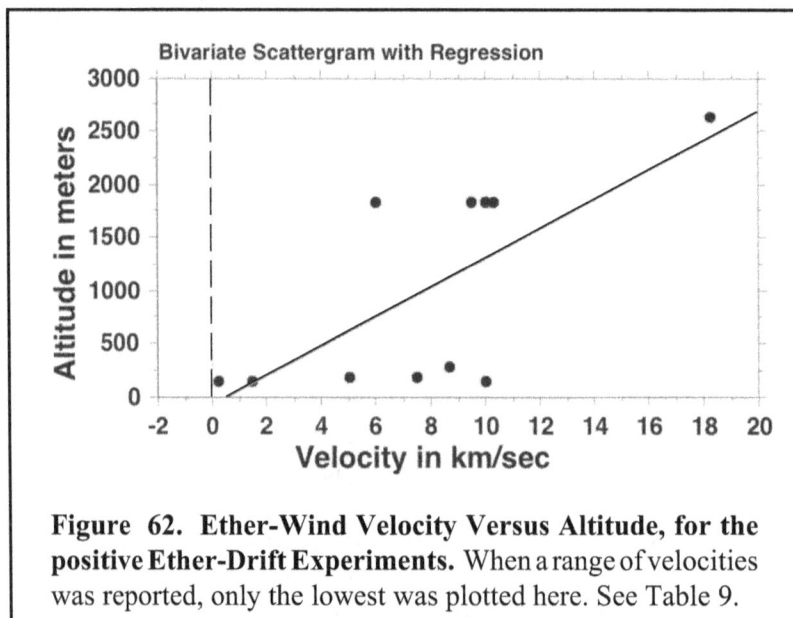

Figure 62. Ether-Wind Velocity Versus Altitude, for the positive Ether-Drift Experiments. When a range of velocities was reported, only the lowest was plotted here. See Table 9.

Table 9: Summary of Successful Ether-Drift Experiments		
Experiment	Approx.Velocity km/sec	Altitude meters
1887 Michelson-Morley Cleveland:	5 -7.5*	199
1902-04 Morley-Miller, Cleveland:	7.5-10 *	199
1905 Morley-Miller, Euclid Heights:	8.7	285
1921 Miller Mt. Wilson preliminary:	10	1850
1924 Miller Mt. Wilson preliminary:	10	1850
1925-26 Miller Mt. Wilson:	10.05	1850
1928 Michelson-P-P, Mt Wilson:	6	1850
1998-99 Galaev radiowaves Kharkiv:	1.5	150
2001 Galaev optics Kharkiv:	0.21	190
2004 Cahill, Adelaid, Australia:	10	150
2009 Munera, Bogotá:	18.25	2640

* Only the lower velocity figures were used in Figure 62.

the cosmos. The similarities of various axes of motional azimuths are presented in a sequential manner, building up the evidence, in Figures 35, 78 and 91, on pages 111, 254 and 289.

The general detection of lower ether-wind velocities at lower altitudes, and higher velocities at higher altitudes, is a direct confirmation of a material ether wind being slowed and entrained at the Earth's surface. These data further explode the widely parroted myth of a Michelson-Morley "null", or that "nobody ever got a positive result for the ether-drift." A great deal of modern astrophysical theory, including Einstein's relativity theory, is *thereby negated.*

Additional but Problematic Ether Experiments

Other ether experiments were undertaken in recent decades which obtained either negative or positive results, but with significant problems in experimental design or serious error in overall theory, frequently with a lack of sufficiently reported detail. Consequently they were not covered in this work. In most cases they embraced the false "null" for Michelson-Morley, rarely mentioning Miller. Or, they blocked their apparatus within metal covers in stone buildings. In such cases, they either obtained no results at all, or questionably high values. Many were either not the usual application of the Michelson optical interferometry, or were rather complicated derivations thereof. And many were run only over very short periods of time. Consequently, it was difficult to directly compare them to the findings already reported. Only a brief account follows, with citations for them in the *Reference* section.

L. Essen 1955: Applied a radio wave resonator inside a steel container at low altitude inside a brick and stone structure, obtaining a 1 km/sec ether-wind velocity, which he dismissed as irrelevant.

Cedarholm, et al. 1958, 1959: Applied dual masers instead of light beams, inside a metallic shield within a massive stone building, inside low-altitude New York City's forest of tall buildings. An ether velocity of 0.03 to 1.5 km/sec was obtained, and was dismissed as irrelevant.

D.C. Champeney, et al. 1963: Applied resonating radioactive sources on a spinning disk with lead shielding inside a metal vacuum chamber, at low elevation inside a heavy brick building. No results were reported, except to proclaim that no ether existed and to affirm Einstein.

Jaseja, et al. 1964: Applied masers similar to Cedarholm, inside a metal shield and stone building at low altitude, obtaining results from 0.03 to 9.2 km/sec. The higher value was summarily dismissed as due to a claimed "magnetostriction", reporting a negative result only.

S. Marinov 1974-1979: Applied a dual rotating mirror apparatus to create interference bands, claiming a questionable positive result of 100 km/sec early in his studies, and 300 km/sec in later efforts, though at cosmic coordinates very far away from those of others previously reported. He did not mention Miller or the issues of altitude or material shielding. His curt dismissal of all prior ether-drift research was suspect, as were his erratic threats to kill himself when *Nature* magazine refused to publish his article. He eventually did commit suicide.

E.W. Silvertooth 1986: Applied an apparatus supposed to overcome alleged "flaws" in the Michelson interferometer, and also said nothing about altitude or shielding. He claimed a detected velocity of 378 km/sec, aligned with the constellation Leo, associated with the 3°K Cosmic Microwave Background Radiation (CMBR). No mention was made of his experimental location, or altitude, or material surroundings. By my estimation, and with a change to an entrained dynamic ether theory, his complex experiment could just as easily be understood as revealing a

~17 hours sidereal vector as obtained by other ether-wind studies, which lays at 90° from Leo. An elaboration of this experiment would require many additional pages, and so must wait for a separate discussion.

R. DeWitte 1991: Identified a phase shift in radio frequencies sent along a 1.5 km copper coaxial cable, oriented north-south. As the Earth turned, he found a sinusoidal pattern with a peak at RA 17 hrs sidereal, in confirmation of prior ether research. However, his claim of a 500 km/sec velocity appears incongruent. While his experiment ran for nearly a half-year, June through December, he never published his data or details except in a few emails and on a website. Only three days of his data were publicly presented, on a graph which could not by itself realistically differentiate between sidereal or civil clock time. Others have reviewed his work with published summaries, but the lack of full details remains a barrier to knowing with certainty the results of his experiment. (see Cahill 2006, Kehr 2002).

Müller, et al. 2003: Applied two opposing laser beams over short distances aiming at cryogenic optical resonators enclosed in a metal shield inside a stone building at low altitude. "Ether" was largely erased from consideration. They claimed to have "measured" a seriously zero result even more precisely zero than the zero results they wrongly asserted was obtained by others. A Big Zero, indeed.

M. Grusenick 2009: Presented a YouTube video of a vertical interferometer experiment, indicating a shifting of about 10 full fringes as the instrument was rotated in a vertically-oriented plane. (WebRef.22) Insufficient detail was presented to make a proper analysis, even while some concerns were obvious, such as a short light path and possible gravitational slumping of the mirror system as it was rotated. Another YouTube video with even fewer details indicated a failure by a second party to replicate this experiment. However, a successful but as yet unpublished replication came to my attention just before going to press with this book. None had sufficient detail by which to judge their merits or present them here. It is nevertheless an important experiment, and should be replicated with significant stress-testing to evaluate for conventional explanations, along with both video and written documentation.

I should also mention, it is very likely there are other more recent ether-drift experiments which I do not know about, and hence could not include in this work. Readers are welcome to inform me about any such neglected experiments, with either positive or negative results, through the publisher, for possible inclusion in newer editions of this book.

These problematic experiments aside, Figure 62 presented a few pages back, as well as Figures 35, 78 and 91, on pages 111, 254 and 289, are the most solid rebuttal to anyone claiming the ether-drift experiments didn't produce positive results, or were faulty in some manner. Those figures not only reflect significant overlapping and reinforcing positive results, but also stand as a confirmation on the material-substantive and motional properties of the cosmic ether, of so-called "empty space".

New properties of the ether have also been identified in this chapter, as with the work of Galaev indicating ether's behavior like a very thin hydrodynamic gas, which interacts with metal shields to create boundary-layer effects. Múnera's team also showed good results, even while – in my opinion – their results are better understood in accordance with a dynamic ether theory. Possible ether-drift signals inside fiber optic or buried copper cables, as with the Cahill and DeWitte experiments, suggest a similarity to how compressional sound waves travel faster in water (~1500 meters/sec) or steel rods (~3200 m/s) than in open air (~350 m/s). Perhaps my hesitations to accept their full results are in error? Beyond the basics of ether velocity and azimuth, which have been repeatedly affirmed, much remains to be studied and determined about the *properties* of the cosmic ether. All these experiments should be replicated at different altitudes within transparent and less-dense structures, as with Miller's work. A renewed era of ether experimentation is clearly necessary, but only if the issues detailed herein are taken into strict consideration.

As will be presented in Part III, experimental works by biologists and chemists have also detected a RA 17 hrs cosmic sidereal signal in the behavior of living creatures, as well as in phase-change physical chemistry. The findings to follow significantly expand our definition of *cosmic ether*, into the realm of *cosmic life-energy functions*. And in the process, additional new theory is provided to better understand many astronomical mysteries. This adds to the controversy, but also provides a great deal of clarification about just what the cosmic ether is, and how it behaves and moves.

Firstly, Part II lies ahead, a discussion of Einstein's relativity theory, and the foundational reasons why it should be rejected.

Part II:

The Empire Strikes Back: Erasure, Mystification, and Falsification of History

There is something fascinating about science. One gets such wholesale returns of conjecture out of such a trifling investment of fact.

– Mark Twain

Einstein Rising

"My opinion about Miller's experiments is the following... Should the positive result be confirmed, then the special theory of relativity and with it the general theory of relativity, in its current form, would be invalid. Experimentum summus judex. Only the equivalence of inertia and gravitation would remain, however, they would have to lead to a significantly different theory."
— Albert Einstein, letter to Edwin Slosson, 8 July 1925 (Hebrew U. Archive)

The Rise of Einstein's Theory of Relativity

In 1905, Albert Einstein published several research papers that are considered to be cornerstones of modern physics and astronomy. Upon first reading his works decades ago, I found his relativity theory to be deeply mystical, referencing unseen forces such as "curved space-time". He ignored measured real-world cosmic motions that affected the velocity of light, and conjured up cosmic motions in a space-time unreality, which still remains as sheer speculation, heavy with maths but never convincingly

Albert Einstein
(1879-1955)

affirmed by empirical reality. Today I accept him as a humanitarian, and his ideas on energy-mass equivalence ($E=mc^2$) *as approximations*. However, the proofs of variable light-speed, as from the successful ether-drift experiments, completely destroy a central assumption of Einstein's relativity theory, that of *light-speed constancy*. Above that concern, when the evidence claiming to prove the accuracy of his relativity theory is critically examined from the viewpoint of the successful ether-drift experiments, one cannot confidently assert Einstein's theory is *unequivocal*, meaning that *only* the Einstein theory of relativity can explain those claimed proofs. A substantive dynamic ether in space works just as easily to understand the modern cosmology claimed as proof for the Einstein theory. My statement is modern-day heresy, of course, and so a discussion of the specific evidence is necessary. For those claimed proofs developed while Einstein was still alive, specifically the issues of the bending of starlight as seen during solar eclipses, and the shifting perihelion of Mercury, counter-arguments are presented below. Other criticisms appear in Part III. Firstly, an introduction to Galilean versus Einsteinian relativity is in order.

The Natural Relativity of Galileo

Viewed by ordinary pre-Einstein maths and logic, in what I call the *natural relativity* of Galileo which governs the real world – the relationships between moving light waves and the velocities of their sources and receivers, and the velocity of the medium in which the waves are waving – is highly relevant, and can be viewed as follows:

$$c + v > c - v$$

This first equation states that light speed "c" plus the velocity "v" of the emitter is *greater than* light speed "c" minus the same velocity "v". An example would be trying to catch a ball thrown towards you by someone driving by in a fast car, versus the same person and car moving away from you and throwing the ball "backwards" after they passed by. In the real world, the velocity of the car adds to or subtracts from the velocity of the thrown ball, depending upon its direction and speed of motion. This is what I call *natural relativity*, which governs thrown balls, rockets, railway trains, airplanes, and all other motions, including light waves (or light particles). The equation above is based upon logic and direct observation. If someone gently tosses a hard baseball at you from a standing position, at a speed of 1 kilometer per hour, the ball's velocity will be much less than if they gently toss the same baseball at you from the open window of a car driving by at 150 km/hr, which propels the ball in an additive manner, at 151 km/hr (~95 mph). In the first case, you can catch the ball with ease. In the second case, you might break the bones your hand.

This is the long-known conventional Galilean 3-dimensional *natural relativity*, and it is one reason why space rockets from Cape Canaveral in Florida, aiming to achieve orbital velocity in the quickest amount of time, using the least amount of fuel, always launch towards the east, so as to gain the Earth's west-to-east rotational velocity at that latitude, of ~1500 km/hr (or 0.42 km/sec) added to the rocket thrust aiming to put a satellite or space ship into orbit.

Einstein's theory of relativity demands that the above common sense *natural relativity* should be ignored, and claims it does not hold true for light waves or light particles. By his theory, light waves always have a constant velocity in all directions, without reference to the motions of light emitters and receivers, or to the densities or motions of the medium upon which the light waves are carried. By Einstein's theory and using the prior thrown-ball analogy, all balls thrown by all people, moving in any and all directions, would depart from the hands of those throwing them at the same identical speed. And all balls would arrive at their destinations at the same speed, always, no matter how hard or slow they were thrown or emitted, and no matter if the receiver was also moving either toward or away from the ball thrower. It is a surreal landscape impossible to imagine by ordinary logic or reasoned analogy. It is also a landscape for which absolutely *no observed evidence exists.*

Einstein's theory therefore *demands* a "null" result from all the ether-drift experiments, prohibiting all variations in the speed of light. His theory insists that space be "empty" of anything that might influence the speed or motions of light, be that the cosmic ether or any other factor. And, as an historical fact, while Einstein's theory gained in popularity, the growing evidence for a real ether drift with variable light speed was increasingly ignored or misrepresented. The universe was declared, *ex-cathedra,* to possess supernatural and deeply mystical, illogical, and never-observed properties.

Here, by contrast is how the Einstein universe commands light and reality to behave: $\mathbf{c + v = c - v}$

This second equation states that light speed plus velocity *equals* light speed minus velocity. Where problems arose in the Einstein theory, regarding velocity differences between light emitters and receivers, it was all neatly taken care of by isolating them into two separate "frames of reference", each residing within its own separate space-time reality. The two realities were then reconnected by mathematical equations which demanded supernatural properties be invoked that had no proofs whatsoever at the time they were proposed, and which subsequently have never been unequivocally proven to exist As such, every major theoretical departure Einstein made from classical physics, every postulate governing the Einstein relativistic universe, exists only in the minds of the theoretical physicists who embrace his doctrine.

From what has been presented in the preceding chapters, it is clear that the Michelson-Morley 1887 experiment documented a slight variation in light speed, approaching ~5 - 7.5 km/sec, a fact which was chronically ignored or misrepresented as a "null / zero" effect. This is so, even if Michelson-Morley and Morley-Miller occasionally thereafter lapsed into use of the "null/zero" language themselves, apparently due in large measure to sheer weight of academic peer-pressure. Later on, there was Miller's own independent work in Cleveland and atop Mount Wilson, from 1921-1926. Miller provided ever more positive results for a cosmological ether of ~10 km/sec with a seasonal variability, throughout the Einstein years. This, too, was ignored or misrepresented.

The ether's velocity and axis of motion in space had been determined, indicating variations in the speed of light along preferred cosmic directions. Add to this the additional ether-detection work of Michelson-Gale, Sagnac and Michelson-Pease-Pearson, which was also chronically misinterpreted and ignored, as if only an exceedingly high velocity of ether wind could be imagined or considered significant.

Over those years of intensive experimentation and public discussion, this large body of evidence for variations in light speed depending upon direction was distorted, erased and ignored by what I see as a growing Einstein cult. Top physicists and astronomers of the early 1900s rushed gladly into the arms of a new metaphysics, rejecting all evidence to the contrary, and declaring Einstein correct. Newer generations of physicists and astronomers assumed what they learned in textbooks was unquestionably true, and rarely consulted the older historical publications. In subsequent decades, a shroud of scientism and toxic ridicule descended over the subject of cosmic ether.

By Occam's Razor, the collective ether-drift experimental findings ought to have, by the late 1920s, settled the matter in favor of the ether, and against Einstein's theory; or at minimum, to create a well-funded continuance of ether-drift research, parallel to other experiments. Why didn't that happen?

Starting around 1906, and amplifying thereafter, a strange compulsion developed within physics and astronomy, to collectively embrace Einstein's relativity theory, even if only 15 years later, as Einstein himself noted in the face of Miller's positive ether-drift results, "...the whole relativity theory collapses like a house of cards". Those words were written by Einstein in 1921, in a letter to Robert Millikan, before the best-ever set of ether-drift data had been collected by Miller over four seasonal epochs atop Mount Wilson.

By 1925, with Miller's new Mount Wilson data getting serious public exposure, Einstein was even more deeply concerned, as expressed in a letter to Edwin Slosson on 8 July 1925, obtained from the Hebrew University Archive in Jerusalem, and quoted at the start of this chapter. It was a truthful admission by Einstein, even if in public he was less forthcoming with his concerns. By Einstein's own theory, light speed had to be constant throughout the universe. Miller's work, more than anyone, had threatened Einstein's very elegant but also deeply mystical relativity imaginings.

In his 1905 paper "On the Electrodynamics of Moving Bodies (Zur Elektrodynamik bewegter Körper)", on the first page of that paper, Einstein wrote:

> "... the *unsuccessful attempts to discover any motion of the earth relatively to the 'light medium,'* suggest that the phenomena of electrodynamics as well as of mechanics possess no properties corresponding to the idea of absolute rest. ...[we must] introduce another postulate, which is only apparently irreconcilable with the [Principle of Relativity], namely, that light is always propagated in empty space with a definite velocity c which is independent of the state of motion of the emitting body. ...The introduction of a "luminiferous ether" will prove to be superfluous inasmuch as the view here to be developed will not require an "absolutely stationary space" provided with special properties..." (Einstein 1905, p.891. Bracket notes and emphasis added.)

Einstein's words revealed a knowledge of the ether-drift experiments, indicating he had read a German translation of the original Michelson-Morley 1887 paper, or was relying on the opinions of other ether-critics. Thus, he set up the artificial conditions of an "empty-space" universe, without an ether having "luminiferous" or other identifiable properties such as substance, motion or gravitational properties. In Einstein's universe, light was "always propagated in empty space with a definite velocity", which had no bearing upon the motions or character of the presumed "empty space" medium, or of the light-emitter or recipient. Einstein therein showed a serious disregard for the ether concept at the very time when ether-drift experiments were continuing to gather better and more convincing evidence. His use of the phrases of "light medium" and "luminiferous ether" in diminishing quote marks, suggest both a knowledge of the ether, along with a contempt against it. Newton's ideas on "absolute space" were also swept aside by Einstein's relativity, as a competing metaphysics.

Even if by 1905 the ether drift could not be measured in velocities greater than Michelson-Morley's ~5 to 7.5 km/sec, similar to at least some of the Morley-Miller results, by contrast Einstein was unable to show *any* direct evidence for his proposed relativistic "space-time" or "gravity wells". By definition, Einstein's theory had no directly observable components, and zero evidence to support its existence at the time of his postulates about it. And because his theory had adopted parts of the FitzGerald-Lorentz contraction theory, which also had never been detected experimentally, his relativistic theory might not ever be directly or independently confirmed. The existence of any evidence for a variable speed of light would pull the rug out from under Einstein's theory, as he well knew.

Einstein's publications on relativity in 1905 followed the writings of FitzGerald and Lorentz, whose metaphysical contraction theory, previously discussed, was gaining in popularity, though not in scientific proofs. Einstein eventually made a series of predictions which also rested upon the assumptions of empty space and a constant light speed in every direction. He eventually achieved scientific popularity when some of those predictions obtained claimed empirical verifications, by exceedingly tiny adjustments to prior Newtonian determinations. However, this early fame and acceptance was driven just as much by favorable newspaper articles, spread globally and quickly by the new inventions of the wireless telegraph and radio broadcasting. This great media attention and the ensuing scientific discourse were accompanied by *a suppression of serious discussions about the ether-drift experiments,* which also might explain the new tiny adjustments to Newtonian theory. Media hype appeared to tip the balance in Einstein's favor.

Einstein's Mystical Ether Concept

In May of 1920, Einstein gave a lecture at Leiden University in the Netherlands, where he made an effort to reconcile his empty-space relativity with a new "ether" concept that, by mere declaration, deprived it of any kind of substance, motion or other detectable or measurable properties. Near the end of his lecture, entitled "Äther und Relativitätstheorie", Einstein stated:

"We may sum up as follows. According to the general theory of relativity, space is endowed with physical qualities; in this sense, therefore, an ether exists. In accordance with the general theory of relativity space without an ether is inconceivable. For in such a space there would not only be no propagation of light, but no possibility of the existence of scales and clocks, and therefore no spatio-temporal distances in the physical sense. *But this ether must not be thought of as endowed with the properties characteristic of ponderable media, as composed of particles the motion of which can be followed; nor may the concept of motion be applied to it.*" (Einstein 1920. Emphasis added.)

In short, Einstein's new "ether" was a *metaphysical dead nothing,* proclaimed into existence and rendered immaterial and irrelevant by the demands of his relativity theory, which otherwise would be totally undermined by a substantive and motional ether. Such theoretical disagreements between the ether-drift scientists and the advocates of Einstein's relativity theory would briefly come to the surface in open scientific debate around 1925, following the announcement of early results from Miller's experimental work at Mount Wilson.

Miller's work still posed an obstacle to the acceptance of Einstein's relativity, even as his interferometer measurements were being ignored as "too small" of a result. And yet, Einstein received tremendous support for his astronomical predictions of *fantastically smaller quantities* – for example, his assertion of having better maths than Newton for predicting starlight bending close to the Sun, as well as for determination of shifts in the perihelion of Mercury. Both phenomena might well be the products of a variable-density cosmic ether, as I discuss below, but no matter. For Einstein, tiny variations in experimental or observational results that supported his claims were no problem! For him, research journals and newspapers made celebrations of exceedingly small findings with a loud Huzzah! By contrast, *an experimentally detected ether wind close to the orbital escape velocity of Earth was ignored.* Such glaring contradictions and biases from the Einstein camp and their media supporters would increase over the years.

The Dynamic Ether of Cosmic Space

It appears that the older Einstein was fully aware of this issue, and was troubled by the empty enthusiasm and vacuous popularity upon which he rode to fame, as well as regarding the growing malignant denigrations of Miller. Late in his life, he wrote:

> "You imagine that I look back on my life's work with calm satisfaction. But from nearby it looks quite different. There is not a single concept of which I am convinced that it will stand firm, and I feel uncertain whether I am in general on the right track." — Albert Einstein, on his 70th birthday, in a letter to Maurice Solovine, 28 March 1949 (in Hoffman 1972, p.328)

It is a pity that Einstein rarely made such self-critical and qualified statements in public, and doubly so that his modern advocates do not have even this small bit of uncertainty or humility. So far as I can determine, none of his biographers have seriously investigated this aspect of Einstein's self-doubt, as they all accept the usual historical falsification of the ether-drift experiments. Einstein died six years later, in 1955, shortly after granting a series of interviews to Robert Shankland, one of Miller's former students who had become an Einstein sycophant. Shankland was at that time also leading a team undertaking a post-mortem reexamination of Miller's work, an effort which I will show, in the next chapter, was an all too typical *academic hatchet-job*.

The Einstein-Supporting Experiments, *Not Unequivocal!*

In 1915, Einstein proposed three tests for his theory of relativity. These were 1) a slight correction to existing Newtonian-based calculations on the perihelion shift of the planet Mercury, 2) another correction to existing Newtonian gravitational starlight bending near to the Sun, and 3) a gravitational cause for the redshifting of galactic light. A number of experiments were undertaken in efforts to test out Einstein's predictions, followed by loud pronouncements of a great success for the Einstein theory over the older theories of both Newton and cosmic ether. Let's review these major assertions objectively.

1. 1916 Precession of the Perihelion of Mercury

In his article from 1915, "Explanation of the Perihelion Motion of Mercury from General Relativity Theory", Einstein offered a more precise calculation on the perihelion shift of Mercury in its path around the Sun. Mercury takes 88 Earth-days to orbit the Sun, with an elliptic orbital eccentricity that brings it to 46 million kilometers distance at perihelion (closest approach to the Sun), then out to 70 million km distance at aphelion (farthest distance from the Sun). However, as early as 1843 the French mathematician Urban Le Verrier determined that the orbit of Mercury possessed rather unusual characteristics, shifting in the forward direction of its orbit at perihelion, by 38 arc-seconds per year. By modern measures, this has been revised upwards to 56 arc-seconds per year, or 5557 arc-seconds per century, a quantity which over 100 years is equal to about 1.53° of a standard 360° circle (or 0.0153° per year). Newtonian mechanics could explain most of this orbital shift as due to the gravitational influences of the other planets in our solar system, all of which rotate around the Sun in the same direction. The Newtonian calculation was nevertheless off a tiny bit from the actual measured quantity of perihelion-shift, by about 43 arc-seconds *per century*, or about 0.008° per year. That slight inaccuracy is about the same angle subtended by a human hair of one-half millimeter diameter at 100 meters distance or more, a very tiny quantity indeed!

In 1916 Einstein claimed his theory of general relativity would explain the 43 arc-sec/century difference. His arguments held sway over astrophysics, but only because the concept of a motional and material cosmic ether in space was ignored. The amount of perihelion advancement, which is a very tiny correction to an already very tiny variance, finds an explanation if we consider the cosmic ether to not only be material and entrainable, but also *in vortex motion around the Sun as it turns. Such a rapidly rotating layer of condensed cosmic ether in the central parts of the solar system, surrounding and rotating with the Sun, and influencing Mercury at the time when it makes its closest perihelion approach, would give Mercury a slight forward boost in velocity, with a consequent slight forward shift in perihelion.*

Conventional astronomy accepts that the motions of planets speed up the closer they are to the Sun. By a dynamic ether theory, this increased velocity is caused and sustained by the faster rotating vortex of ether, which pushes and floats those planets along in their orbits. By this understanding, vortex motion of the Sun's ether envelope would be moving even faster than Mercury's own orbital velocity. Consequently, as Mercury approaches the Sun at perihelion, it is exposed to this region of faster rotational motion, which then exerts a "push" to its orbital velocity during that closest approach. The cosmic dynamic ether

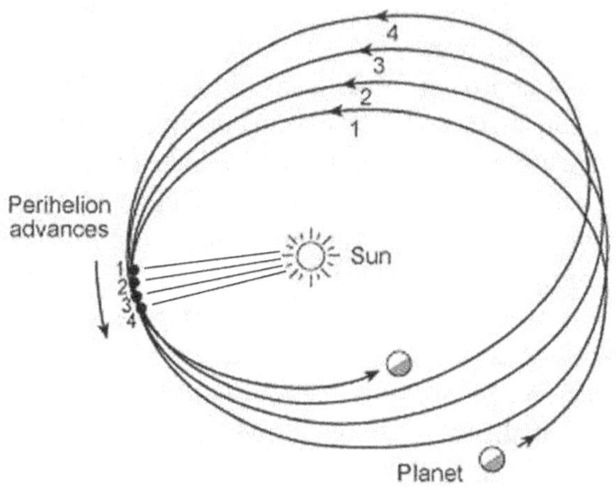

The Dynamic Ether of Cosmic Space

Figure 63. Highly Exaggerated Diagram of Mercury's Orbital Perihelion Advance

thereby provides a straightforward real-worldly understanding of Mercury's perihelion advancement.

By analogy, wintertime mid-latitude cyclonic storm systems, tropical hurricanes and tornadoes increase in both wind speed and rotational angular velocity the closer one gets to their central core regions of lowest pressure. The same appears generally true of water whirlpools. It is also seen in Kepler's equations governing the ecliptic-plane orbital velocity of planetary bodies as a function of their distance from the Sun. Conservation of angular momentum is surely at work, but the raw equations governing planetary motions in empty space don't take into account such things as a possible fast rotating material gravitational ether in the solar system, which would give a slight forward push to orbiting objects as they moved slightly closer towards the Sun.

Mercury moves around the Sun once every 88 Earth days, which is the fastest orbit of all the planets. The Sun's surface makes a full rotation at its equator once every 24.5 Earth days, reflecting the increase in *angular velocity* as one moves closer to the Sun's surface. This is shown in Table 10, on the Orbital Properties of the Planets and Solar Surface. Mercury also has a high eccentricity, its orbital plane pole being tipped off from the rest of the planets in the same general cosmic direction as the Sun's rotational plane pole. This was already pointed out in Figure 28 (p.128), in the chapter *Which Way Drifting*, indicating a coupling of Mercury to the Sun with a shared tipping force closer to the center of the solar system. Mercury remains the only major planet

	Orbital Distance from the Sun (10^6 km)	Eccentricity	Orbital Velocity km/sec. days/yr	
Sun Surface	0	0	2.0	24.5d
Mercury	57.9	0.205	47.4	88d
Venus	108.2	0.007	35.0	225d
Earth	149.6	0.017	29.8	365d
Mars	227.9	0.094	24.1	687d
Jupiter	778.6	0.049	13.1	11.9y
Saturn	1433.5	0.057	9.7	29y
Uranus	2872.5	0.046	6.8	84y
Neptune	4495.1	0.011	5.4	165y
Pluto	5906.4	0.244	4.7	248y

Table 10. Orbital Properties of the Planets and Solar Surface

to have such a high eccentricity, dipping at perihelion closer to the Sun and gaining a slight boost in orbital velocity.

Mercury therefore appears as a special case, given how only in the closer regions to the Sun would the cosmic ether significantly increase in its density and rotational velocity sufficient to propel a planetary object into a slightly advanced orbit. A similar phenomenon of far lesser magnitude may also occur with the other planets, of tiny forward shifting of the sidereal azimuths of all the planetary perihelions, but none are so large, or so pointedly discussed by the astronomers.

2. 1919 Eclipse Photos of Gravitational Bending of Light

Einstein also claimed a more precise measure of starlight bending close to the Sun. The effect was too tiny to be directly observed, but photographs were finally possible during total eclipses when the Sun darkened, making background stars appear in the daytime. During such times, the distant stars behind the Sun have their light paths diverted, ever so slightly, so as to appear a bit farther away from the Sun than they actually are. This light-bending effect is a widely observed fact, and is not in question. Einstein's claim was made in 1916, in his "Foundation of the Generalized Theory of Relativity", in *Annalen der Physik*.

In a famous pair of 1919 expeditions to Africa and Brazil, Arthur Eddington and his associate Edwin Cottingham made solar eclipse photos of stars close to the Sun, and measured their deviations from anticipated positions. They were touted in public as "definitively" supporting Einstein's new theory of relativity over the prior calculations based upon Newtonian gravitational theory. Figure 64 below is taken from a memorial plaque on the Island of Principe, off the coast of Africa, celebrating their expedition, which also included a separate expedition to Brazil by their associates. Both depict starlight bending near to the eclipsed Sun, though in a highly exaggerated manner. The text in the graphic on the facing page admits the small starlight deviations as depicted are *600 times larger* than actually measured. The two items, plus the graphic at top-right from another news report of that time, give a reasonable though highly exaggerated summation of the very tiny effect that Eddington and Cottingham had measured.

The apparent displacement of starlight proposed by Einstein, and documented by Eddington/Cottingham and others since, was equal to an angle of 1.75 arc-seconds, or about 5/10,000ths of one angular degree (more precisely, 0.0004861 degree). That is roughly the angle subtended by holding that same half-millimeter diameter hair out at 2 kilometers distance, certainly a much tinier quantity to ponder than the ~10 km/sec ether drift measured by Dayton Miller and others.

And yet, such repeated ether detections as outlined in prior chapters are routinely scoffed at or ignored in the public and scientific discussions about the meaning of the eclipse observations. Figure 66 gives another highly exaggerated diagram of solar bending of starlight, but now includes a representative layer of denser cosmic ether surrounding the Sun, which could also create starlight bending through more ordinary ether refractive effects. Let's explore this idea a bit further.

Figure 64. Part of a Plaque Celebrating Eddington's Eclipse Observations on the Island of Principe, 29 May 1919.

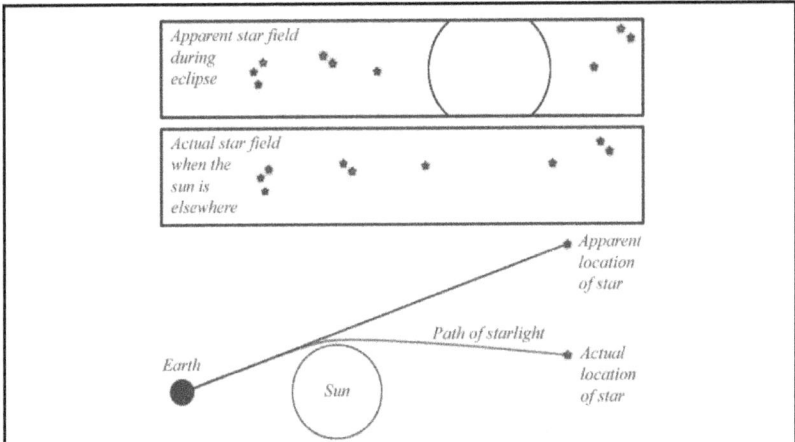

Figure 65. Graphics from Newspaper Reports on the 1919 Eclipse Expeditions of Eddington-Cottingham.

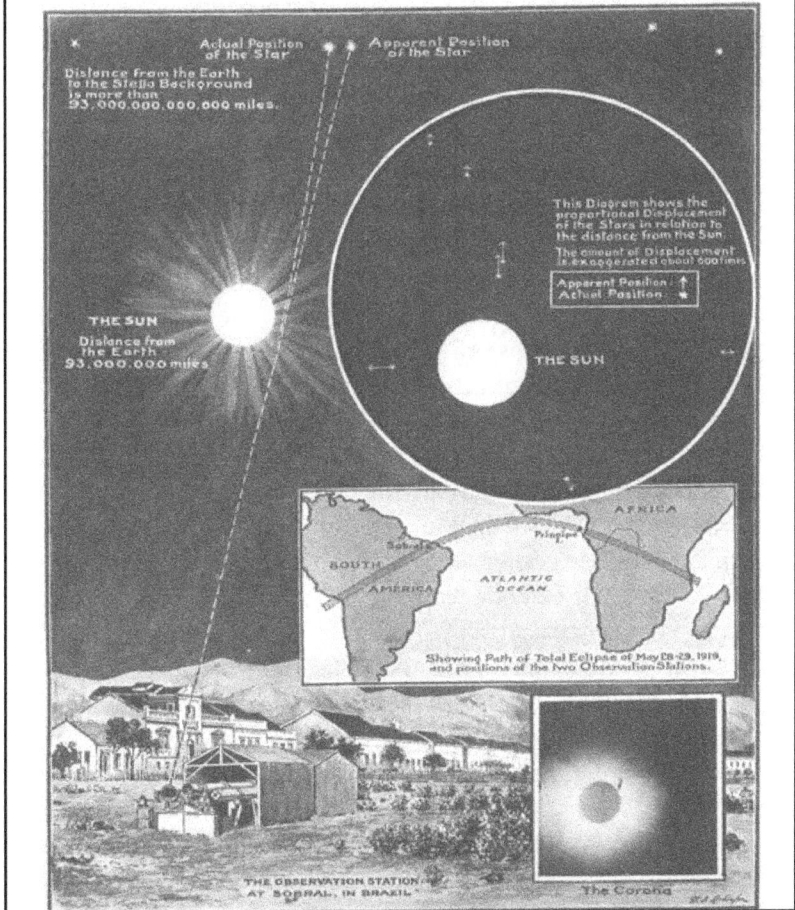

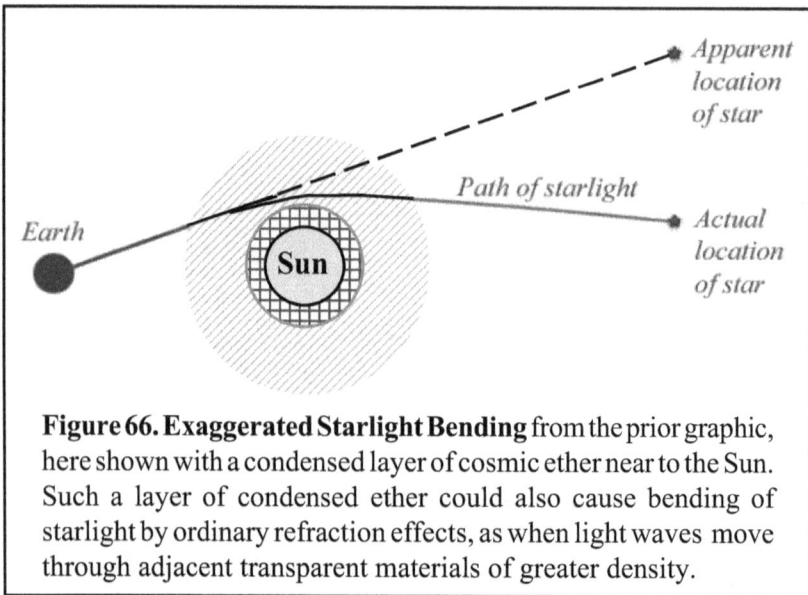

Figure 66. Exaggerated Starlight Bending from the prior graphic, here shown with a condensed layer of cosmic ether near to the Sun. Such a layer of condensed ether could also cause bending of starlight by ordinary refraction effects, as when light waves move through adjacent transparent materials of greater density.

Most of the contemporary photographs of solar eclipses focus primarily upon the thin corona layer immediately adjacent to the Sun, where magnificent solar flares and prominences are visible, but not on the larger visible but less-dense *extended corona* components. The extended corona extends outwards by *at least 7 solar diameters*, to around 5 million kilometers by recent NASA findings, even farther than what is depicted in the above diagram. Figure 67 the facing page shows at the top an early eclipse drawing from 1806 and a 1980 eclipse photograph. They are contrast-enhanced to reveal the corona components extending outwards by about two or three solar diameters. The larger image at bottom is a similar contrast-enhanced "eclipse" view of the Sun made from the SOHO satellite in synchronous solar orbit. The black circle substitutes for the Moon and obscures the Sun. The interior parts of the extended corona are apparent in all three cases, though not to its full extent. It extends well beyond the much brighter interior corona immediately adjacent to the Sun. The extended corona reacts to the solar wind and coronal mass ejections, but is not the same thing. This suggests a condensed ether-like composition, where the inward motion of the cosmic ether compresses and confines much of the outward-bound solar radiation, wind and ejecta, to yield the enigma of a hotter solar surface layer compared to just below that layer. The gravitational ether thereby deflects all but the mightiest of solar coronal flares and mass ejections rising up to breach its barrier. It is also

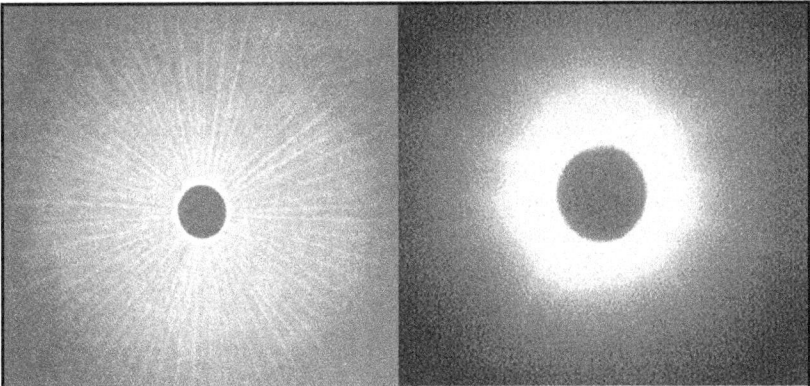

Figure 67. The Solar Corona at Larger Distances
The top-left is a drawing from a June 1806 eclipse, while the top-right image is from a photograph of an April 1980 eclipse. At bottom is a composite contrast-enhanced image from August 2017, emphasizing the larger solar corona reaching out many solar diameters, taken by the SOHO satellite in synchronous orbit around the Sun. The black disk substitutes for the Moon.

reasonable to believe all planetary bodies, the stars and our Sun included, would also possess a substantial envelope of condensed cosmic ether, which plays a similar role of atmospheric containment and gravitation. As for example with the Earth's *plasmasphere*, shown in Figure 86 (p.275).

The idea that the Sun or massive planets like Jupiter could bend starlight was anticipated to exist prior to Einstein and the Eddington-Cottingham observations, according to Newton's theory of gravitation. However, Newton postulated light as a "shower of particles" rather than wave motions in the ether, accepting only a static and immobile, nonmaterial ether, defined according to his "absolute space". By Newton's theory, and by the early 1900s, the gravitational pull of the Sun was calculated to bend passing starlight by an amount of 0.875 arc-seconds. In 1913, Einstein predicted a similar quantity of light bending near to the Sun, but in 1916 revised his calculation upwards to 1.75 arc-seconds, doubling the Newtonian expectation. Eddington and Cotting-ham confirmed the slight bending of starlight by eclipse photographs, at an angular deflection closer to Einstein's predictions than to those of Newton. This was the first major breakthrough for Einstein's accep-tance, and as mentioned previously, was accompanied by a large amount of worldwide media attention. Global news media covered the May 1919 eclipse voyages in a breathless manner, as something heroic and exotic, like the Victorian-era explorations of the Congo or Nile Rivers. The claimed verification of Einstein's prediction also was somewhat restorative for a tarnished German civilization and science, a big change from the depressing news reports of the First World War which had ended only six months earlier.

Unmentioned in all the media hoopla was that the Sun's starlight-bending effects certainly could also be due to a light-refracting layer of cosmic ether enveloping the Sun, being more condensed closer to the Sun's surface, just as is postulated for the Earth, where a layer of cosmic ether was then considered as a cause for stellar aberration. The bending of starlight was never at issue, only the diverging theoretical explana-tions for it. Unfortunately, the subject of cosmic ether as a competing explanation was not given mention in the Einstein media blitz, except for the false claim that the ether was somehow defeated by the eclipse observations. Nor was it mentioned by Einstein and his followers, all of whom continued to fully reject the growing ether evidence of a ~10 km/sec velocity. This was a substantial quantity, far more so than the exceedingly tiny amounts of angular deviation being argued over by the respective followers of Newton and Einstein. Both theoretical viewpoints rejected a material entrained ether.

Other scientists besides Miller remained skeptical of the loud claims about the eclipses, and continued to embrace the concept of a material ether. Oliver Lodge, aware of all that has been so far summarized in this book, was unconvinced that the eclipse data from Africa and Brazil had made any inroad against ether theory, and continued to decry the negative interpretations of the Michelson-Morley and Morley-Miller experiments. He wrote two papers for *Nature* magazine, whose editors apparently also felt unsettled by the Einstein theory: "Aether and Matter: Being Remarks On Inertia and On Radiation and On the Possible Structure of Atoms"(1919), and "Relativity: The Geometrisation of Physics and its Supposed Basis on the Michelson-Morley Experiment" (1921). All such ether-based arguments were simply ignored by Einstein and his followers, except for their continued false statements about the claimed "null" result of Michelson-Morley.

Campbell and Trumper (1923) replicated the Eddington-Cottingham experiment in 1922, finding the displacements of stars during the 1922 eclipse were rather chaotic, being in every which-way, including to move closer towards the Sun. Their observations are more congruent with a *turbulent* cosmic ether rather than a smoothed-out kind of time-space gravitational effect. In more recent years, Collins and Pinch (2012) reviewed the Eddington-Cottingham photos in archive, and claim the photographic evidence does not so clearly support Einstein. By their analysis, Eddington cherry-picked only the photos that affirmed Einstein, ignoring the ones that didn't, and made large assumptions throughout. A search of internet shows a few claims of successful replication, with a few dissenting voices. Where's the truth?

The eclipses and debates notwithstanding, over April and December of 1921, Miller obtained his first results from atop Mount Wilson, with another 1045 turns of his large interferometer, showing an average ether-wind signal of ~10 km/sec. It was at this time, after learning of Miller's preliminary results from the April experiments, that Einstein made his deep anxiety known to fellow scientist Robert Millikan in a letter, writing as previously noted that his relativity theory was at risk of collapsing.

In more recent years, the Laser Astrometric Test of Relativity instrument has been placed on the International Space Station (ISS) to better document the effects of starlight bending near the Sun. This apparatus uses a type of Michelson interferometer. However, the results of this and similar experiments, no matter how exacting are their measures, do not refute the possible role of the cosmic ether, which is simply ignored and presumed to "not exist". Stellar aberration, a phenomenon once considered to be caused by the Earth's ether field, shows stellar deviations of about 20 arc-seconds. That is more than 11 times greater than observed starlight bending near the Sun during eclipses. At best, *neither Newtonian gravitation theory, nor Einstein's space-time calculations stand as unequivocal explanations.* They are no better or worse than my assertion of a condensed layer of cosmic ether near to the surfaces of planets and stars, producing optical refraction and atmospheric containment effects. The same seems apparent with the powerful energetic fields surrounding galaxies, today understood as being far more extensive than previously considered.

On this latter issue, the bending of starlight near to the Sun as seen during eclipses bears a relationship, it appears certain, to how light is similarly bent gravitationally by dense ether fields surrounding far off galaxies, to create impressive and strange warped shapes such as bent galaxies and "Einstein Rings". This is not too different in how the Earth's atmosphere distorts the shape of the Moon when it stands at the horizon. Those parts of the Moon which remain below the horizon are slightly bent up and over the Earth making it look somewhat like a squashed pumpkin on the lower side. Any dense layer of transparent cosmic ether surrounding star or quasar object would also deform the shapes of luminous objects at a distance behind them. Even exceedingly tiny fractions of an angle, bending galactic light over and around an obscuring central object with its own powerful ether field, might easily appear as an arc or circular distortion, bent by the more commonly understood principles of light refraction. An *ether lens* effect?

The astronomers often put forth astrophotos of very distorted cosmic objects, generally with a bright bluish object in the center of the distortion, and around which bending of starlight or galactic light can be seen. In these cases, one cannot justifiably rule out the possibility of a strongly condensed and broad halo of cosmic ether around those centrally-located cosmic objects, as a competing explanation. The same mechanism of a dense layer surrounding our Sun and Earth, can be invoked for starlight bending. However, with the explanation of cosmic ether, the bending of light waves is a direct here-and-now form of cosmic lensing. One does not have to reference Einstein's surreal space-time warps or "gravitational wells", and possibly not even Newtonian gravitation, to find a general theory.

3. Gravitational Redshifts

In 1911 Einstein also postulated that a gravitational redshift would exist for massive stars, where light photons might not escape their "gravitational well" of space-time distortion without undergoing some kind of change or transformation. The idea of gravitational effects on starlight predated Einstein, as already explained, based upon Newtonian theory. However, Newton's particulate theory of light was, like his other optical theories, based upon a static ether. A return to the wave theory of light in subsequent centuries led to a questioning of Newton's ideas. In the end, however, the unjustified discarding of positive evidence for ether and ether drift favored only Einstein's metaphysics. For Einstein, curved space-time could lead to a slowing of starlight as it "struggled out of the space-time gravity well", without reference to a possible highly-condensed layer of ether around a star creating its own redshift effects. As with the eclipse photos, the shifting perihelion of Mercury, and likewise with "Einstein Rings", we can anticipate an exceptionally thick and dense layer of cosmic ether surrounding quasar and galactic masses. If sufficiently thick and dense, that would also affect emitted light waves, slowing and redshifting their light without reference to Doppler effects or space-time gravity wells. I will discuss this matter more directly in Part III, referencing the exceptional work of astronomer Halton Arp and his findings on intrinsically high cosmological redshifts in young hot quasars. An ether envelope surrounding every quasar or galaxy, its density proportional to its age, mass and gravitation, may well create an intrinsic redshifting of its light, beyond whatever Doppler effects might exist due to conventional motional factors. Again, Einstein and space-time are *not unequivocal* on this, nor on any other issue.

From the above, one can see how the cosmic ether of space, if granted a slight material property and variable density in the proximity of large massive bodies such as our Sun and planets, along with a vortex or spiral motion, could bend starlight by refractive effects, and also slightly modify the orbits of eccentric inner planets such as Mercury. Cosmic ether could, in those cases, substitute for the tiny additions which exist beyond standard Newtonian predictions, thereby standing as a competing theory to Einstein's relativity. Neither would be unequivocal, but the ether theory has the advantage of empirical proof of its existence, without the need to reference never-observed and mystical, other-worldly dimensions or un-realities.

One additional point needs to be re-emphasized. The Michelson-Morley-Miller experiment was rejected, not due to the absence of any measured results, but because those results were considered too small to satisfy the anticipated static-ether velocities of hundreds of km/sec. And yet, the measured 5 to 20 km/sec as detailed among the various experimenters of prior chapters was substantial, even without the "k-factor" upward-calculations to higher velocities. The lower velocity figures were a clearly "ponderable quantity" of real-world ether wind. By contrast, the two main experiments which propelled Einstein's relativity theory to fame were exceedingly tiny quantities, of fractional arc-seconds of angular displacements of sunlight bending or perihelion shifting. This important point bears repeating: *On the one hand, a significant result validating a material ether, but dismissed as "too small". On the other hand, an exceedingly tiny improvement of an already-known quantity, used to promote a totally mystical theory, getting major global media acclaim.*

In Part III I will review other astronomical phenomena where cosmic ether can provide an explanation just as good as or better than the various modern "empty space" theories, such as the redshifting of galactic light, the 3°K cosmic background radiation, the theory of "big-bang" creationism, invisible "black holes" and the new LIGO experiments, all of which rose to significance in the decades following the premature discarding of the cosmic ether evidence and theory.

The Shankland, et al.
Hatchet Job on Miller *

Over the years after Miller's death in 1941, his Mount Wilson ether-drift results continued to trouble Einstein and his followers. A post-mortem of Miller's work was finally undertaken in the early 1950s by a team from Case School, led by Robert Shankland, and with "extensive consultations" with Einstein. As one might anticipate, the new evaluation of Miller's findings made all the wrong assumptions about the cosmic ether as previously exposed in prior chapters, with a clear bias to "disprove Miller". Given how Einstein's supporters continue to place a high value on the Shankland, et al. study, I will go into some detail to expose its serious flaws and biases.

Shankland in fact was Miller's graduate student for many years, and only emerged to become a professional advocate of Einstein's relativity after the death of Miller in 1941. His early career as a scientist got off to a rocky start, in his first published paper (1936) "An Apparent Failure of the Photon Theory of Scattering". In that paper, Shankland

---

Robert S. Shankland, former student of Dayton Miller and later Chairman of the Physics Department at Case Western Reserve University in Cleveland Ohio. Shankland's academic career soared following publication of several widely read interviews with Einstein, and after he organized a post-mortem on Miller's work in cooperation with Einstein, pronouncing Miller's work as worthless. Shankland subsequently became a bureaucrat within the Atomic Energy Commission.

---

* This chapter was originally presented to a meeting of the Natural Philosophy Alliance, in Berkeley, California, May 2000, titled "Critical Review of the Shankland, et al. Analysis of Dayton Miller's Ether-Drift Experiments".

reported negative results on the question of Compton x-ray scattering, an experiment that is today a cornerstone of the particle theory of electromagnetic radiation. Shankland's research on that question was reasonable and competently undertaken, and in fact cited others who also failed to reproduce the Compton effect. However, this "incorrect" result set Shankland against the growing trend towards resurrection of the particle theory of light. A few years later, Shankland veered more decidedly into the mainstream of physics, away from all things ether and waves.

Shankland became Chairman of the Physics Department at Case School following Miller's retirement and death, and aside from his 1936 paper, his original contributions to science were basically "null-zero". However, as the apparatus and archives for both the Michelson-Morley and the Miller experiments resided at Case School, Shankland maintained a correspondence with Einstein, and fielded many inquiries from scholars around the world, asking about Miller's experiments and results. Such was the background leading to the renewed investigation of Miller's data in the early 1950s.

With encouragement from Einstein, and working with a team of Einstein's followers at Case School, the Shankland, et al. study "New Analysis of the Interferometer Observations of Dayton C. Miller" was undertaken and published in *Reviews of Modern Physics* in 1955. The title of the Shankland paper suggested the authors had made a serious review of "the interferometer observations" of Miller, to include some kind of a new thorough-going and serious evaluation. Such was *not* the case.

The very first sentence in the Shankland team's 1955 paper re-peated the widely parroted falsehood, that the Michelson-Morley experiment had a "null" result. The third sentence in the Shankland paper was similarly false, claiming that "All trials of this experiment except those carried out at Mount Wilson by Dayton C. Miller yielded a null result within the accuracy of the observations." This kind of chronic misrepresentation of the positive results of many interferom-eter experimenters, including Michelson-Morley, Morley-Miller, Sag-nac and Michelson-Pease-Pearson, exposed an extreme bias and delib-erate misrepresentation. The fact that this is a very popular prejudice does not excuse it. By redefining all the positive results observed by what may in fact have been the majority of ether-drift researchers, as mere expressions of "observational inaccuracy", Shankland and friends narrowed their task considerably.

The Shankland team misrepresented Miller's work in numerous unsupported statements, to the point that it was hard to determine what was based upon severe bias, versus an expression of their profound ignorance on the subject of ether-drift experimental and analytical procedures. They not merely tore down Miller's findings in transparently inaccurate ways, but also uplifted the Michelson-Morley experiments with the usual "null" falsehood, as if it were the most solid evidence on the question – which by itself implied that Miller, unlike the well-respected Michelson, was some kind of incompetent.

There were two basic approaches used in the Shankland team's analysis: 1) a search for random errors or statistical fluctuations in Miller's data, and 2) a review of selected data sets which they declared demonstrated significant thermal artifacts in the data. We can critically review these claims.

1) Claimed Statistical Fluctuations in Millers Data
The Shankland paper did present a statistical analysis of a portion of Miller's published 1925-1926 Mt. Wilson data, concluding that his observations "...cannot be attributed entirely to random effects, but that systematic effects are present to an appreciable degree" and that "the periodic effects observed by Miller cannot be accounted for entirely by random statistical fluctuations in the basic data" (Shankand 1955, p.170). Also, the Shankland team admitted they "...did not embark on a statistically sound recomputation of the cosmic solution, but rather [looked for]...local disturbances such as may be caused by mechanical effects or by nonuniform temperature distributions in the observational hut." (p.172)[5] In short, they conceded the sidereal patterns in Miller's data could not be due to any systematic measurement error, nor result from any mechanical flaws in the interferometer apparatus itself – while simultaneously admitting a disinterest in recomputing a potentially validating ether-drift axis ("cosmic solution") from his data. These were important admissions, suggesting that, unless they could find some other fatal flaw in his data, Miller had really got it right, and had accurately measured a real Earth-entrained ether drift.

This statistical analysis, perhaps the most centrally important thing in the Shankland team's paper, was not undertaken by any of the four team members listed as authors of the paper. It was instead carried out by Case School Physics student Robert L. Stearns, for his Master's

---

5. This stand-alone page number, and all similar page number citations reference to Shankland, et al 1955, unless otherwise indicated.

Thesis (Stearns 1952). Stearns was given only a footnote credit in the Shankland paper. In his thesis, Stearns noted the large amount of data gathered by Miller, of "316 sets of data...by Miller in 1925-26" for the most significant Mount Wilson experiments (Stearns 1952, p.15-17). While already covered, it is important to repeat this vast compilation of ether-drift data, so as to know what a miserable picayune job was delivered by the Shankland team.

Each of Miller's data sets was composed of 20 turns of the interferometer, with sixteen data points per turn (a total of 320 data points per data set). Altogether, Miller conducted over 6000 turns of his interferometer. His work at Mount Wilson was undertaken at four different seasonal epochs, each of which encompassed a period of around ten days, centered on the following dates: April 1st, August 1st, and September 15th, 1925, and February 8th, 1926 (Miller 1928, 1933a). It must be kept in mind, that these Mount Wilson data from 1925 and 1926 provided the most conclusive and foundational obser- vations for Miller's ether-drift calculations and conclusions, as pre- sented most clearly in his 1933 paper. As detailed below, the Shankland team did mention these Mount Wilson data, but in a manner that confused them with his earlier and less significant efforts, including various control experiments conducted at Case School. The signifi- cance of this obfuscation of dates will be highlighted below.

2) Shankland Team's Assertion of Temperature Artifacts

Regarding possible temperature artifacts in Miller's data, this objection was raised early on in the history of ether-drift interferom- etry, and specifically rebutted by Miller when he was still alive. A letter exchange between Miller and Joos from a 1934 issue of *Physical Review* (see pages 159-162, this book) records part of this debate, and appears to be one of the few published criticisms on the temperature issue Miller ever received while still alive. Miller had this to say about the problem:

"When Morley and Miller designed their interferometer in 1904 they were fully cognizant of this...and it has never since been neglected. Elaborate tests have been made under natural conditions and especially with artificial heating, for the devel- opment of methods which would be free from this [thermal] effect". (Joos/Miller, 1934)

In their search for possible thermal artifacts, the Shankland team failed to evaluate any large set of Miller's data, in contrast to the reasonably comprehensive statistical part of their study, as undertaken by graduate student Stearns (1952). Instead, they appear to have "gone fishing" in Miller's data for something by which they could simply dismiss him. For example, they discussed Miller's own 1923 temperature-control experiments, where radiant parabolic heaters were used to artificially create a general doubling of the size of interference fringes. Miller describes these experiments:

> "Several electric heaters were used, of the type having a heated coil near the focus of a concave reflector. Inequalities in the temperature of the room caused a slow but steady drifting of the fringe system to one side, but caused no periodic displacements. Even when two of the heaters, placed at a distance of three feet from the interferometer as it rotated, were adjusted to throw the heat directly on the uncovered steel frame, there was no periodic effect that was measurable. When the heaters were directed to the air in the light path which had a covering of glass, a periodic effect could be obtained only when the glass was partly covered with opaque material in a very nonsymmetrical manner, as when one arm of the interferometer was completely protected by a covering of corrugated paper-board while the other arms were unprotected. These experiments proved that under the conditions of actual observation, the periodic displacements could not possibly be produced by temperature effects." (Miller 1933a, p.220)

Perhaps without intending to do so, after examining Miller's laboratory notes for the Cleveland temperature control experiments, the Shankland team basically repeated Miller's words on this point, without, of course, doing any new experiments themselves:

> "In the experiments where the air in the optical paths was directly exposed to heat, large second harmonics (0.35 fringe for one heater, and about twice this value for two heaters) were always observed in the fringe displacements, and with the expected phase. Shifting the heaters to a different azimuth produced a corresponding change in the phase of the second harmonics. *When the optical paths and mirror supports were*

> *thermally insulated, the second harmonics were greatly reduced to about 0.07 fringe.*" (Shankland et al. 1955, p.174. Emphasis added)

This statement confirmed the wisdom of Miller's approach. The added insulation reduced the thermal effects from a nearby radiant heater to only 20% of the un-insulated readings. I have an ordinary commercially-available electric radiant parabolic heater at my home, and it gets so hot you cannot stand closer than 12" without burning yourself, or possibly catching your clothing on fire. If Miller had used a parabolic heater even half as strong as this, it would certainly have been a source of heat seriously stronger than anything present in his Mount Wilson experiments. This would be so particularly at night, during foggy or overcast conditions, and when the entire interferometer house was sheltered with a tent, with the apparatus and light-beam path covered with cork, glass and paper insulation.

Consider a radiant heater at several hundred degrees C, creating a steep thermal gradient but only a 0.07 fringe shift in the insulated interferometer. How much less of an effect would be produced by a human body, or even from the inside of a daytime shaded wall? Assuming an environmental thermal effect only one-tenth that seen with the parabolic heater, fringe shifts of merely ~0.007 would have been produced, well below observational detection. Miller's data sheets, for example, recorded observations "in units of a tenth of a fringe width", though readings down to hundredths of a fringe were possible with care. Overall accuracy of the ether-drift measurements approached a hundredth of a fringe after mathematical averages of many readings were extracted.

The Shankland paper nevertheless used these control experiments as a weapon against Miller, claiming without evidence that heater-type effects *might* have occurred in his Mt. Wilson experiments, even where no such heater or similar heat source was present. But why would the Shankland team shy from undertaking a more systematic evaluation for temperature artifacts? They could have evaluated only Miller's daytime interferometer experiments, and looked for a thermal effect from the southerly wall of the structure during the optimal solar heating hours, of 10 am to 2 pm by the civil clock time. If they could have shown an effect present in daytime data which was not present at night, it would have possibly proved their case. However, this obvious analytic procedure was either not undertaken, or not reported. Nor would such a heating effect be anticipated, as Miller *did* organize his data on both civil-clock and sidereal hour graphs, which showed no such daytime-heating peak for the civil-clock organization. (See Figure 33 on page 106, this book) *A sinusoidal pattern of ether-drift appeared in Miller's data only when it was organized by sidereal time.*

The Shankland paper also resurrected the temperature criticisms by Joos (1934), but unethically so, without reference to Miller's rebuttal in the same published exchange. If the periodic effects observed by Miller were the product of temperature variations, as was claimed by Shankland and Joos, then why would that variation systematically point to the same set of azimuth coordinates along the cosmic sidereal clock, but not to any single terrestrial coordinate linked to civil time? Miller repeatedly asked this question of his critics, who had no answer for it. The Shankland team erased Miller's precise rebuttal to Joos.

It is clear Miller had been deeply engaged on the problem of temperature effects, and worked hard to know exactly how thermally-produced errors might develop, and how to eliminate them. The Shankland team seized upon Miller's open acknowledgment of fringe-shifts from air heating by powerful radiant heaters during control experiments, and a few other sentences written in his lab book, and abused his words to claim thermal anomalies were probably the source of whatever periodic effects were subsequently measured by Miller at Mount Wilson – where no radiant heaters were used, and the empirically-developed thermal insulation features and procedures were put into place. Without some kind of independent experimental evidence to support such a claim of a thermal influence, the Shankland team's arguments were illogical, to say the least. At worst, they were a *lying dirty-trick.*

The Shankland paper also went through a series of arguments about the interferometer house, how the wall materials, roof angles, interferometer glass housing, etc., *might* result in a definable effect upon the air temperature in the light beam path, concluding only they could not rule out such an influence. However, *neither could they rule it in.* They stated that it "...is not in quantitative contradiction with the physical conditions of the experiment" (p.175). Given their ignoring of the sidereal nature of the periodicities, this statement could hardly be taken seriously, and certainly did not constitute a rebuttal of Miller's data.

The Shankland paper finally attempted to correlate several selected daytime interferometer runs with temperature measurements made at the same time, by cherry-picking a few data sheets to support their allegations. They nevertheless acknowledged difficulty in correlating low fringe-shift values with low temperature differentials. However, they found one set of high fringe-shift values correlated with slightly higher temperatures, even while noting another set where high fringe-shift values correlated with lower temperatures. Finally, they complained that "...no temperature data are available to reveal thermal conditions at the roof, which may be responsible for the large fringe displacements at the times of highest altitudes of the Sun" (p.176). Were they even aware of the large tent shade Miller had erected over the measuring house? If this sounds like a confused and unconvincing rebuttal of Miller, a reading of the full original text provides little clarification. In any event, this latter claim by the Shankland team cannot be true, as Miller did graph his data by civil time coordinates, as a control feature against his more telling plot by sidereal hour. *The civil time graph did not show any kind of solar heating effect around the noon or afternoon hours.*

Failing to find any damning evidence in the daytime data sets, when temperature gradients inside the interferometer house might be expected to be at a maximum, they turned their focus to nighttime data sets. Once again, only a few of Miller's data sheets were cherry-picked to "prove their case". Data from two nights (30 Aug. 1927 and 23 Sept.1925) with stable air temperatures were reviewed. These nights showed very clear and systematic fringe variations (p.176), but because the azimuth of the fringes changed by an unstated "minimal" amount over the approximate 5 hours of observation, the critics complained "it would be extremely unlikely if the fringe shifts were due to any cosmic effect" (p.177). Apparently, the Shankland team was so locked into the older "static ether" assumptions of the original Michelson-Morley experiment, that they were unclear about what they should have seen in Miller's data. In 1927, at a Conference on the Michelson-Morley Experiment held at Mt. Wilson Observatory, where Michelson, Lorentz, Miller and others made presentations and engaged in open debate, Miller addressed this question: "Observations were made for verifying these [static ether] predictions ...but it did not point successively to all points of the compass, that is, it did not point in directions 90° apart at intervals of six hours. Instead of this, the direction merely oscillated back and forth through an angle of about 60°..." (Miller 1928, p.356-357). The reason for this is, Miller's detected axis of ether drift is oriented reasonably close (within ~30°) to both the Earth's axis of rotation and the axis of the plane of the ecliptic.

More importantly, not all of the interferometer data sheets for a given date – which presumably would have had similar weather and temperature conditions – were included by the Shankland team for critical review. They selected only those data sets which appeared to support their argument of a claimed thermal artifact. For example, they chose "ten sets of observations, Nos. 31 to 40 inclusive, made in the hut on the Case School campus between midnight and 5:00 AM on August 30, 1927" and "...runs 75 to 83 inclusive taken from 12:18 AM to 6:00 AM on September 23" from the 1925 Mount Wilson experiments (p.176-177). Other than proclaiming these selected data gave them the *impression* of being the result of temperature errors, they had no stated criterion for bringing them into discussion. This biased cherry-picked data-selection, or rather *data-exclusion* procedure, forces one to ask: What about data sets No.1 to 30, and runs 1 to 74? Similar unexplained data selections or data exclusions occurred throughout the Shankland paper, leaving one to wonder if the excluded data, which constitute the overwhelming majority, simply could not provide support for their ad hoc criticisms and biased *a priori* conclusions. One can imagine the howl of protest which would have arisen if Miller had taken this approach, arbitrarily excluding data from his calculations which super-ficially suggested something other than a real ether drift. Well, Miller never did that, but the Shankland team did!

A third data set from 30 July 1925 was highlighted by the Shank-land team as it contained one extremely large peak where Miller noted "Sun shines on interferometer". This data *is included* as part of Miller's published Mt. Wilson analysis. However, the Shankland team ex-tracted only "observations Nos. 21 to 28 inclusive, made between 1:43 AM and 6:04 AM on July 30, 1925." Obviously, at around 6:00 AM the sun rose and caught Miller and his assistant off-guard. What about observations Nos. 1 to 20, or other early-morning data, where the Sun didn't shine on the interferometer? These other data were not brought into discussion, except they did note the runs prior to the sunshine incident demonstrated "...an extremely erratic behavior...we have no ready explanation for this apparent departure..." Here, the Shankland team basically confesses their grab-bag of ready explanations was empty, and how the idea that those data were expressing a real ether drift was simply *too positive and "impossible"* for them to consider. The fact that Miller wrote the note about the sunlight on that particular data sheet and included it in his final analysis, speaks to his honesty.

The Shankland team also selected data sets Nos.56-58 from 8 July 1924, from Miller's control tests made in a basement location at Case Physics laboratory — the temperatures were very stable and the fringe oscillations quite small, so they argued these data were proof for thermal effects on the apparatus for all other occasions. However, it was this very problem of basement and dense surrounding materials which led Miller on the path to use the apparatus in locations not subject to significant ether shielding or Earth entrainment.

The Shankland team's paper concluded its temperature criticisms by discussing a few additional data sets: Nos. 113-118 from 2 April 1925, Nos. 88-93 from 8 August 1925, and Nos. 84-91 from 11 February 1926 (p.177). In this case, the data sheets were taken from Miller's Mount Wilson experiments, where the amplitudes and phases were claimed to have been "nearly alike", suggesting they were merely criticizing a repetitive sidereal ether-drift signal in his data. Insufficient detail was given to allow a review of the critic's assertions, however, it appeared they were once again incorrectly misinterpreting Miller's data along the lines of static ether assumptions.

As in almost all the cases given above, none of these data were analyzed systematically, nor were they presented in such a manner that the Shankland team's criticisms could be factually justified. I had the impression, they simply scanned through a pile of Miller's data sheets, and not knowing what they were looking at, picked and pointed to a few selected parts which appeared to show minimum results, then dismissed it all as the product of thermal artifacts. Miller's detailed control experiments to exclude thermal artifacts were ignored or misrepresented, as was the overall sidereal-cosmic component of his results.

For the casual reader ignorant of Miller's original experiments, the Shankland team's paper might appear to present a reasoned argument. However, they obfuscated and concealed from the reader most of the central facts about what Miller actually did, and in any case it was so unsystematic and biased in its approach, excluding from discussion ~90% or more of Miller's extensive Mount Wilson data, as to render its conclusions *meaningless*.

From the above recounting, I find it impossible to view Shankland, and other members of his team, as anything better than *extremely biased reviewers of Miller's work*. At worst, they produced a *lying dirty-trick* and *academic hatchet-job*. And yet, while their published paper was filled with negative suppositions, what-ifs and maybes, and could not solidly establish any kind of serious evidence against Miller, it often contained admissions they had no evidence upon which to base their too-easy dismissal of him. For example, they wrote the following:

"In this case, we must admit that a direct and general quantitative correlation between amplitude and phase of the second harmonic observance on the one hand and the thermal conditions on the other hand could not be established. The reason for this failure lies in the inherent inadequacy, for our purpose, of the temperature data available." (Shankland 1955 p.177)

It appears Einstein was a part of this incompetency or deception. According to Shankland's 11 Dec. 1954 interview with Einstein, (Shankland 1963), Einstein was pleased to have already read a pre-publication copy of their forthcoming paper. He was 76 years old at the time the Shankland paper was published, and died only four months later on 18 April 1955. I therefore cannot believe Einstein was ignorant of what Shankland and friends had done, or were preparing to do – in which case, it might have deeply troubled him, and contributed to his passing away so coincidentally. I must wonder, if Einstein kept a diary during this period, and wrote something expressing serious concerns about the *post-mortem, auto-da-fe* of Miller's life's-work, that was about to take place in his name?

Following publication of these falsely-derived conclusions about Miller's work, Einstein was dead and Shankland's reputation rose. He went on to publish a series of widely-read interviews with Einstein, based upon their meetings over the prior years (Shankland 1963, 1964, 1973a, 1973b). In Shankland's interviews, Miller's ether-drift experiments were rarely discussed, except only in passing and in a dismissive manner, quite agreeable to Einstein's theoretical views. This wasn't always so. As a student and later as a colleague, Shankland was quite appreciative of Miller's person and work, writing a glowing letter to him in 1936 upon Miller's appointment as Honorary Professor (similar to modern Emeritus status). Shankland wrote:

"...may I add my own congratulations and voice my deepest thanks for all that your influence has meant in shaping my life, in exalting my ideals, and in enlarging my intellectual and cultural horizons.... With the greatest esteem for your works, and the deepest affection for you..." (Shankland letter to Miller, 1 June 1936.)

Such appreciative sentiments were not to last. By October of 1941, only a few months after Miller's death, Shankland wrote a 10-page memoriam article devoted to Miller's life and work, with photos and great detail, for the *American Journal of Physics*. However, Miller's years of work with Morley on the ether-drift question were reduced to one passing sentence regarding their meeting with Lord Kelvin in Europe. Miller's independent ether-drift experiments in Cleveland and at Mount Wilson were reduced to one paragraph. In neither case were the results of the ether-drift experiments even mentioned, nor the excitement and debate his work had created in at least a few scientific journals and the newspapers. More space was devoted to naming the various operas attended by Miller and his wife, and to the phonodeik device which has no more relevance today than buggy whips. Miller's lectures, prizes and awards were mentioned, but not one of the primary reasons for them – his ether-drift experiments. The description was more of a Professor of Music, than of a physicist who had earned a PhD in astronomy at Princeton for calculating the orbits of comets from his own telescope observations, and who later constructed the first working American high vacuum tubes, x-ray machinery and x-ray photographs. The ether-drift subject was basically erased from Shankland's memoriam.

As I stated at the beginning of this chapter, the Shankland et al. evaluation of Miller's work was conducted with serious ignorance and bias against him, dissecting Miller without serious concerns, and certainly without consulting any advocate of ether theory in the process. By the 1950s, there was probably nobody surviving in any physics or astronomy department who could fill Miller's shoes to make an adequate defense of his work. Ether theory was then being compared to "the search for perpetual-motion machines" (Swenson 1972, p.239). This had a silencing effect upon the subject, across the entire fields of physics and astronomy. It still does, as indicated in my *Introduction.*

Swenson also suggested that, during his later years, Miller was ignored and isolated. This appears to be correct. Shortly before Miller's death in 1941, he had given all of his interferometer data sheets – hundreds of pages of measurements – to his one-time student Shankland, with the bitter statement that Shankland should "either analyze the data, or burn it" (Kimball 1981, p.2). Shankland also blamed Miller's findings for having blocked the awarding of a Nobel Prize in Physics to Einstein for his relativity theory – this also appears to be correct, but in my opinion, Miller should have received the Nobel.

As a final note, during my visit to the Case Western Reserve University (CWRU) in 2001, I met with physics professor William Fickinger, who stated rather bitterly that "Dayton Miller had set back Case Physics by 50 years". In my view, however, Miller had made a consistent attempt to move Case Physics in the right direction – away from Einstein's metaphysics. That today so many in the Academy have hungrily devoured the Einstein theories, without caution or critical review, and know so very little about the actual history and experimental facts of the ether-drift experiments, is a testament to the powers of academic censorship, mystical thinking, media-driven popularity contests, Nobel-Prize chasing, and biased awarding of grant money. It is not due to any greater scientific legitimacy of Einstein's theory over the many experimental confirmations of a tangible cosmic ether and variable light speed.

While at Case, I was shown a large closet filled with all kinds of apparatus created and used by Miller, and the CWRU Archive department allowed my review of their many boxes of materials on Miller, Michelson, Morley, and Shankland. However, copies of Miller's laboratory notebooks and data sheets could not be located.

After returning home, I wrote to Fickinger and the archivist, asking them to redouble their efforts to find the missing notebooks and data sheets, which today I can report have indeed been found. Copies are now available at the Case Western Reserve University Archive Department, for scholars to study and review. However, my opinion is that new experiments on the question of ether drift, organized according to the factors and protocols described in this book, would be a more fruitful approach than further dissection of the Miller data. His findings should be accepted at face value, as positive evidence for cosmic ether, ether drift and variable light speed, just as we accept the findings of Sagnac, Michelson, and similar experimental results. New efforts should now be made towards independent confirmations, following a strict protocol that embraces, rather than denigrates, the idea of a motional and material, entrained ether. The current unethical and scientifically destructive "court-jester / skeptic club" attitude has no place in science, and must end.

Part III:

Into New Territory:

Additional Evidence

for a

Material, Motional

and Dynamic Ether

Ether as Cosmic Life-Energy

> "There is no such thing as 'empty space'.
> There exists no 'vacuum'. Space reveals
> definite physical qualities [which] can be
> observed and demonstrated. Some can be
> reproduced experimentally..."
> – Wilhelm Reich,
> *Ether, God and Devil,* 1948, p.111

In the years after the historic ether-drift experiments were concluded, and figuratively "driven into exile", multiple converging lines of evidence from other scientific disciplines indicated the discovery of an interconnecting self-organizing cosmic medium, a *cosmic life-energy* with ether-like dynamic and plasmatic properties. The discovered life-energy functioned within living systems, influencing chemistry and biology, and could change the physical structure of water. It also existed as a background medium filling the atmosphere and vacuum of space. Experimenters such as Jacques Benveniste (memory of water), Frank Brown (external biological clock mechanisms), Harold Burr (electrodynamic fields), Björn Nordenström (bioelectrical circuits), Giorgio Piccardi (physical-chemical fields), Wilhelm Reich (orgone energy) and Viktor Schauberger (living water) independently documented different aspects of this phenomenon. Entire bodies of scientific work and literature have been developed over the years by these and similar scientists, far too large to review here. For some, I can only give a general citation to their work in the *References*. For those in the 20th Century up to c.1995, an annotated bibliography was developed by John Burns (1997), *Cosmic Influences on Humans, Animals and Plants.* Science journals such as *Cycles* and the *Interdisciplinary Journal of Cycle Research* published numerous papers on these subjects. Today their journals have nearly vanished, their leading scientific luminaries passed away. When alive, most were subjected to public "skeptic" attacks, academic misrepresentations, and unethical erasure.

229

The Dynamic Ether of Cosmic Space

Regrettably, nothing of the most insightful and productive of the above list of life-energy scientists, Wilhelm Reich, is found in Burns annotations, and little of fact about him is found elsewhere in mainstream science, pop-media or internet. This was the result of a deadly 20th Century's slander and book-burning campaign directed against him in the 1940s and thereafter, as discussed in the *Introduction.* (WebRef.1) He published his findings in his own institute's journals and books, which were eventually reprinted in the 1970s, after the book-burning epoch.

In this chapter I will survey the facts on Reich's experimental findings, speak to my own positive replications of his experiments over the last decades, and end with a short discussion on Piccardi and Brown. These latter two scientists identified specific cosmic components in their investigations which, I will show, are agreeable in the details with Reich and Miller. Taken together and merged with the prior findings on cosmic ether under discussion, these studies collectively document a *major scientific breakthrough, the discovery of a unitary cosmic-atmospheric-biological energy*, ignored, suppressed and dismissed prematurely during the 20th Century.

Wilhelm Reich's Dynamic Ether-Like *Orgone Energy*

From 1934 to 1957, Wilhelm Reich produced a series of experimental reports documenting the existence of a unique form of energy, called the *orgone*, or *orgone energy*. Reich's line of research began with the clinical and experimental investigation of Freudian libido theory, including a milestone study on the bioelectric nature of emotions, somatic impulses, sexual excitation and sensory perception. Reich's research proceeded also into microbiology, with a study of motility and impulse-creation within simple microbes such as the ameba, which has no brain, nerves or muscle tissue by which to move its protoplasm towards food or away from irritating influences. His studies (Reich 1934, 1938) identified bioelectric commonalities in the motions of raw protoplasm in ameba, to nervous and muscular impulses in humans.

Wilhelm Reich
1897-1957

The work by Seifriz (1936) and others on motile slime molds suggests related findings. Slime molds are a large single cell of protoplasm with multiple cell nuclei and a common outer cell membrane. They can measure several centimeters in diameter, and can move about and take different forms. They show bioelectrical and purposeful motile characteristics similar to what Reich identified in single ameba, and yet they also do not have any brain or central nervous system, nor muscular tissues by which their motility and capability to change form could be understood. Reich concluded the tiny bioelectric signals detected in such cellular forms were only a superficial expression of a more powerful biological energy, which he later demonstrated experimentally as the orgone energy. (Reich 1942, 1948)

Starting around 1939, Reich created a special ferromagnetic-dielectrical enclosure to aid in capture and study of the orgone radiation. These enclosures resembled a Faraday cage, though I consider them more accurately described as a large *hollow capacitor,* a box-like enclosure with multiple layers of conductors and insulators in the walls, but with a final exterior dielectric insulation and interior ferromagnetic composition. Reich's layered enclosures were found to attract and accumulate this unusual biological-atmospheric energy, which could be felt on the inside as a penetrating revitalizing energy. It was named the *orgone energy accumulator* (ORAC).

By his early determinations, orgone energy charged the tissues of living organisms, but also existed in water and soil, and played a fundamental role in life processes. It entered organisms by food, water, breath and direct skin absorption, the latter effects of which could be amplified in the ORAC enclosure. He identified the orgone energy in a freely-moving dynamic form within the atmosphere, and also within high-vacuum tubes, implying a cosmic component.

Reich viewed the orgone as a ubiquitous medium similar to the cosmic ether which filled the vacuum of space; by his time, ether had been mostly discarded. The properties of Reich's orgone energy, as I have experimentally investigated and confirmed over nearly 50 years of experimental study, are in many respects quite similar to the Miller type of cosmic ether, *so long as ether is extended to include water-reactive and biologically life-enhancing properties.* In this context, we may ask, if one accepts the existence of cosmic ether, in any form, would it not also play a role in weather, chemistry, atomic processes, biochemistry and living organisms?

ORAC devices concentrate the orgone in a manner allowing for more detailed study. Controlled experiments indicate the ORAC can boost plant growth, increase tissue regeneration and biological healing, with immune-system boosting effects. These controversial claims have been verified in multiple controlled experiments with plants, with cancer mice, and with human subjects suffering from various low-energy or immune system disorders. There is an extensive published literature on these questions (WebRef.1). The ORAC can increase the energetic charge and vitality of the entire organism, with a general stimulation of the parasympathetic nervous system. The bioelectrical *zeta-potential* of red blood cells, a critical indicator of disease resistance, or immunity, tends to elevate with ORAC treatment, opposing trends towards disintegration of blood and other tissues. (Bauer 1987, Kavouras 2005, Müschenich 1986, 1995, Reich 1948).

The ORAC also shows various anomalous physical effects such as spontaneous heat production and an increased electrical density within its interior; and the ability to induce biologically-significant spectral absorption and fluorescence signatures into water so charged up inside it. It can suppress the rates of spontaneous "natural leak" discharge from electroscopes and create anomalous ionization effects within orgone-charged high-vacuum and Geiger tubes. Reich was an MD, and so applied his ORAC to human illness, finding it had a strong life-positive benefit for low-energy conditions, as well as the ability to disintegrate or slow tumor development. Reich and his associates backed up these claims with published clinical reports and controlled experimental evidence, gaining support from other physicians and scientists, which made him a considerable threat to mechanistic pharmaceutical medicine and the various "empty-space/dead universe" theories of his day. (Reich 1944, 1945, 1948. DeMeo 2011, 2018)

Through Reich's experiments and observations, and the verification studies by others including myself, the basic properties of the orgone energy can be described. The orgone is a *ubiquitous continuum* which fills all space, much like the cosmic ether, but is in constant *lawful* motion, with flowing, streaming, pulsating and spiral-form motions. This is confirmed by direct observations and experiments, including with high-vacuum tubes charged up inside the orgone energy accumulator (ORAC). Such evidence indicates the orgone exists also in cosmic space. It has a variable density, can shift its concentration from one place to another, and also exhibits the capacity to spontaneously expand and contract, with *pulsation*. The orgone can penetrate matter in a *mass-free* form, but in other contexts or conditions, it can weakly or strongly interact with matter. It is attracted to and charges all matter to a certain maximum capacity level. Metals strongly attract but

Figure 68. The Orgone Energy Accumulator

Resembling a *Faraday Cage* or *Hollow Capacitor*, the Orgone Accumulator (ORAC) acquires a charge of a previously unknown energy directly from the atmosphere and cosmic space. The interior charge is documented in many biological experiments, as with the stimulated growth of seedlings, speeded healing of animal tissues, stimulus of the human parasympathetic nervous system and boost in immune system functions. Anomalous physical effects develop inside the ORAC, such as a slightly higher temperature and an increase in electrical charge density. The ORAC can also increase the count-rates of certain nuclear radiation detectors and produce "ionization" effects within deep-vacuum tubes. The spectrographic absorption of ORAC-charged water increases in the far-UV frequencies with a compliment near-UV/blue fluorescence. The photo below shows two human-sized experimental orgone accumulators inside a larger orgone energy darkroom at the author's laboratory.

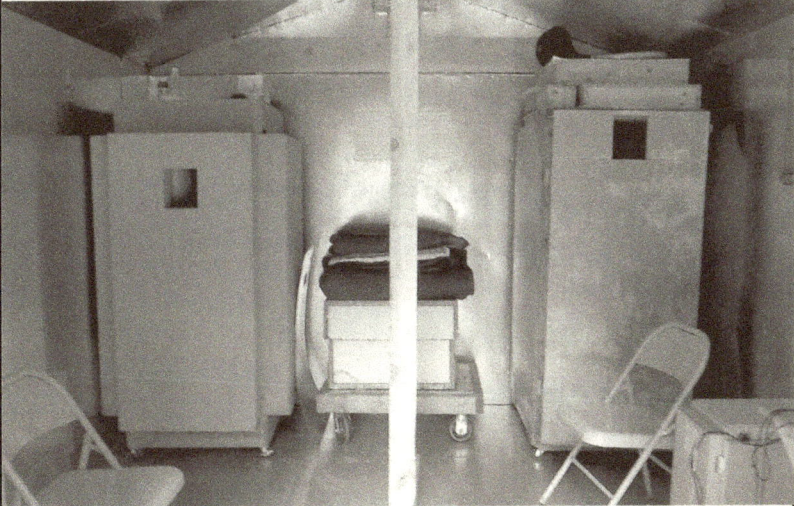

Reich's orgone energy and the accumulator are *world-class discoveries*, confirmed many times by independent scientists and physicians. However, their findings on the accumulator continue to be slandered and misrepresented by irrational critics and hostile media, who ignore published experimental confirmations, and deliberately lie about nearly everything in Reich's science and biography.

DeMeo, *Orgone Accumulator Handbook*, 2010
www.academia.edu/4211927

Figure 69. The Orgone Accumulator Thermal Anomaly (To-T)
The temperature inside a strong orgone accumulator (To) will be
slightly higher than inside a thermally-balanced control enclosure
(T), by a few tenths of a degree or higher. The effect has diurnal
patterns *unrelated to daily high or low temperatures*. To-T is higher
on low humidity sunny days when the accumulator effect is strongest.
The author and others have successfully replicated and verified the
To-T effect in controlled experiments.

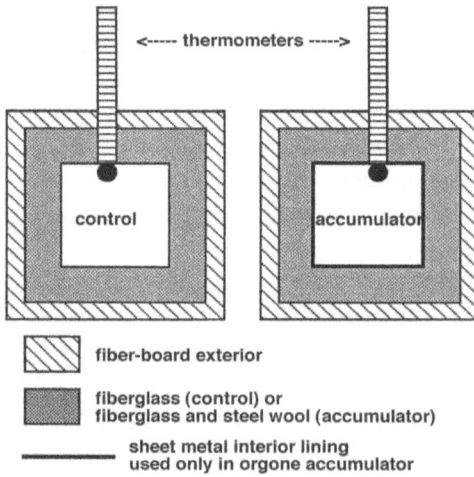

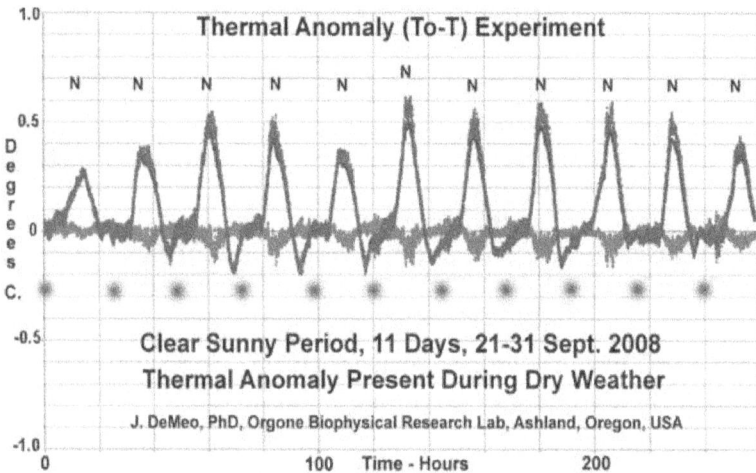

DeMeo, *Confirmation of Thermal Anomaly*, 2009
www.academia.edu/3677742

Figure 70. Biological Effects of the Orgone Accumulator

Reich and his associates undertook many clinical and controlled experimental studies proving the orgone radiation could benefit the growth of plants and the health of humans and other animals. The studies are myriad and beyond the scope of this publication to review in detail (see WebRef.1). Below is a typical result from one short run of a controlled mung bean seed-sprouting experiment at the author's laboratory. The left-side group of seeds was sprouted inside the orgone accumulator, as compared to a control group on the right, kept in a non-accumulating enclosure with all other factors being the same. A 3-year study of this effect showed an average ~34% boost in growth over controls, with high statistical significance (p<0.0001).

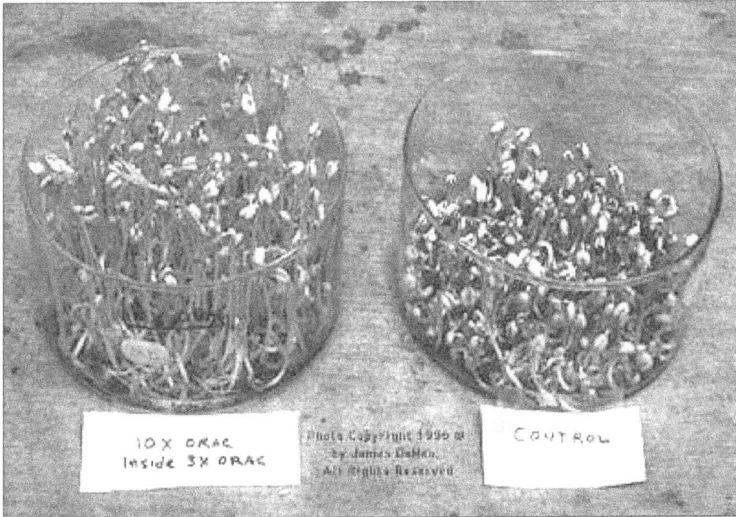

TABLE 1.	Control Groups	Orgone-Charged Groups	Percent Change
Average Seedling Lengths	149 mm	200 mm	+ 34%
Germination	95.8%	97.3%	+ 1.6%
Weight Increase	49.0 gram	53.2 gram	+ 8.6%
Average Water Consumed	109.9 ml	118.3 ml	+ 7.6%
Refractive Index (%Brix)	6.3	5.1	– 19%

DeMeo, *ORAC Stimulation of Sprouting Beans*, 2010
www.academia.edu/3677850

Reich's Discovery of the Biological, Atmospheric and Cosmic Orgone Energy

- First discovered as a radiation from blue-glowing *sand bion microbes*, isolated within an insulated metal *Faraday*-type enclosure, which was later developed into the *orgone energy accumulator*.
- Experiments proved orgone energy is an excitable and mass-free energy continuum, filling all space, similar to the cosmic ether, also being motile, pulsatile, excitable and reactive to matter.
- Orgone energy has a negative entropy and will concentrate to higher levels where possible. The orgone accumulator can charge up objects to yield thermal, electrostatic, humidity and other anomalies. It also has healing effects upon burns or wounds, with symptomatic benefits for biopathic, degenerative health conditions.
- Reich's theory of *cosmic superimposition* postulates spiraling and merging orgone energy streams as the basis of cyclonic rainstorms, hurricanes and galaxy formations. Energetic superimposition also functions at the microscopic, cellular and organism levels, in the creation of matter and governing sexual attraction and procreation.
- The energy flows according to spiral-wave, rotational characteristics, the *spinning wave or Kreiselwelle,* as Reich observed.
- The affinity of orgone energy towards water underlays its regulatory function within clouds, weather dynamics and the atmosphere, a fact which is proven by experiments with the cloudbuster. Reich provided some of the first scientific discussions on acid rain, forest death, and hazy drought/desert atmospheres.
- Reich's ideas accord with much of the older theory of cosmic ether, and both predated and anticipated such physical concepts as the *neutrino sea, cosmic plasmas, interstellar medium,* and *dark matter* – many of which are identified by similar blue-glowing and ubiquitous characteristics. However, Reich is rarely cited for his scientific priority. A modern "taboo" surrounds his name and work.
- The moving and streaming *cosmic orgone* was described by Reich as the *cosmic prime mover* in the Galilean sense. Reich's orgone theory is in harmony with the *empirical* foundations of modern astronomy, but challenges the varied metaphysical theories of astrophysics such as Einstein's space-time relativity, big-bang creationism, and fanciful quantum magic.

quickly discharge, reflect or radiate away the orgone energy. Dielectric insulators (wool, fiberglass) attract and hold the orgone, creating both an *orgonotic charge* and electrostatic effects in the process. (Reich 1949, 1951)

The orgone is also excitable and luminous. When exposed to atomic radiation or sparking electricity, it becomes highly excited and irritated, much as living protoplasm. Reich considered the aurora to be a direct expression of orgone energy in excitation and movement, based upon its motional similarities to living protoplasm. The *orgone charged vacuum tubes (vacor)* can also glow with a deep bluish color when sufficiently soaked inside the ORAC, and then excited by DC or high-frequency electricity. As I discovered, the vacor tubes would yield soft blue flashes of light merely by stroking with the hand, which carries no more than a few hundred millivolts.

Controlled experiments with ORAC-charged distilled water reveal a strong spectroscopic *absorption* of far-ultraviolet (UV) light frequencies well beyond a control distilled sample, with a related *fluorescence* in the near-UV and bluish frequencies. The *Cerenkov radiation* seen in nuclear reactor pools appears related, as is the intensive blue glow of highly-charged waters in certain hot-springs, natural lakes, deep oceans and glacial ice. (DeMeo 2018).

Based upon such observations, Reich postulated the Earth had a solar-excited, blue-glowing *orgone energy envelope,* which moved west to east around the planet, faster than Earth's rotation. The blue-glowing daytime atmosphere, by his thinking, was a local phenomenon created by solar excitation of the Earth's orgone envelope. His theory opened up new lines of inquiry not previously considered, suggestive of the older theory of a *luminating or luminiferous* cosmic ether. Reich expanded upon the findings of orgone motility more generally to include cosmic motions of all kinds. He postulated the existence of large spiraling streams of orgone energy in cosmic space, setting the solar system and spiral galaxies into motion.

Reich identified streams of orgone energy in the wind patterns on Earth, one aligned with the plane of the solar system ecliptic yielding a W to E atmospheric motion, and a second stream moving SW to NE, curling around the planet as an expression of the plane of the Milky Way Galaxy. He argued, these two streams of cosmic energy move down to Earth from space, permeating through the atmosphere, setting both Earth and atmosphere into motion. In their combined interaction, he argued the two energetic streams not only produced the Earth's 23.5°

axial tilt, but also created spiraling atmospheric motions, to include tropical hurricanes. The orgone energy's water-attracting and nega-tively-entropic properties drive the above interactions. Reich made these observations and postulates in the years prior to public knowledge of the jet streams and before satellite images, which is remarkable in itself. Mathematically ordered spiral and vortex structures in living creatures come from similar energetic motions and processes, as seen in embryology, sea shells and plant-growth patterns.

Reich also noted certain subjective light impressions as seen at night, in darkrooms after the eyes had adjusted for ~20 minutes, and specifically within a darkened room-sized ORAC, the *orgone energy darkroom.* He observed that orgone energy under higher charge and excitation had two basic visible expressions. One was a *fog-like* form, of a slow shifting and pulsing nature. This form I have observed and documented on videotape as a slow moving pulsation, using a dispersed laser light inside a strong ORAC. The other is a *pointed* form, of numerous pin-points of light which move in and out of existence, with a lifetime of less than 1 second. They emerge from the background orgone ocean, move mostly randomly, then quickly sink back into the same energetic medium from which they came. On occasion of higher energy situations, they last longer and inscribe specific spiral-forms which may attract each other and converge, before ebbing away. The pointed form is easily observed, in either darkrooms or out in the open sky, but has so far proven more difficult to record on photo or video.

These pin-points of light are also somewhat confirmed in the more sensitive image intensifiers and CCD video cameras, which reveal motile light flashes and particulate motion in completely darkened rooms. Engineers call them "random photons" or "cosmic rays", but their behavior is not so well understood from orthodox perspectives, as seen in the shifting, vague explanations given to them. A YouTube search of "gen.3 image intensifiers - astronomy" will bring up many examples of this phenomenon, which isn't exactly what one can see with the eyes, but is the closest objective verification I have so far identified. *This phenomenon must however be separated from the documented observation of blood-corpuscles in the eye,* which move in a patterned manner, retracing the same pathway relative to your retina and cornea, and also move together with bursts of speed, in coordina-tion with your pulse or heartbeat. Orgone units do not, and are random in their motion and appearance. Not everyone has the ability to see them; I estimate around half the population can do so.

Reich came to view these pin-points of moving light as *orgone units*, expressions of the orgone energy moving from the more quiet fog-like state into an excited pointed expression. He argued, similar orgone functions were at work within both microscopic and larger macroscopic atmospheric and cosmic scales, as a basic universal principle. For these and many other reasons, I believe Reich's orgone energy and the *dynamic cosmic ether* are functionally identical phenomena.

By Reich's theory, cosmic orgone units attract each other, superimpose centripetally in a spiral form and thereby condense to create new matter out of the same cosmic energy substrate. Reich described one expression of these spiral waveforms with the German term *kreiselwelle* (*spinning wave*), and came to believe they underlay various biological, atmospheric and cosmic motions. By Reich's theory of *Cosmic Superimposition* (the title of his 1952 book), the streaming, spiraling movements of microbes, of certain plants and molluscs, of hurricanes, of the rotation of planets on their axes, of moons around planets, and the revolution of planets around their suns, are all products of superimposing streams of cosmic energy.

There are other phenomena associated with Reich's orgone units, or *packets of energy*, which may be related to Planck's *quanta* in some unknown manner. Reich discovered that a Geiger tube as used for radiation particle counting, when charged up in an orgone accumulator, would initially yield erratic counts or go "dead", but later produce very high counts for ordinary background radiation. This was confirmed at my laboratory over many years, with an orgone-charged neutron counter (Ludlum model 12-4), which originally gave only 1-2 counts per minute. After soaking in the orgone darkroom for about a year and turned on only occasionally, one day it surprised everyone by giving very high counts over many hours. It was then set up with a data recorder, and continued with high counts permanently thereafter. It would also react to the presence of human beings (with their own life-energy fields) who entered the darkroom, as well as to periods of large sunspot and solar flare events, yielding up to ~4000 cpm. This was a saturation value for the recording system, so the pulses were probably much higher. This behavior continued over the period of high sunspots, but then declined to "only" ~200 cpm, then down to ~50 cpm over the current extended period of mostly zero sunspots. Reich also noted this reaction of his orgone-charged Geiger tubes to sunspots. Spanish physicist Victor Milian and his associates have also investigated this issue, obtaining positive indications supporting Reich's prior observations on radioactive anomalies in the ORAC. (Milian 2002, 2007)

The issues involved in this particular experiment go to the heart of cosmic factors and radioactive decay mechanisms, and lead to open questions about what is being counted inside the ordinary nuclear vacuum-type radiation detectors. The anomalously higher reactions in the ORAC radiation experiments must be due to other factors than natural background radiation. For example, the ORAC may produce an increase in the energetic potential and ionization properties within the Geiger tube. So it yields a higher count or pulse rate for a given voltage setting. Whatever the reason, *it is a phenomenon we have also seen disrupting video cameras and tube-type sensors, sometimes with permanent damage to their electronics.* Reich's most informative experiments on this subject focused upon orgone-charged high vacuum (*vacor*) tubes, and on the process of *oranur*, a highly-excited state of the orgone energy when exposed to irritating radioactivity, sparking devices, and electromagnetic fields. Reich's publication on the *Oranur Experiment* (1952c) presents the details.

From multiple lines of argument and experimental evidence, the orgone thereby fulfills the requirements of a dynamic and motional, cosmological luminiferous ether, but one also with specific identifiable properties which expand our definitions of cosmic ether into the realms of biology, meteorology, and nuclear processes. Certainly, it is clear that Reich's findings and theory are *not* compatible with "empty space" or with the older static or *stagnant and immobile ether concepts*. Nor does his work agree with the Einstein theory of relativity, or with other theories requiring the *absence* of an energetic and substantive nature to cosmic space. However, his theory and findings are compatible with a *motional and substantive dynamic ether*, something which would also fulfill the role of a *cosmic prime mover*. Reich further noted the orgone energy possessed a negative entropy, as it would spontaneously concentrate to higher levels of charge until reaching a maximum capacity level. Reich's orgone energy thereby stands as the long-sought *self-organizing principle in nature,* which must exist to oppose the degenerating and destructive properties of mechanical entropy.

While Reich was familiar with the cosmic ether as a concept, by the time of his writings on the subject, in his 1949 book *Ether, God and Devil*, he was confronted with a world where the false "null" interpretation of Michelson-Morley was well established. He apparently knew nothing about the work of Miller, and so he wrote:

Figure 71.

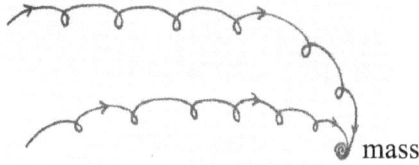

Reich's *Kreiselwelle,* or "Spinning Wave" orgone units as seen in the metal-lined orgone energy darkroom. Here, two units join and are postulated to create new matter which retains the momentum and structure of the cosmic energy.

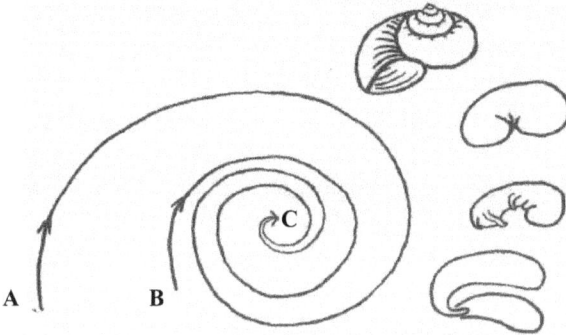

Cosmic Superimposition in Living Systems, basic form, two energy streams A and B converge on C. The imprint of such superimposing spirals are found in seashells, beans, embryos and in the orgonotic fusion of male and female organisms.

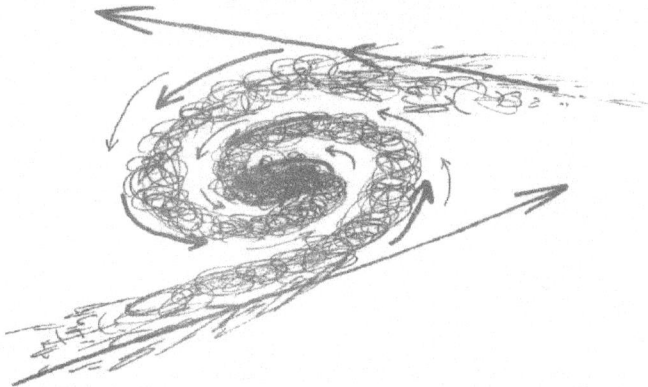

Cosmic Superimposition in Cyclonic Storms, Hurricanes and Galaxies, where two streams of energy approach and converge to form larger structures.

241

Figure 72. The <u>OR</u>gone <u>A</u>nti-<u>NU</u>clear <u>R</u>adiation Effect: ORANUR and VACOR (Orgone Charged Vacuum)

Reich's experiments indicated the orgone was a cosmic energy with life-like properties, filling all space but previously undetected by physics or biology. He tested the orgone radiation with various kinds of standard energy detectors, finding that most would not register anything. However, Geiger-Müller (GM) tubes would react if allowed to soak in the orgone accumulator for an extended period. This led to experiments with orgone-charged high vacuum tubes at around 0.5 micron pressure, which showed anomalous reactions not anticipated by conventional physics. These experiments indicated there is no "empty space". Orgone research, like that into "dark matter", "neutrino sea", "cosmic ether", "cosmic plasma" or "zero-point vacuum" proved the background of open cosmic space is *energy-rich*. In the historic *Oranur Experiment*, Reich also observed changes in the decay-rates of radioactive isotopes. (Reich 1951c)

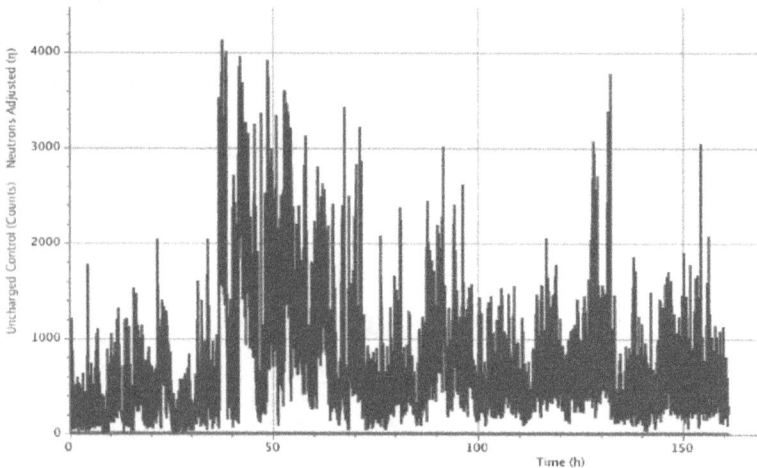

Very high count rates from a special thick-walled GM detector tube, soaked for a year inside a strong orgone accumulator at the author's laboratory. No radioactive materials were present, only natural background. Normally the device yields about 2 cpm, but with orgone charging it produced up to 4000 cpm during high sunspot years, declining to ~50-200 cpm during low sunspot years.

" 'Objective experiments', such as the Michelson-Morley light experiment which did away with the ether, are catastrophic events in scientific research." (Reich 1949, p.8)

His grasp of the principles of ether dynamics was nevertheless sound:

" 4. The negative result of the Michelson-Morley experiment, which was designed to demonstrate the ether, must be comprehended. The premises that led to the performance of the Michelson-Morley experiment rest on incorrect assumptions. ... Though I must leave a thorough critical evaluation of this experiment to the physicists, who are at home in the realm of its premises, the following remarks may be justified on the basis of some observations in orgone physics:

a) One of the premises of the Michelson experiment was the assumption that the ether is at rest; the earth, accordingly, moves through a *stationary ether.* This assumption is clearly proven incorrect by observation of the atmospheric orgone. If the 'ether' represents a concept pertaining to the cosmic orgone energy, it is *not stationary, but moves more rapidly than the globe of the earth.* The relation of the earth's sphere to the surrounding cosmic orgone ocean is not that of a rubber ball rolling on stagnant water, *but of a rubber ball rolling on progressing water waves.*" (Reich, 1949, p.112-113)

Reich's orgone, a real and objectively measured energy, connects life, atmosphere and cosmos into one bold theory, with expressions in both the living and non-living worlds. His theory of *cosmic superimposition,* later published in a book of the same title (1951a), encompasses the phenomena of cell growth and division, sexual excitation and attraction, emotions, cloud dynamics, atmospheric circulation patterns, and finally, planetary movements and galactic structure. Matter is not only created in the universe by streams of flowing and pulsing cosmic orgone energy, but this same energy also acts to move the planets and suns along on their paths in the heavens. Reich described the process as follows:

"The sun and the planets move in the same plane and revolve in the same direction *due to the movement and direction of the cosmic orgone energy stream in the galaxy.* Thus, the sun does not 'attract' anything at all. It is merely the biggest brother of the whole group. ...

Both moon and Earth spin along in space, with their respective open (not closed) pathways mutually approaching and separating again. Therefore *it is not the gravitational masses, but the PATHWAYS of the gravitational masses, which meet.*

The moon does not 'circle around the Earth', since the lines of movements are open, spiraling curves. ...

The cosmic orgone energy flow that carries both moon and earth along in the same direction, in the same plane, and in perfect coordination of their speeds, is the true agent of the gravitational free fall. ...

The function of gravitation is real. It is, however, not the result of mass attraction but of the converging movements of two orgone energy streams. From these converging streams the 'attracting' and 'gravitational' masses once emerged and they are still carried along in the universe by the same streams in an integrated, unitary fashion as expressed in their common direction of movement, their common planes of motion, the mutual approach of their centers, and the mutually coordinated speed of their spinning motion." (Reich 1951, p. 191, 274, 276.)

Reich further compared the spinning wave to the line described in space by a point near the rim of a turning wheel, or by a rotating top, and as seen in the pulsating behavior of pendulums. Figure 74 provides graphical views of Reich's analogies.

"The 'Swing' can be easily visualized as the line described in space by a point on the rim of a wheel rotating forward. In relation to the ground, this point on the rim though rotating with even speed in itself, describes a movement of alternating acceleration and deceleration. In other words, its motion expands and contracts alternatingly. On the forward turn, the point moves faster. On the backward turn, it moves slower. The ratio of speed change depends, of course, on the basic speed of rotation: The faster the rotation, the shorter the contraction with respect to the forward motion.

A spinning top shows the same basic function of speed contraction and expansion. The top will move in a more or less

Figure 73.

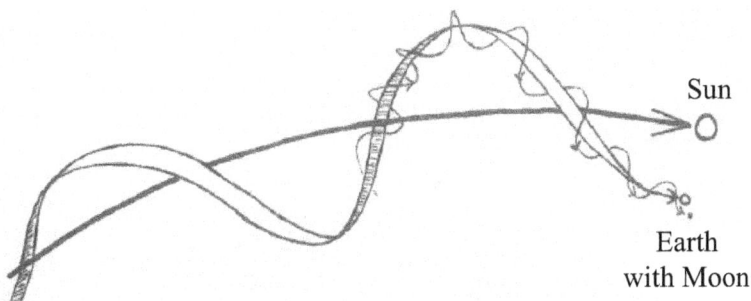

Planetary Motions by Reich, not in Newtonian circles nor Keplerian ellipses, but in forward-moving open-ended spirals. The *kreiselwelle* or spinning wave of the orgone energy is a basic motional property in both the microscopic and macroscopic realms. The Sun and planets are "passengers" riding along on superimposing, spiralling streams of cosmic energy.

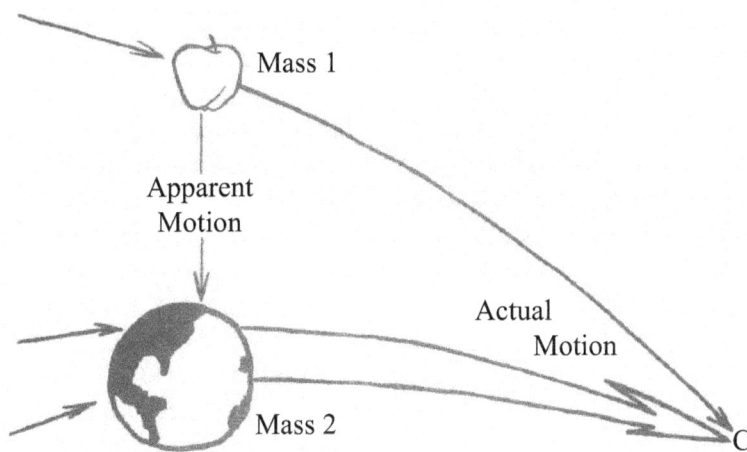

Gravitational Superimposition: Mass 1 approaches Mass 2 following *curved* lines of the superimposing orgone medium, not straight lines. Matter is thereby moved towards a common center-point at C. Space is filled with a spiralling cosmic energy that sets all matter on curved pathways of mutual approach, as the gravitating force.

245

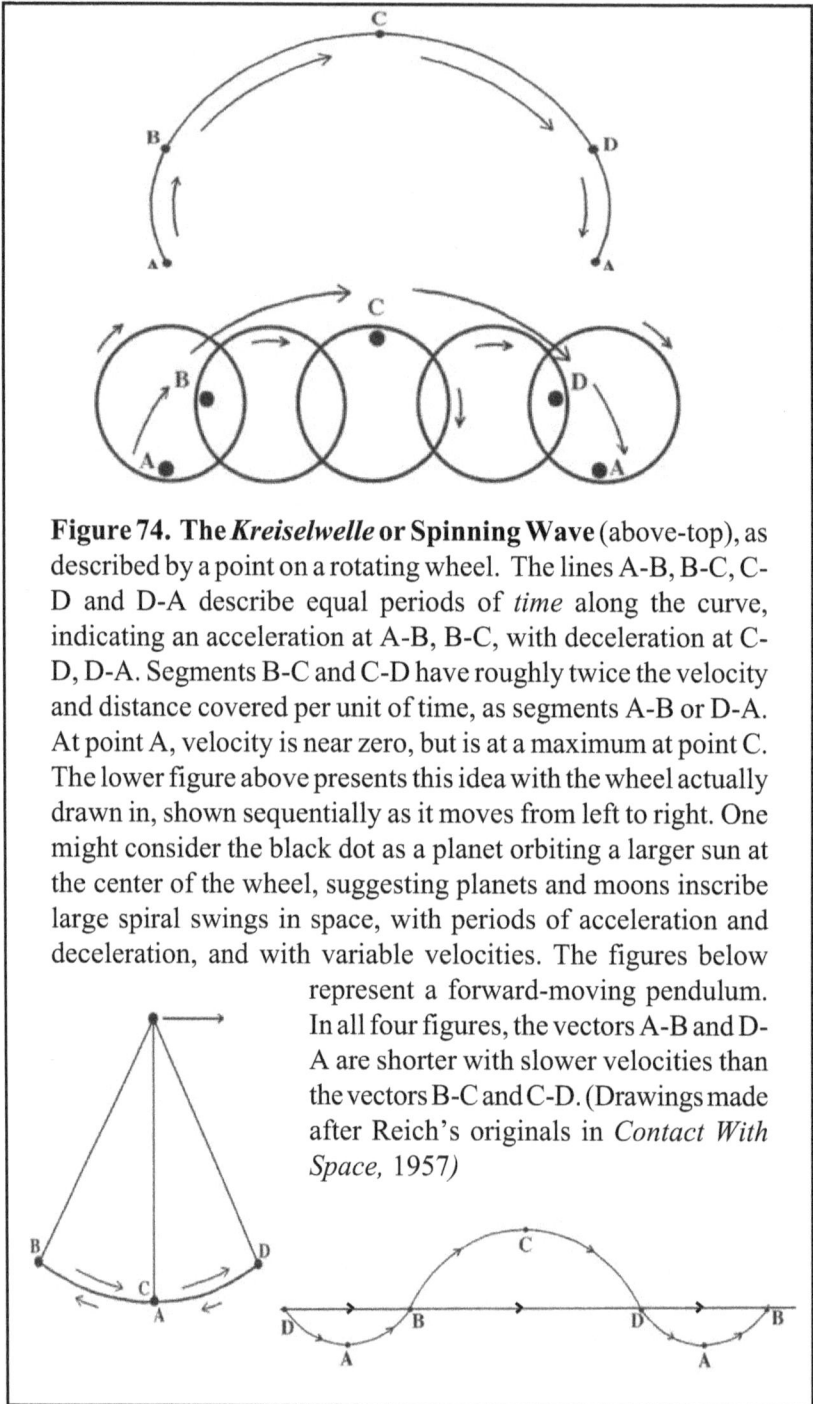

Figure 74. The *Kreiselwelle* or Spinning Wave (above-top), as described by a point on a rotating wheel. The lines A-B, B-C, C-D and D-A describe equal periods of *time* along the curve, indicating an acceleration at A-B, B-C, with deceleration at C-D, D-A. Segments B-C and C-D have roughly twice the velocity and distance covered per unit of time, as segments A-B or D-A. At point A, velocity is near zero, but is at a maximum at point C. The lower figure above presents this idea with the wheel actually drawn in, shown sequentially as it moves from left to right. One might consider the black dot as a planet orbiting a larger sun at the center of the wheel, suggesting planets and moons inscribe large spiral swings in space, with periods of acceleration and deceleration, and with variable velocities. The figures below represent a forward-moving pendulum. In all four figures, the vectors A-B and D-A are shorter with slower velocities than the vectors B-C and C-D. (Drawings made after Reich's originals in *Contact With Space,* 1957)

curved line at high speed. The line of motion forward will be more even the greater the speed.

At a lower speed of rotation, the pin on which the top rotates will clearly describe a spinning wave, a KRW (Kreisel-welle) and swings with alternating acceleration and deceleration...

Alternating expansion and contraction of forward motion may also be easily observed in the movement of swinging pendulums under the condition that the point of suspension moves onward in space, while the pendulum body swings." (Reich, 1957, p.95-110)

These and other descriptions in Reich's writings suggest very real and testable hypotheses regarding planetary movements, some of which already appear to be accepted at a basic level by modern astronomy, though for completely different reasons, and certainly without the multi-disciplinary significance given by Reich. For example, it is superficially acknowledged that the orbits of planets around our Sun inscribe large spiral-forms in space. However, *no special emphasis is placed upon this fact, given the assumption of empty space.* Only a few textbooks make mention of it.

Miller's Ether-Drift Velocities Confirm Reich's Open Spirals

Reich was keenly aware of how his theory of cosmic superimposition was in partial conflict with astrophysical theory, and wrote:

"The path of the planets is neither a Copernican circle nor a Keplerian ellipse. It is of necessity *open*, and not closed, since there is a *forward* motion in space of sun and earth which never returns into its own path. Correlation of the classical astrophysical calculations which use the circle and the ellipse with the orgonomic '*open* path' of the course of the planets, becomes now a major task of natural science. Since the paths of the planets are necessarily a spinning wave, the coordination of classical and orgonomic astrophysical observations will have to deal with the integration of the Keplerian ellipse with the spinning wave." (Reich 1951, p.82)

Table 11. Miller's 4-Epoch Averages of Ether Drift Velocity

Epoch Center Date	Ether Velocity Average
1926 February 8th	9.3 km/sec
1925 April 1st	10.1 km/sec
1925 August 1st	11.2 km/sec
1925 September 15	9.6 km/sec
4-Epoch Average:	10.05 km/sec

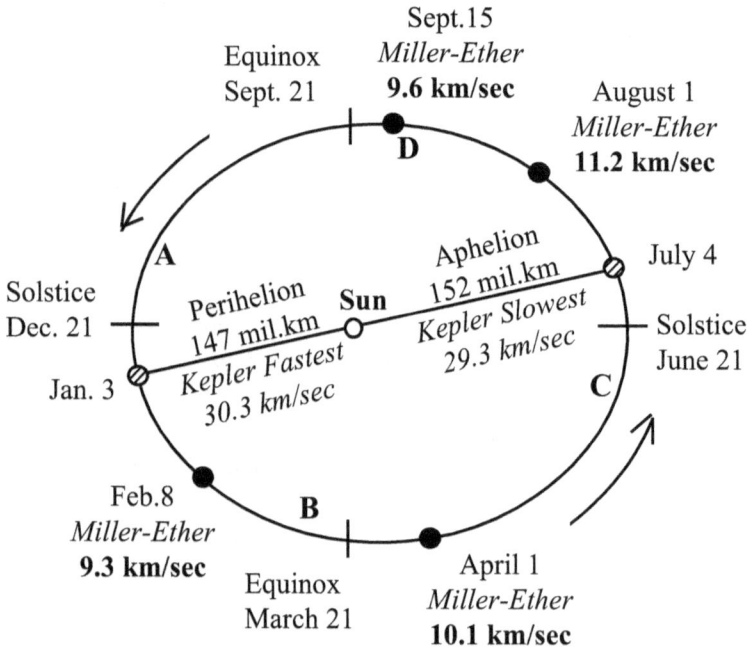

Figure 75. The Keplerian 2-D View of Earth's Orbit is Not Congruent with Miller's Ether-Wind Velocities, nor with Reich's Cosmic Spiral Motions. By Keplerian theory, planets move fastest at perihelion and slowest at aphelion. Miller's ether-drift velocities appear incongruent in this flat 2-dimensional presentation, but fit well when the same motions are viewed in spiral form (see next figure). The letters A, B, C, D can be matched up to the same letters in Figures 74 and 76. *Point "A" is where Earth's 3-Dimensional spiral motion in the background of space is at a minimum velocity, while point "C" is the location of maximum velocity.*

With the work of Miller, there is support for Reich's cosmic superimposition theory of open spiral motions of the Earth around the Sun. This will require a review of Miller's ether velocity determinations, given in Table 11, comparing them to both Kepler and Reich.

By Kepler, velocities of planets through open space are determined in relationship to the Sun, and hence *increase* as they move towards perihelion, closest to the Sun. The Earth's perihelion in space is on January 3rd, about two weeks after the Winter Solstice of December 21st. Keplerian solar-orbital determinations also show the velocities of planets *decrease* through open space as they approach aphelion, their farthest distance from the Sun. Earth's aphelion is around July 4th, about two weeks after the June 21st Summer Solstice. However, Miller's ether-drift velocities contradict Keplerian formulations.

When viewed according to Kepler on a flat 2-dimensional (2-D) plane, the Earth's orbit speeds up at perihelion and slows at aphelion. However, when Miller's individual epoch ether velocity determinations are plotted on that same flat surface, by conventional theory they are *not congruent with 2-D Keplerian theory.* Miller's slowest ether velocity of 9.3 km/sec, on February 8th, is closest to perihelion, a date when by Kepler, the speed of the Earth's orbit across the flat 2-D plane is the fastest. And in the part of the flat plane where Earth's speed should be slowest by Kepler, closest to aphelion on August 1st, Miller's ether velocity was the fastest, at 11.2 km/sec. This is presented in Figure 75.

The fastest ether velocity is found closest to aphelion, where Keplerian velocities should be slowest, while the slowest ether velocity is found closest to perihelion, where Keplerian velocities should be fastest. Conventional 2-D flat-surface astronomy calculates an empty-space velocity of Earth at perihelion to be 30.3 km/sec, and at aphelion to be 29.3 km/sec. Miller's findings contradict that determination. *This incongruity is resolved and shown to be understandable, when the same data are viewed in their real-world 3-dimensional (3-D) form of an open-ended spiral*, as per Reich's theory, seen in Figures 76 and 77.

Relative to the background of space, Earth moves a greater distance during the period March-September (B-C and C-D) than during the period September-March (D-A and A-B). The spiral velocity of Earth in space is then congruent with Miller's ether-drift data, and by the letters A-B-C-D, also with Reich's *kreiselwelle* diagram in Figure 74.

If conventional astronomy is correct that Earth is moving generally towards Vega, then the northern axis of the ecliptic plane is ~28° off from the direction of the Sun-Earth spiral motions through space. This motion towards Vega would be a *cambered, off-center and open-ended spiral, leaning towards right ascension (RA) 17 hrs sidereal by around 62° up from the plane of the ecliptic.* This has important implications, *but only if cosmic space has energy, substance and motility.*

Notably, this is not something affecting only the Earth. *The velocities of all planetary motions in the background of space, as they are orbiting the Sun, are significantly variable over the course of a year, deviating from the Keplerian expectations when viewed in their 3-D spiral motions.* This spiral variability is also greater than the variations within the Keplerian flat-surface, 2-D ellipse velocities.

By Miller's measurements, given on Table 11 and Figures 75 and 76, the Earth's velocity variance along a spiral trajectory ranges from 9.3 to 11.2 km/sec over the course of the year, a difference in velocities of ~17%. The Keplerian velocity variations for Earth are from 29.3 to 30.3 km/sec, or about 1 km/sec difference between perihelion and aphelion; that is about a 3.3% variation. Miller's ether velocity variations are

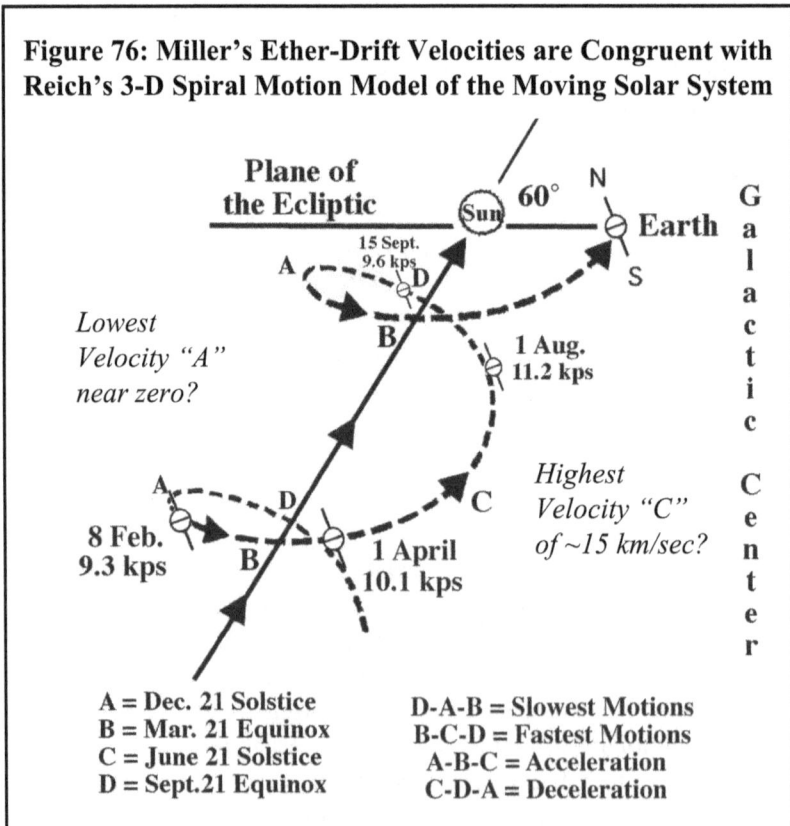

Figure 76: Miller's Ether-Drift Velocities are Congruent with Reich's 3-D Spiral Motion Model of the Moving Solar System

A = Dec. 21 Solstice
B = Mar. 21 Equinox
C = June 21 Solstice
D = Sept.21 Equinox

D-A-B = Slowest Motions
B-C-D = Fastest Motions
A-B-C = Acceleration
C-D-A = Deceleration

therefore about five times larger than Kepler's variations, in addition to their respective maxium and minimum velocities being inverted.

Is the True Axis of Ether Motion Closer Towards Vega?

Recall from an earlier chapter (p.96) how, after consulting with unnamed Mt. Wilson astronomers, Miller did not undertake his Mount Wilson ether-drift measures in the December period as he was convinced that "...tests during [that period] should give a resultant value for Earth motion near zero..." (Swenson 1970, p.65) A very low ether velocity is predicted for December by spiral-motion dynamic ether determinations, when the Earth dramatically slows in motion against the background of cosmic space. Swenson did not clarify how this conclusion was drawn, but from our viewpoint of Earth's spiral-form motion in space, December is a period of minimal ether velocity.

By the 3-D spiral motions, Earth moves at a very slow velocity at point "A" on my figures, but then increases velocity as it moves past "B" towards "C", where a maximum velocity would exist. This is fully oppositional to the 2-D Keplerian concepts. As noted, Miller regrettably never undertook systematic ether-drift measurements in either December or June. With this in mind, we can ask, how would a theoretical June maximal ether-wind of around 15 km/sec, and a December theoretical ether wind of only a few km/sec, affect Miller's determination of Earth's azimuthal motion in the cosmos? If velocity measures had been made on those "A" and "C" dates, with those results, it would have significantly pulled his final azimuthal determination towards a *lower declination*, while keeping along the same 17 or 18 hrs sidereal vector. The ether wind would then aim closer towards the star Vega, in better agreement with the modern "Sun's Way".

We can understand why Miller may not have desired to undertake an epoch of ether measurements during a period of predicted low ether velocity. But why avoid making measurements at the June maximum? We do not know.

An identical apparent contradiction in celestial velocities was also raised in the "dark matter wind" phenomenon detected in recent years by the DAMA (DArk MAtter) project led by Rita Bernabei, presented in more detail towards the end of the next chapter. Piccardi came to similar conclusions about the variable velocities of Earth's orbit in a spiral-form manner, given how this factor showed up in his chemical phase-change experiments. So far as I can determine, only Reich,

Piccardi and Bernabei (the latter two discussed momentarily) wrote about this issue of variable speed, spiraling orbits in space, moving through a cosmic energy. *They also got it right.*

Reich was fully correct, the Keplerian determinations of highest velocity at perihelion and slowest at aphelion are precise, but only for 2-dimensional flat-plane astronomy. *For 3-dimensional spiral trajectories, Kepler's equations are incomplete,* lacking the added spiral-form velocities, even though they retain practical value for orbital motions on a flat surface. Calculations for sending rockets to the Moon or Mars adhere to the 2-D Keplerian model, as all objects remain within the same inertial framework. Moving from the "flatland" world of 2-dimensional planetary motions, into the 3-dimensional reality will, as Reich noted, require new mathematical formulations. This might only become obvious when investigating possible short-cuts or fuel-savings in space voyages, by navigating through a spiral-form cosmos, much as jets and ships save time and fuel today by using great-circle routes.

Figure 77 on the next page presents the material-motional dynamic ether theory with additional stellar and galactic features included. The Figure 78 northern sky chart on the page thereafter is updated from the Figure 35 (p.111) in the chapter on *Dayton Miller's Experiments.* Figure 78 now includes all the prior vectors plus the *average* azimuth location of the numerous studies on Earth's movement through the cosmos which Miller originally accepted for his northerly solution, and published as a table in 1931. Further added are the ether-drift determinations from Galaev, Munera and Cahill, as previously discussed. Vectors are also included for Piccardi and Brown, as will be discussed in the pages immediately thereafter. One more update of this same Figure 78 will be seen at the end of the next chapter, plotting additional independent vectors of Earth's cosmic motion in space.

Again, this particular set of velocities and motions would not have any serious repercussions for conventional theory if space were an empty void. However, as space is filled with a material and motional ether medium, it becomes an important consideration, doubly so if the cosmic ether wind is a realization of the older concept of a *prime mover.*

My diagrams, based as they are on the actual movements of the Earth and Sun, with Miller's ether-drift velocity findings added in, not only provide a solution in agreement with conventional astronomy, and with Miller's original determination of a northerly apex, but also are in harmony with the overall existence of a material and motional, dynamic cosmic ether, indicating measured variable speeds of light in

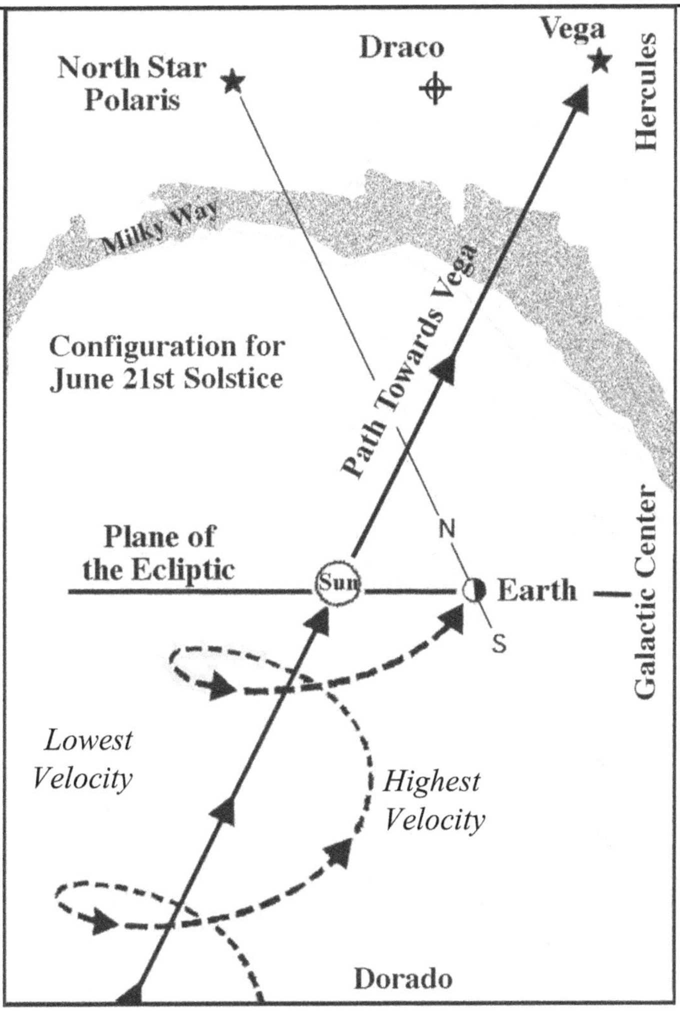

Figure 77. The Earth's Spiral Path Through the Cosmos.
Neither a flat circle nor ellipse, nor even a symmetrical spiral
as in a screw thread, Earth's motion is a cambered, off-center
spiral, which imparts *variable* velocities to the Earth and other
planets over the course of their yearly orbit around the Sun. The
northern pole of the ecliptic plane is identified at the '⊕' mark
near the top center, while the Earth moves more closely
towards the star Vega. The diagram shows the Earth at the June
21st solstice position, a general period of maximal spiral
velocity through the cosmos. Actual velocity determinations
are given on the preceding Table 11 and Figure 75.

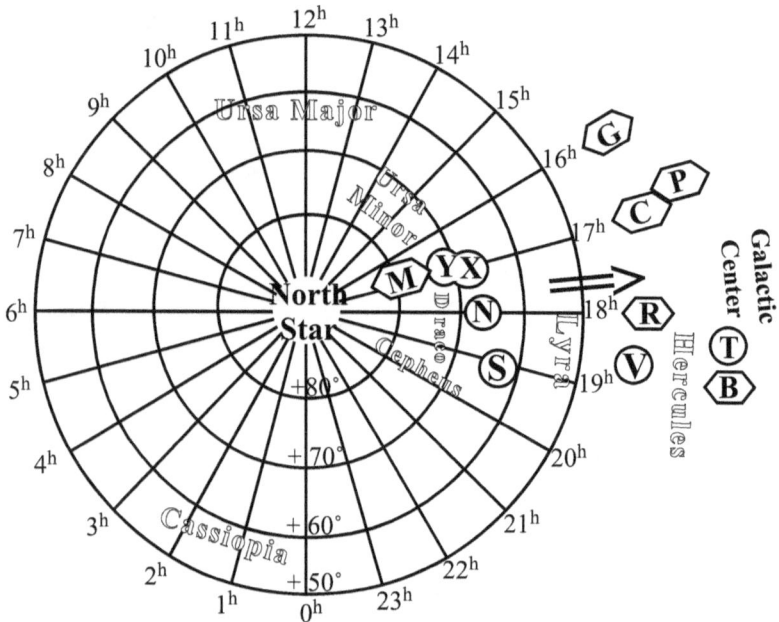

Figure 78. Northern Sky Map of Cosmic Vectors, Update 1

Looking North and Up to Polaris (Pole Star)
from Earth's Northern Hemisphere

Northern Pole of the Solar System Ecliptic = Ⓝ
RA 18:00h Dec +66.5° North
Modern Sun's Way Towards Vega = Ⓥ
RA 18:36h Dec +39° North
Sun's Rotational Axis Northern Pole = Ⓢ
RA 19.4h Dec +64° North
Center of the Milky Way Galaxy RA ~17:45h ⟹
Miller's Northern Axes of Ether Drift
1928: RA 17h Dec +68° North = Ⓧ
1933: RA 16.9h Dec +70.5° North = Ⓨ
Reich's "Point C" Cosmic Velocity Max = ⟨Ⓡ⟩
Piccardi ⟨Ⓟ⟩ **& Brown** ⟨Ⓑ⟩ **Chemical & Biological Vectors**
Galaev ⟨Ⓖ⟩ **16h and Cahill** ⟨Ⓒ⟩ **17h Ether Vectors**
Munera (Dynamic) Ether Vector 17.5h Dec +79° N = ⟨Ⓜ⟩
Miller's 1931 Average, Table of Earth's Motion = Ⓣ
RA 18.2h Dec +28° North

different directions. They are also congruent with Reich's theory of cosmic superimposition. This clarification makes logical sense when the solar system is placed into its *real-world,* spiral-form vortex motion, moving towards a northerly apex. *Miller's and Reich's findings are thereby congruent with contemporary empirical astronomy, but not with contemporary theoretical astrophysics.*

If the ether-drift experiments are the consequence of some strange set of errors or bad assumptions, then why is there such a strong correlation of observed astronomical motions gathered together on such an axis, aligned generally towards RA ~17 to ~18 hrs sidereal?

Reich's Meeting With Einstein on the ORAC Thermal Anomaly

One of Reich's proofs for the orgone energy and the ORAC as a remarkable and functioning device – aside from the documented biological growth and immune-boosting properties, its electrostatic charging and other physical effects – was the *To–T experiment*, previously discussed and summarized in Figure 69, p.234. The orgone accumulator was found to create a slightly warmer temperature within its interior (To) than a thermally-balanced control enclosure (T). The difference, To-T, maintained an average positive thermal anomaly of a few tenths of a degree C, with variations up to several degrees C. Reich developed an experimental apparatus to measure the specific To-T effect, using sensitive tenth-degree mercury thermometers.

It is a simple experiment conceptually, though not so easy to undertake in practical terms, if one wishes to rule out all possible conventional thermodynamic influences. When properly done, it stands as a refutation of the 2nd Law of Thermodynamics, and also as a proof of the negative-entropic nature of the orgone energy.

After a preliminary letter to Einstein in late December 1940, describing his findings, Einstein agreed to meet with Reich at his home in Princeton, on 13 January 1940. The two men spent the evening discussing Reich's discovery, as well as their respective times in Berlin, and escapes from the Nazis to the USA. The meeting went on for five hours. Reich brought with him a *To-T* demonstration apparatus, as well as an *orgonoscope*, a device for making visual observation of orgone energy units and motions in the atmosphere. During that meeting, Einstein confirmed the visual phenomenon and also witnessed the thermal increase inside the small orgone accumulator, declaring that "should it be true, it would be a great bomb" in physics. Reich departed the meeting on friendly terms with Einstein, leaving him with the To-T and orgonoscope devices for further study.

In his subsequent letter of 7 February to Reich, Einstein *confirmed having replicated the To-T effect.* Unfortunately, one of Einstein's assistants offered a conventional explanation, that the thermal anomaly was an artifact of heat convection in the room. Einstein accepted that explanation without further investigation, and communicated his opinion to Reich in the same letter. Reich wrote back to Einstein on 20 February, outlining various control experiments which, if undertaken, would refute the idea of simple convection effects. However, Einstein never replied to Reich on this issue of control experiments, and there is no evidence he undertook further investigations.

Today we know Reich's orgone energy, which is functionally identical to the cosmic ether, would have undermined Einstein's theory of relativity no less than did the positive ether-drift evidence. Only 15 years earlier, Einstein had been confronted with experimental evidence of a similar cosmic ether continuum by Dayton Miller, as already discussed. Einstein never embraced the Miller results, given how it would destroy his theory of relativity. He had dismissed Miller's findings *ex-cathedra* as thermal artifacts. The same "explanation" was resurrected by Einstein in 1941, to reject Reich's findings, which also suggested the discovery of a *ponderable medium* – the orgone energy in Reich's case.

Einstein's response to Reich after their initial meeting and letter exchange was total silence. Rumors eventually came to Reich's attention that "Einstein has disproven Reich", which was not the case. *Einstein had confirmed the To-T effect,* but then dismissed it as due to an unproven, ad-hoc objection of "thermal convection". He then refused to undertake the control experiments, and did not further respond to Reich. To defend himself against rumors, Reich published the correspondence between the two men in a small booklet entitled *The Einstein Affair* (Reich 1953). Their relationship cold, Reich demanded and obtained a return of his demonstration apparatus. They had no further contact.

Reich and Miller

Reich never cited Miller's work, though he knew about and referenced the work of Michelson-Morley, as previously quoted. Similar to Miller, Reich noted orgone energy was more active at higher altitudes, and he identified specific times of year when orgone energy exhibited higher versus lower states of excitation. My own experiments with orgone-charged neutron counters verified Reich exactly on this point. The work of Giorgio Piccardi, discussed below, also found similar cosmic effects in phase-change and chemical reactions in water, occurring simultaneously in both northern and southern hemispheres. It was not something due only to seasonal temperature variations, or related mechanical factors.

Reich's findings and theory, as presented in the book *Cosmic Superimposition,* agree with much of acknowledged astronomy, in that moving stars and orbiting planets describe large open spiral-forms in space. From this, Reich worked out his own special functional equations of gravitation and pendulum behavior, based upon his insights into the spinning wave, and space being filled with an energy-rich substrate. His findings are compatible with the concept of a dynamic ether, which would also fulfill the role of being a *cosmic prime mover*.

However, Reich's ideas are *not* compatible with the concept of a *static, stagnant or immobile ether*, and only *partially compatible* with Miller's *passive Earth-entrained ether*, through which Earth is assumed to be forcibly pushing. For Miller, ether motion was unrelated to gravitation. By contrast, Reich's universe is animated by streams of flowing, pulsing and superimposing cosmic orgone energy, which moves the planets and suns along on their paths in the cosmos. Streams of superimposing orgone energy create matter, move the planets in their orbits, and form the spiral galaxies, just as they give rise to the hurricanes, cyclonic storms, and even the spiral twists often found in plant and animal growth and form. Reich's views thereby unite astronomical determinations with biology, chemistry, atmospheric motions, light transmission and gravitation. His ideas are unique for cosmology, even while having similarities to the older ideas of Descartes on whirlpools of dynamic ether, and also to Michelson's rarely-expressed interest in a vortexing cosmic ether. They are deserving of our attention, as much as do Miller and the other ether researchers, if science is to break free of immobilizing and illusory mystical concepts.

Had Miller lived long enough to review his work from the standpoint of Earth's spiral motions, I believe he might have been led to reconsider his average azimuth of ether drift, in accordance with my Figures 75-78, derived as they are from both his own and Reich's findings.

Other Evidence for a Dynamic Ether Energy

Giorgio Piccardi's Chemical Tests

The Italian chemist Giorgio Piccardi independently discovered a set of cosmic motions, similar to Miller's and Reich's, over several decades of study of anomalous variations in laboratory phase-change experiments. These included changes in precipitation rates of bismuth chloride from solution, and in the freezing temperature of supercooled water, as related to sunspots and other cosmic factors. The anomalies appeared in the same months at separate laboratories in both northern and southern

**Giorgio Piccardi
1895-1972**

hemispheres, indicating a cosmic factor affecting the entire Earth simultaneously. Use of metal enclosures or shields, suggestive of Reich's orgone energy accumulator, produced a higher rate of reactivity of solutions, as did use of stirring devices containing a small amount of mercury sealed inside a partially-evacuated glass tube. These were once sold commercially for de-scaling of boilers, such as the *scalebuoy* device, which emit a subtle silver-blue light when shaken, inducing energetic changes in the physico-chemical solvent properties of water.

From his laboratory results, Piccardi theorized that the helicoidal movement of the Earth around the Sun, with changing velocities over the course of the year, was a factor which imparted variations within his tightly-controlled laboratory physical chemistry experiments. It was not something related to ordinary seasonal temperature or humidity, which would affect the northern and southern hemispheres in an opposing, out of phase manner. (Faigl 1990, Piccardi 1962, 1965, 1966, 1968, 1972)

As with Miller and Reich, Piccardi's work was also subjected to denigration and academic erasure. His contemporary advocates experienced similar censorship after his death, though for a short period of time in the 1950s and 60s, Piccardi's work did enjoy research support through the UN *International Geophysical Year* science programs. Piccardi and his associates founded the International Society for Biometeorology, an organization which later banished those associates (as noted in the *Introduction*). Piccardi's spiral-helicoidal planetary movements were illustrated in his primary English publication (Piccardi 1965); he also made a 2-D graphic and 3-D working model. These are presented in Figures 79 and 80. Piccardi wrote:

> "The Earth is displaced with the Northern hemisphere leading... If space were empty, of fields of matter and inactive, a consideration of this type would be of no importance. But today, we know instead that both matter and fields exist in space. For this reason, *the displacement of a body such as the Earth in one direction or another is not inconsequential. Its general physical conditions must vary in the course of a year.*" (Piccardi 1965, p.97-98. Emphasis added.)

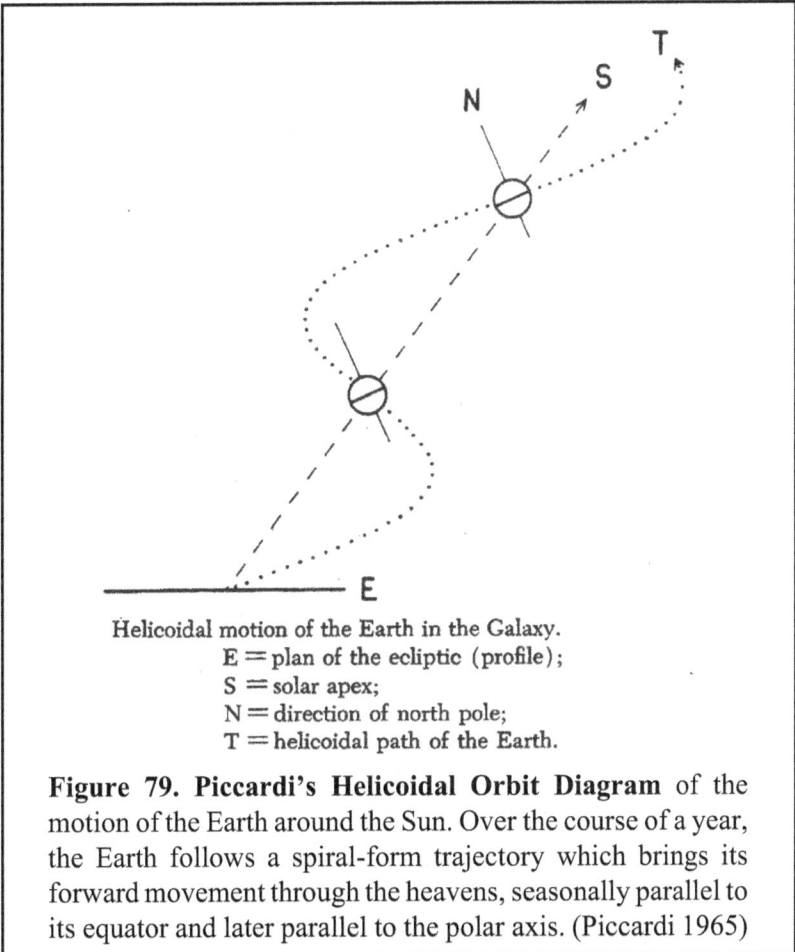

Helicoidal motion of the Earth in the Galaxy.
E = plan of the ecliptic (profile);
S = solar apex;
N = direction of north pole;
T = helicoidal path of the Earth.

Figure 79. Piccardi's Helicoidal Orbit Diagram of the motion of the Earth around the Sun. Over the course of a year, the Earth follows a spiral-form trajectory which brings its forward movement through the heavens, seasonally parallel to its equator and later parallel to the polar axis. (Piccardi 1965)

Piccardi's experiments suggested a cosmic mechanism similar to that observed by both Miller, Galaev and Reich. Like them, he observed the reactivity of his chemical tests would increase at higher altitude locations, both above sea level and height above the local ground surface. His calculations of the Earth's net motion through the background of space agree with the variations presented in my figures, with the Earth moving in the northerly direction along a cambered, offset spiral manner. We therefore have converging lines of evidence from three different sources, Miller, Reich and Piccardi, in the latter case coming from the field of chemistry. It does not appear that Reich, Piccardi or Miller knew of each other's work. Their findings were all developed independently, and are in good general agreement on the basic points regarding the substantive-reactive properties of cosmic space, and the importance of cosmic motions and spatial, 3-D geometry in the understanding of experimental results, and in formulation of theory.

Figure 80: Piccardi's Animated Model of the Helicoidal Motion of the Earth Around the Sun, as presented at the Brussels World Fair in 1958. The Earth moves faster through the cosmos in June than during December. (Piccardi 1965)

Frank Brown's Cosmic External Biological Clock Mechanism
From 1959 to 1977, the biologist Frank Brown worked at Northwestern University in Chicago, and later at Wood's Hole Institute in Massachusetts, documenting cosmic sidereal day, lunar and sunspot cycles in the biological clocks of a variety of plants and animals. The organisms used in his studies were always maintained under constant light, temperature, humidity and also often barometric pressure environmental conditions. For example, Figure 81 shows the *sidereal day variation* in the oxygen consumption of potatoes kept inside a hermetically-sealed tank, with all other environmental conditions held constant. The experimental data were gathered over an 11 year period. The

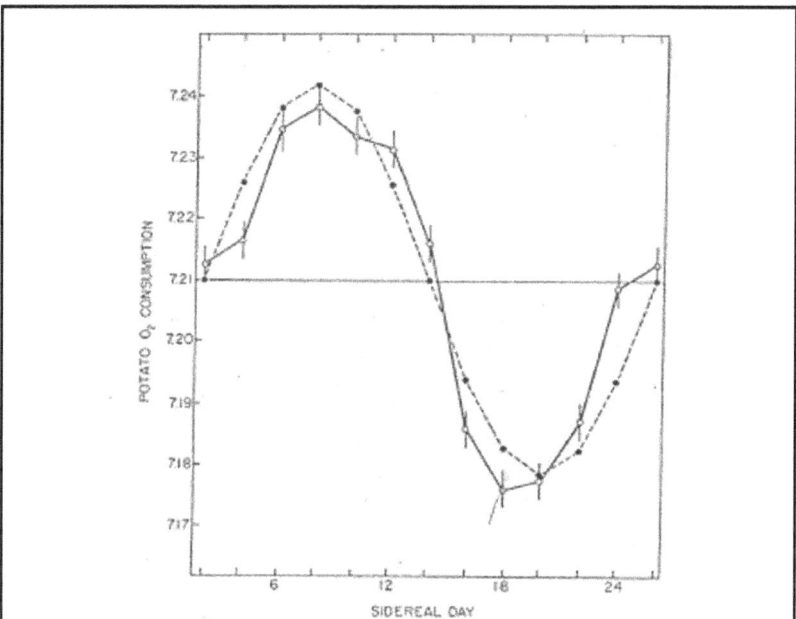

Figure 81: Sidereal-Hour Cosmic Variations in Brown's External Biological Clock Experiment, showing the changes in oxygen uptake of potatoes, kept inside a hermetically-sealed container with constant conditions of light, temperature and humidity, averaged from an 11-year study by Frank Brown (1988). It demonstrated the potatoes reacted to a cosmic factor with a maximum at 6-7 hrs and minimum at 18-19 hrs, close to the 5-hr and 17-hr sidereal variations in the Miller ether wind.

sidereal-day average of maximum potato oxygen consumption reveal a burst of plant growth at about 6-7 hrs sidereal, with a low point in oxygen consumption at 18-19 hrs sidereal. These identified sidereal patterns in potato metabolism lag about one hour behind the 5 hr and 17 hr sidereal times of highest and lowest ether-wind velocities, as moving through Brown's laboratories in Chicago and/or Woods Hole, Massachusetts, both near latitude 42° N. Figure 82 shows a similar graphic, but for the *seasonal variations* over the year, for a variety of life forms maintained in constant, controlled environmental conditions.

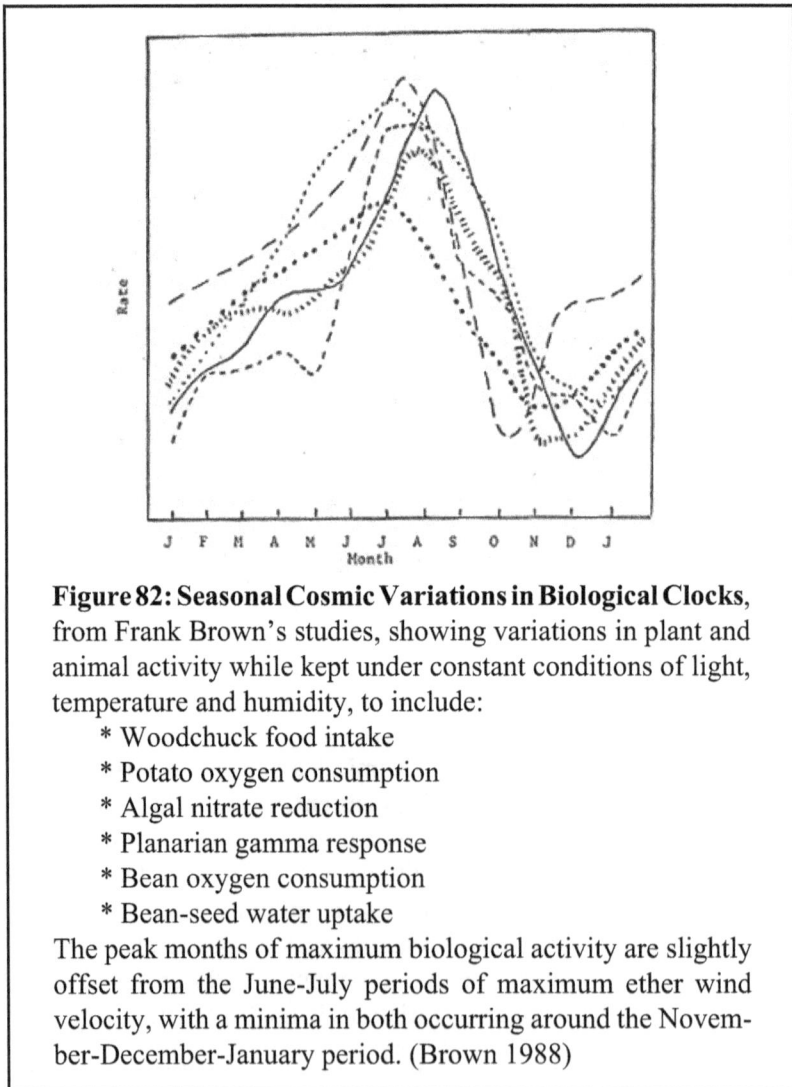

Figure 82: Seasonal Cosmic Variations in Biological Clocks, from Frank Brown's studies, showing variations in plant and animal activity while kept under constant conditions of light, temperature and humidity, to include:
   * Woodchuck food intake
   * Potato oxygen consumption
   * Algal nitrate reduction
   * Planarian gamma response
   * Bean oxygen consumption
   * Bean-seed water uptake
The peak months of maximum biological activity are slightly offset from the June-July periods of maximum ether wind velocity, with a minima in both occurring around the November-December-January period. (Brown 1988)

The biological reactions in Brown's experiments are at maximum approximately at the same months as Earth's highest velocity of motion through the background of space in June-July. The periods of biological minimums occur at the lowest velocity of Earth's motion through space, around December-January. While superficially these data curves may appear to reflect normal summer-winter thermal and growing season effects, the fact that they reveal a cycle in plants, amphibians and mammals *within the controlled constant conditions in his lab*, refutes such a simple explanation. Brown's experiments do not show biological reactions to simple thermal boosts or stress, but instead indicate a response to external cosmic factors from outside the laboratory which could not be controlled for. The fact that Brown's results match up with the cosmic ether-wind experiments undertaken a half-century earlier, as well as with the findings of Piccardi and others, is nothing less than astonishing, *indicating common cosmic causal factors at work.*

Other Similar Discoveries of Cosmic Energy
Unfortunately, space does not allow discussion of significant related work by other notable scientists. These include: Burr (1971) on *electrodynamic fields*, Gurwitsch (1932, see Beloussova 2000, Burns 1997) on *mitogenic radiation*, Kervran (1971, 1972) on *biological transmutations*, Shnoll (1979, 2009, WebRef.4) on *anomalously coordinated but distant radioactive decay histograms*, and Tchijevski (see Beloussova 2000, Burns 1997) and Wheeler (1943) on sunspot influences upon biology and weather, as well as upon cycles of human energy and wars. References are provided for each of them. Also noteworthy are the CIFA organization (www.cifafondation.org), and the Gerald Pollack water research group (www.waterjournal.org) the latter of which continues to host conferences and publications on similar lines of investigation.

All of these latter investigators, in my opinion, are making experimental findings which can best be understood from the standpoint of a unitary cosmic-biological ether/life-energy, in keeping with all what is contained in this book.

The Dynamic Ether of Cosmic Space

Direct Evidence
For a Dynamic Ether

Motional, Dynamic, Spiralling, Luminiferous,
Variable Density, Matter-Forming, Substantive

A Review of What We Know

Let's start this chapter by reviewing the specific nature and properties of the cosmic ether as learned from the different experiments already recounted in this work.

From Michelson-Morley 1887, we learned a cosmic ether wind with an upper value of ~5 to 7.5 km/sec was detected, able to partially penetrate through the stone basement building in which the light-beam interferometer experiment was conducted. Their results were a much lower velocity than the ~200-300 km/sec anticipated from Newtonian static ether "absolute space" assumptions. While the 36 turns of their interferometer were minimal, over only a few days, their results were never "null" or "zero". They stated the experiment would have to be repeated again at a higher altitude over intervals of three months. This repetition was never conducted by them.

From Morley-Miller 1898 to 1906, we learned that light speed is not affected by a strong magnetic field. They later constructed a larger and more sensitive light-beam interferometer, used for experiments over several years, with *nearly a thousand individual turns of the instrument over different months.* They experimentally tested their interferometer for the postulated "matter contraction" of FitzGerald-Lorentz, which was never confirmed. This was accomplished by mounting the interferometer optical components on a base of different density materials, such as wood, concrete or steel, and comparing that to the sandstone base used in the Michelson-Morley experiment. However, in the process, Morley-Miller repeatedly confirmed a real ether drift of ~7.5 to ~9 km/sec. The highest ether velocity was obtained when the interferometer was moved out of the basement at Case School into a small open-window hut on top of a nearby hill at Euclid Heights. This suggested a slightly but not fully entrained or dragged ether, which was slowed when coming into contact with dense matter. An average velocity of 9.2 km/sec was obtained overall by Morley-Miller, who did not then attempt to calculate the azimuth of Earth's net motion in the galaxy. Their results nevertheless violated Einstein's assumption of light-speed constancy, refuting his theory of relativity.

From Sagnac 1913, we learned that two light beams, when projected in opposing circular directions on a rotating turntable, would reveal a variance in the velocities of the two light beams. Sagnac's work further refuted Einstein's relativity assumption of light-speed constancy.

From Miller 1921 to 1926, we learned that he improved the operation and sensitivity of the large Morley-Miller interferometer, and moved it to a high altitude, near to the Mount Wilson Observatory. From that location, Miller obtained an average ~10 km/sec ether wind velocity. His work from 1925 to 1926 acquired data on ether drift over four seasonal epochs, which allowed a more precise determination of sidereal vectors for maximum and minimum ether-wind velocity, as well as the net motion of the Earth in space. The maximum velocity of ether wind was at 5 hrs sidereal, with the Earth moving in a 17 hrs sidereal net direction, generally towards the northern apex of the solar system ecliptic. Miller's work, which proceeded over many years with thousands of turns of the interferometer, confirmed the existence of an entrained ether close to the Earth's surface, along with a clear variation in light speed, dependent upon direction, altitude and season.

From Michelson-Gale 1925, we learned that two light beams projected in opposing directions around a large circuit laid out in an open field inside ether-blocking metal tubes, produced data with minimal average results indicating little effect of Earth rotation on light-speed. However, the raw data had a great variation suggesting significant ether-drift influences upon light velocity. This aspect confirmed Sagnac's endorsement of the luminiferous ether and further challenged Einstein.

From Michelson-Pease-Pearson, third experiment of 1928, we learned they confirmed an ether-wind velocity with an upper value of ~6 km/sec, using a new interferometer not as sensitive as Miller's, inside the concrete base and metal dome at the Mount Wilson observatory. However, they ignored their own results due to a bias favoring Newtonian "static-ether" expectations of a ~300 km/sec velocity.

From Michelson-Pease-Pearson 1930-1931, we learned that even while this experiment was aimed at calculating the absolute speed of light, without reference to any ether wind, the experiment nevertheless yielded variations of around 11 km/sec within one standard deviation.

From Kennedy 1926, Piccard-Stahel 1926-1927, and Joos 1930, we learned that an ether-drift experiment undertaken within stone buildings, basements, and/or under heavy metal shielding leads to a blocking of the ether wind, in some cases down to nearly zero.

From Kennedy-Thorndike 1932, we learned that a heavily barricaded ether-drift experiment of novel design, contained in a metal chamber, inside buildings within a large city at low elevation, could nevertheless produce an ether wind result of ~10 to 24 km/sec. This significant ether velocity was ignored and called "null" by Kennedy-Thorndike, however, as they anticipated only an extreme Newtonian static-ether velocity of "thousands of kilometers per second".

From Galaev 1998-2003, we learned of newer experiments with a radiofrequency link that allowed determination of Earth's net motion in space along ~15 hrs sidereal. A novel first-order optical interferometer design was also developed that took into account and used the ether-blocking nature of a metal pipe shield. This experiment detected ether wind velocity variations within relatively small changes in altitude above the ground surface, with the maximal ether wind at 5 hrs sidereal, and the Earth in a net motion along the ~17 hrs sidereal vector. Galaev's experiments, by his statements, confirmed Miller "down to the details", finding a direct relationship between ether velocity and altitude. His experiments further indicate the ether behaves much like a hydrodynamic gas of exceedingly low density, interactive with matter.

From Múnera 1998+, we learned of an interferometer experiment undertaken at the highest altitude location to date, in Bogotá Colombia. An ether velocity was measured, towards the northern ecliptic apex, in agreement with prior ether-wind azimuths made by others. By static-ether theory, as used by the Múnera team, the motion was along RA 5.4 hrs sidereal, Dec +79°N at 365 km/sec. By my interpretation *according to a dynamic, material and entrainable ether*, their RA results of ~5 hrs sidereal may be seen as a pushing force that moves the Earth towards RA ~17 hrs sidereal, at 18.25 km/sec, but with the same Dec +79° N. This dynamic direction of motion is in close agreement with Miller's determinations of 1928, and with those of Galaev, further indicating an altitude/ether velocity correlation.

The Dynamic Ether of Cosmic Space

From Cahill 2004+, we learned how a novel fiber-optic cable method of ether-detection at low-altitude Australia could also show a positive result of ~10 km/sec at 17 hrs sidereal, as recalculated downwards according to Miller's "*k*" altitude/velocity coefficient.

From Reich 1939-1957, we learned of new evidence for a life-energy with cosmic dimensions. While Reich was misinformed (as were we all) that the Michelson-Morley experiment produced a negative result, he nevertheless objectively identified his life-energy as having properties functionally identical to the cosmic ether of prior decades. Reich understood the true nature of planetary orbits as being open-ended spirals, and not circles or ellipses. From that, he challenged the old Keplerian theory, which nevertheless retained its mathematical validity within its own theoretical flat-surface orbital viewpoint. By applying Miller's measured ether-drift velocity determinations to the Keplerian model of planetary motions, contradictions appeared which vanished when the same Miller data was viewed within a Reichian spiral-moving solar system. Reich also determined many properties for his orgone life-energy that can help in our understanding of the cosmic ether of space. These include: A negative entropy, electrostatic and radiological properties, a variable density and motility, a prime mover with gravitational properties, and an energetic substrate out of which new matter can form. The cosmic orgone life-energy thereby serves as a *self-organizing principle* throughout all of nature. Both orgone energy and cosmic ether were shown to be similar dynamic phenomena.

From Piccardi and Brown 1950-1980, we learned of biological clock activity and chemical tests which revealed cosmic influences related to weather, lunar cycles, sunspot cycles, and sidereal coordinates. The velocity vectors along ~17-18 hrs sidereal were identified in both of their respective experimental findings.

From all the above research efforts, we can further distill out the experimentally determined properties of the cosmic ether/life-energy:

1) The cosmic ether/life-energy is optically detectable at the Earth's surface in velocities ranging from a few to ~18 km/sec, depending upon altitude, latitude, season of year, material environment and direction.

2) Cosmic ether is interactive with matter, such as Earth's crustal surface materials and perhaps the oceans as well. The ether is entrained and slowed down the closer one moves to sea level. Higher altitudes and elevations above the ground surface yield higher ether velocities.

3) Stone buildings or basement locations also inhibit the ether wind, generally in proportion to their density, in keeping with a slightly material ether which can be entrained or slowed by dense matter.

4) Less dense or transparent materials, such as canvas, cardboard or thin window glass, do not significantly block the ether flow.

5) The ether's flow can be substantially blocked within the interior of a metal pipe laid perpendicular to its motion, but allowed to flow *through* the same metal pipe when it is oriented parallel to ether flow. This creates a change in the speed of light moving through the same pipe, which is dependent upon its orientation, as discovered by Galaev. A recovery time may be necessary to allow for restoration of motion as the pipe is moved from the perpendicular to the parallel orientation, and vice versa. This indicates the ether behaves much like a hydrodynamic gas, albeit one of exceedingly low density.

6) The averaged maximum velocity of ether wind in the northern hemisphere is observed at around 5 hours sidereal. This ether wind maximum appears to blow from a generally south to north direction, carrying the solar system along in the same direction as its own motion, towards a ~17 hours sidereal destination, close to the star Vega.

7) The repeatedly observed RA of ~17 hrs sidereal is along the same general direction as the modern determination of the "Sun's Way" towards the star Vega, which is additionally along a meridian roughly bisecting the center of the Milky Way Galaxy. This further indicates our solar system is being propelled by a motional and substantive, entrainable ether wind, rather than being dragged through a stagnant "absolute space" static ether by a second mystery force.

8) The view of the solar system as a flat disk composed of elliptical orbits of planets is shown to be incomplete, as the real-world motions of planets are open-ended spirals around a moving Sun.

9) The cosmic ether as life-energy plays a more fundamental role in the creation, motility and behavior of living material, in the creation of matter and gravitation, and in cosmic motions than previously considered. Ether as life-energy is not merely a satisfactory creative and self-organizing principle, but is also the cosmic prime mover.

Aside from the immense detail presented in the preceding chapters, a few additional issues will now be raised which provide further affirmation for a dynamic ether/life-energy, and by which certain anomalous cosmic factors find a reasonable, good explanation.

"Planet 9" as an Ether-Drift Gravitational Anomaly?

Planet 9 is a postulate which has not yet been observed or confirmed, and I doubt if it ever will be. Its existence is nevertheless rooted in a rational consideration, of a gravitational anomaly which has powerful influences upon several Kuiper-Belt planetoids orbiting the Sun at great distances.

The Kuiper Belt exists far beyond the orbits of our solar system's outer planets, such as Neptune which lies at 30 astronomical units from the Sun. One Earth distance from the Sun is termed one astronomical unit, or AU. Neptune orbits 30 times farther from the Sun than Earth. The Kuiper Belt objects lay at 200 to 400 AU, but a few of them are pulled inwards towards the center of the solar system, at high velocities with very eccentric orbits. There is no explanation for these erratic Kuiper planetoids unless one postulates a cosmic gravitational anomaly which pulled them out of the larger population of other Kuiper planetoids, as shown in Figures 83 and 84.

Since modern astrophysics proclaims an "empty space" with no gravitational properties, an invisible "Planet 9" or something similar had to be postulated to provide the missing gravitational pull. Interestingly, the gravitational anomaly has been computed to lie along an average direction of RA 16 hrs sidereal, and Dec of +40°. *That location is within the same small section of cosmic vectors previously presented on Figure 78 (last chapter), associated with ether wind and planetary-solar motions.* By empty-space theory, these vectors should not exist nor have any relationship whatsoever to Kuiper Belt planetoid motions. However, the direction of motion of the Kuiper planetoid erratics is in full agreement with the theory of a motional material ether drift in cosmic space, moving along that same cosmic vector and acting as an ether-wind gravitational pushing force. I therefore propose that *"Planet 9" does not exist, that the vector of its location is approximate, and that the same ether wind that pushes the solar system along on its path towards Vega and the center of the Milky Way galaxy, is the factor that has dislodged a few planetoids from the Kuiper Belt,* pushing or floating them along on a similar trajectory. (WebRef.5)

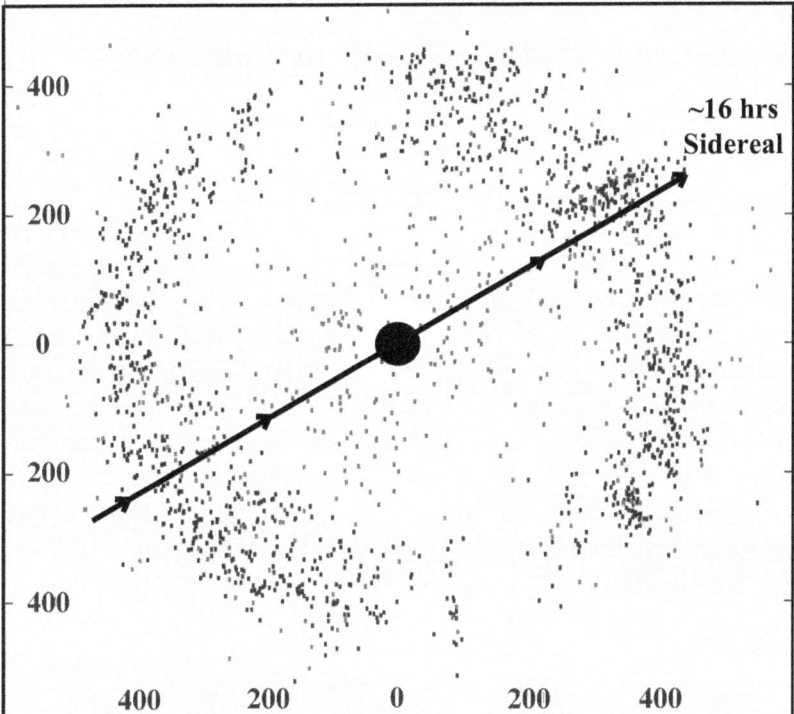

Figure 83. Kuiper Belt Planetoids scattered in a ring around the solar system, at 200-400 Astronomical Units (AU) from the Sun. The black dot in the center identifies the width of our solar system containing all planets (plus the recently evicted Pluto).

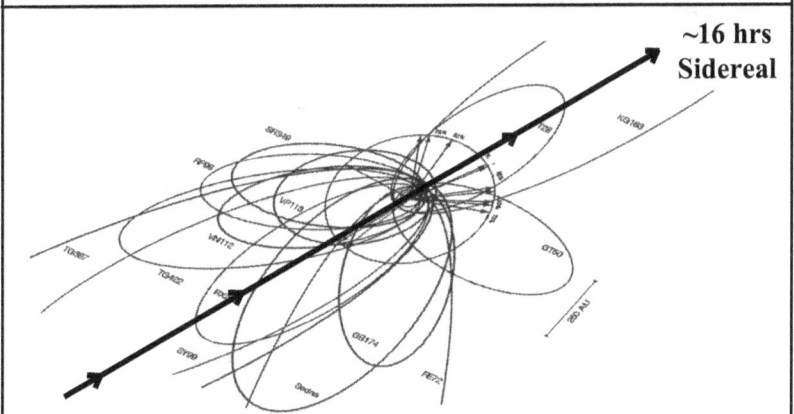

Figure 84. Planet 9 Gravitational Anomaly centered on the Sun, pulls many Kuiper Belt objects into highly eccentric orbits, being moved in the same direction of the ether wind.(WebRef.5)

Galactic Core Tilt and the Sun's Rotational Tilt: Related?

A view of nearby Andromeda Galaxy on the next page shows its remarkable spiral form, but presented with a slight horizontal compression to exaggerate its rotational structure. The image reveals a tilting of the rotational plane of its core region of stars, about 25° off from the plane of the galactic arm band rotation. One observes a flat plane of galactic arm bands spiraling around and moving towards the core region. The core, however, thickens to appear somewhat like a squashed ball, with a rotational plane and axis that is tipped significantly away from the plane and axis of the peripheral spiral arm bands. Cosmic motions in the outer parts of the galactic structure do not match up with the inner core motions, for some unknown percentage of spiral galaxies. This might be a precessional effect, as in gyroscopic motion, where the abundant gathering of stars in the core begins to behave as a separate gravitational mass, with a different rotational shape and characteristics than the peripheral galactic arm bands. Such differences are not anything new; they form a part of the arguments regarding the claimed necessity for "dark matter", as will be discussed shortly.

Our Milky Way Galaxy is also a spiral galaxy, but we are so deeply immersed in it that photographic evidence only shows the thickening of the core region, without any clear tipping of the core rotation. A slight systematic tilt has been observed in Milky Way gamma-ray jets, however, whose emissions from the core of the galaxy are roughly correlated in a very diffuse but identifiable 15° tilt off from the central galactic axis (WebRef.6). Such galactic core tilts are likely variable, and may express a condition which all spiral galaxies develop at some point in their formation. I would postulate, however, that this may reveal a more systematic function at work in other large cosmic spirals, including within our own solar system. Consider, for example, the 7° tilt of the Sun's (and nearby Mercury's) northern pole of rotation, away from the northern axis of the plane of the ecliptic, already mentioned in the prior chapter on *Which Way Drifting* (see Figure 38, p.128). That also appears as a separation of spiral-core gravitational motions from the rest of the orbiting planets, just as the core stars of Andromeda are gravitationally separated from its own stars in the spiral arm bands.

An open question is raised by such phenomena, which are documented but not explained in terms of their patterned structure or mechanism, revealing a difference between the core and peripheral aspects. Could this also be at work in the observed differences between the Miller northern axis pole of ether drift, the northern pole of the plane of the ecliptic, and the Sun's Way through the Milky Way? If for example we were to measure the plane of galactic rotation with reference only to the motions of the stellar arm bands, we would conclude a different plane and axis of galactic rotation than if we referenced only the galactic core region.

Is it possible, through mechanisms not yet understood, that ether wind determinations made from planets in the peripheral areas of our solar system, from Venus outwards, would be different than if measured on Mercury or close to the Sun? Ether measurements on other planets are a current impossibility of course, but this tipping of the Sun's orbital plane off from the axis of the plane of the ecliptic by ~7° is a fact, similar to the tipping seen in Andromeda and other spiral galaxies, though at different angles and directions. Future research could clarify these interesting patterns, as better space telescopes are constructed to study a greater number of spiral galaxies.

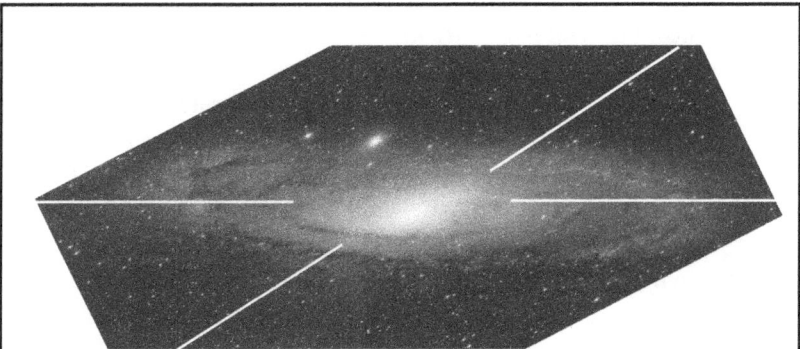

Figure 85. Andromeda Galaxy: Tilting of the Core Mass
Above is an image of Andromeda Galaxy, M31, compressed horizontally so as to exaggerate and better see the difference in the plane of rotation of the peripheral stellar arm bands, versus the tipped plane of rotation of the core region of stars. The image compression of Andromeda makes its core appear more like a squashed ball. The two planes are off from each other by around 25°, apparently due to a change in the gravitational dynamics of the core, relative to the peripheral spiraling galactic arm bands. So separated, the core may begin to precess independently, somewhat like a gyroscope.

The Plasmasphere, Rotating Faster than Earth?

In the preceding chapters, evidence was presented that the ether tends to condense and slow closer to the Earth's surface. This creates an ether layer with a *residual motion* of around 5-18 km/sec as revealed in the ether-drift experiments. Reich identified a similar thick energetic layer, calling it the Earth's *orgone envelope*. Modern research on what is called the magnetosphere indicates a similar phenomenon, of a thick layer surrounding the Earth, very different from open space at greater distances, and emitting light in the ultraviolet (UV) frequencies. When the NASA *IMAGE* satellite was placed into high Earth orbit around the year 2000, it showed the expected UV from the solar-lit side of the Earth, and from the auroral oval. However it also revealed a wide UV-emitting energy field extending outwards by around 1/2 of the Earth's diameter, with other odd features. These included a streaming motion coming to Earth from deep space, wrapping itself around the planet. This wrapping motion occurs in the same west to east direction as the Earth rotates, but *with a slightly faster rotation*. This phenomenon is now conventionally termed the *plasmasphere*.

I saved most of the movies from the IMAGE satellite's website, comparing the rotational speed of the Earth versus the rotating UV-emitting layer. Only a few were of sufficient clarity to make a determination of its velocity, but of the two I analyzed, one moved 3.6% faster than the Earth rotated, while the other moved 9% faster. Figure 86 presents two images from one such evaluation, showing the UV-emitting energy field on the 179th day of the year 2000.

When I wrote to the NASA staff in charge of that satellite and told them of my determinations, they replied that the UV-emitting field was slower in motion, not faster. I disagreed and asked how they made their determination, but got no reply back. Did they just concoct a ready-made, "off the cuff" explanation, in agreement with existing theory? Or did they actually make their own investigation? I cannot know, but they did change their website. Their on-line satellite archive, with color images and movies, today states with emphasis that "the plasmasphere DOES NOT corotate with the Earth" (original emphasis, WebRef.7).

I actually agree with that NASA statement, but determined *the plasmasphere rotates faster than the Earth*, at least in the several NASA movie clips I analyzed, something which is suggested in another line where NASA wrote "the Earth actually spins faster than the

2000/179/06:46
range: 8.16 RE   S/C latitude = 52.06

2000/179/06:46
range: 8.16 RE   S/C latitude = 52.06

2000/179/06:46 = 6:46 AM GMT
range: 8.16 RE   S/C latitude =   52.06

Figure 86. Satellite Images of Earth's Plasmasphere in negative and positive for clarity, looking down from an angle towards Earth's northern pole. A dynamic Earth energy-field envelope, extending outwards by around half the Earth's diameter, is immediately apparent. The inserted circle shows the size of the Earth within the UV-emitting envelope. The rotation is counter-clockwise to the viewer, where the tail comes in from space at the top, then moves towards and wraps around the Earth, much like a snail-shell. (NASA WebRef.7)

plasmasphere *on average*" (emphasis added) suggesting that, *at some times the plasmasphere rotates equal to or faster than Earth.* The moving plasmasphere images do not conform to expected theory, irrespective of who is correct on the rotational velocities. No astronomer or physicist at NASA predicted such a structured energy field would appear on their images, as nothing in conventional theory anticipated it. However, the theory of a condensed layer of cosmic ether, and the theory of Reich on the orgone energy envelope did anticipate it, although the specific structure of the "plasmasphere tail" (which "wags the dog" in this case) and its outward extent was not predicted by anyone. I have already written elsewhere on the Earth's major wind belt's variable response to a rotational ether/life-energy wind moving west to east faster than the Earth's rotational velocity. (DeMeo 2002)

Stratospheric Winds, Created by Cosmic Energy Streams?

Another factor giving support to Reich's theoretical model, is the more recent discovery of the upper stratospheric winds, which move at a velocity up to several hundred kilometers per hour, at very high altitudes. This is something fully unexpected by conventional solar-heating theory. These winds have now been well documented and are known to have a *downward convection* of their energy and velocity, into the lower stratosphere and upper troposphere. (Kodera, et al 1990, Labitzke 1997, 2001) Such winds may drive the jet streams, which exert a powerful control over Earth's weather. Investigations of space weather, above the entire depth of the Earth's atmosphere, may eventually be identified as the driver of the stratospheric winds, confirming Reich's assertion that Earth's weather is influenced by specific cosmic energy streams, aligned with the plane of the ecliptic, and with the plane of the Milky Way Galaxy. (Reich 1951a, DeMeo 2002f)

This concept of cosmic energy streams creating atmospheric winds gains additional support when reviewing the dynamics of other planetary atmospheres. The two most distant planets, Uranus and Neptune, have very high velocity clouds moving in their upper atmospheres, 10 to 15 times faster than Earth's stratospheric winds. Uranus has winds up to 900 km/hr (580 mph), while Neptune's winds can reach up to 2400 km/hr (1100 mph), or 1.5 times the speed of sound. Solar heating alone, especially at such great distances from the Sun, cannot explain these winds, though a fast-moving stream of ether-energy might do so.

Substitute Names for Ether Wind: *Interstellar Wind? Cosmic Ray Wind? Neutrino Wind? Dark Matter Wind? Higgs Field?*

Over my entire professional life, I have observed scientific investigations resting on empty-space assumptions invariably returning to postulates of specific particles which, due to their supposed fantastic abundance and invisibility, and difficulty in measuring them, collectively appear very much like a cosmic ether. The examples of interstellar wind, cosmic ray wind, neutrino wind, and dark-matter wind, all appear as theoretical substitutes for the cosmic ether wind. They *are* the cosmic ether, in all but name, though of course using modern detector technologies quite different from the methods of light-beam interferometry. The empty space universe of conventional modern theory could not predict or understand the subtle properties of radioactivity, or the unusual variable rotational characteristics of spiral galaxies, or the other anomalies already discussed, without reference to some kind of a ubiquitous cosmic energy to make them comprehendible. Without the cosmic ether, new postulates of mysterious, other-worldly particles and forces arose, with increasingly complicated efforts to explain never-ending contradictions. The substantive and motional cosmic ether with life-energetic expressions had been prematurely discarded, decades earlier, in spite of the evidence in favor of it. And still today, such a material-motional and vortexing cosmic ether/life-energy provides a single comprehensive understanding of what is going on in the outer reaches of cosmic space, as well as on Earth, right under our noses.

Interstellar Medium and Wind

The terms "interstellar medium" or "intergalactic medium" have come into widespread usage, as a convenience to acknowledge substance and energy out in cosmic space. But they dare not use the forbidden word "ether". Sometimes "cosmic plasma" as from Hannes Alfven (1981) is used, or "cosmic ions", which is partly what composes a plasma. Certainly a look at cosmic objects from Hubble and other telescopes reveals an organized structure to the background "medium of space", not too different from Earthly aurora or cloud forms, with streaming and boiling, or gentle undulations and great frothing turbulence within a transparent ocean. As noted in the *Introduction*, one sees cosmic objects moving as if through a resisting fluid medium, billowing or leaving a frictional debris trail behind them, with similar impressions of fantastic movement and energy of an almost *protoplasmatic* nature. However, all these phenomena are frozen in time, given the very short duration of our life span compared to the infinity of cosmic time.

A paper by Frisch et al. (2013, WebRef.8) presented a 40-year summary of data on the directions and intensity of the interstellar winds, gathered by the IBEX (Interstellar Boundary Explorer) satellite and 10 other satellites. Their project focused upon *interstellar helium* as a proxy for the interstellar medium as "it was the easiest to measure", given how it emits ultraviolet (UV) light and also carries a presumed electrical charge. The flow of this "interstellar wind" is along an axis aligned at ~5 hr to 17 hr sidereal, the latter end of which aims towards the star Vega and the centerline of the Milky Way Galaxy. The velocity of this interstellar wind is around 50,000 mph, or 22 km/sec. Sound familiar? Figure 87 provides an image of the clouds of interstellar wind within 20 light-years of our solar system, from the original NASA press release. From the descriptions provided, it seems clear that what is termed the "interstellar medium" is identified by an emission of UV light with a given motion. This suggests, the "interstellar wind" is in fact a proxy for *ether wind*. Since we have no satellites that have gone so far out into space to collect particulate material and return it to Earth, the UV light is today the only firm empirical phenomenon by which the interstellar medium is being defined. This point bears an interesting relationship to the section on "Cosmic UV-Blues" at the end of this chapter. Empty-space theory blocks astronomers from considering such *outside the box* heresy, of course.

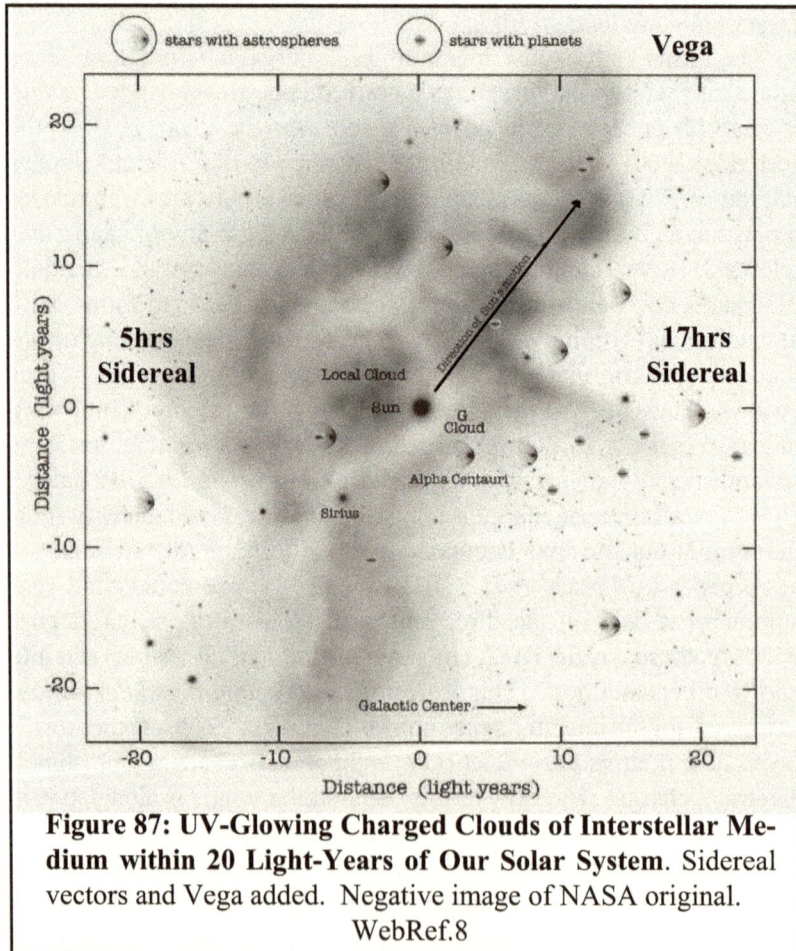

Figure 87: UV-Glowing Charged Clouds of Interstellar Medium within 20 Light-Years of Our Solar System. Sidereal vectors and Vega added. Negative image of NASA original.
WebRef.8

Cosmic Ray Wind or Anisotropy

Cosmic rays as detected by Geiger-Müller (GM) tubes or scintillation counters, have been known for around a century, and have been most widely studied. They are abundant within the atmosphere, and increase the higher one rises in altitude. Cosmic ray detectors, shielded with lead except in specific preferred directions towards which they can be aimed, have found them most strongly in the core region of the Milky Way Galaxy, being attributed to pulsars, stars emitting gamma rays, and supernova remnants. At Earth's surface, while they are mostly chaotic in motions, there is a west to east bias of around 10% of the total cosmic ray muon abundance. Muons are conventionally understood as the break-down products of more highly energetic cosmic rays. These vectors are agreeable with Reich's view of the rotating orgone envelope, moving west to east slightly faster than the Earth's rotation. Marett (2002) experimentally confirmed this West to East bias of cosmic ray muons, stating "...the west-east asymmetry of cosmic rays and the wave-like west-east phenomena of Reich betray an energetic motion leading the earth in its rotation."

A 2017 study by Díaz-Vélez, et al. (WebRef.9), combining data from the northern polar HWAC and southern polar IceCube cosmic ray detector arrays, shows an anisotropy in cosmic rays with a maximum at RA ~60° (4 hrs) sidereal and minimum at ~225° (15 hrs) sidereal, an axis aimed generally towards the center of the Milky Way. This suggests a pushing wind at the ~4 hr sidereal vector, moving the solar system towards the ~15 hr, Milky Way central direction. Another study in 2019 by Erylkin et al. (WebRef.9) found similar but antithetical data, with a peak emission of EAS cosmic rays (Extensive Air Showers), a more episodic but higher energy phenomena, close to the Vela star cluster near the center of the Milky Way. The maximum emissions near Vela are found at RA 277° (18.4 hrs) sidereal, with its antipode minimum being at 97° (~6.4 hrs) sidereal, an axis similar to the above study by Díaz-Vélez, et al. Both these measured vectors correlate with a pushing flow of cosmic ether, possibly carrying cosmic rays within it, or creating energetic pulses as it moves from the ~4-7 hr sidereal vector, moving the Earth and solar system along towards the Galactic Center at ~15 to ~18 hrs sidereal. The occasional strong comic ray burst then comes from the Milky Way galactic center itself, from the general direction Earth is moving towards.

Neutrino Sea and Wind

The elusive neutrino is perhaps the most mysterious of all the cosmic "particle winds". By conventional theory, an incredible quantity of neutrinos occupies every tiny bit of space in the universe, including our living and breathing space. By that view, the Sun theoretically produces 18×10^{37} neutrinos per second, a very large number indeed (18 followed by 37 zeros). Of these, the Earth intercepts 8×10^{28} neutrinos every second. The Earth also purportedly gives off a cousin particle, the anti-neutrino, at a rate of around 1.75×10^{26} per second. Of these gigantic numbers, the average human body receives about 3 trillion natural cosmic neutrinos every second, theoretically only from Earth and Sun, racing through our bodies. A large nuclear reactor also puts out anti-neutrinos, at a rate of about 10^{18} neutrinos per second. (Asimov 1966) By this theory, we live within a virtual neutrino ocean, which has nothing to do with cosmic rays, dark matter or the interstellar medium. They are all considered to be separate entities.

Neutrinos were originally proposed as a "book-keeping" particle, said to be created whenever there is radioactive decay creating beta or gamma particles. They were postulated as a way to understand why some radioactive particle decay events were strong, and others weak, even while the parent unstable radioactive atom which ejects that variable energy particle always lost the same quantity of mass and energy. In theory, a strong radioactive event would be accompanied by a weak neutrino, and a weak event would partner with a strong neutrino. By the neutrino theory, the "accounting books" were thereby balanced, so it was said. But neutrinos could not be observed or detected when they were first postulated by Wolfgang Pauli in 1930. It was several decades before their existence was allegedly proven.

In yet another metaphysical postulate of modern physics, neutrinos are described as almost "other worldly", like ghosts. They are presumed to be of such an exceedingly low mass, without electrical or magnetic properties, that they can pass through millions of miles thickness of solid lead before reacting with one of the lead atoms. Such ideas become all the more mind-boggling when one asks, what happened to all the neutrinos and antineutrinos created since the beginning of time, from all the nuclear radiation decay-events within the universe, even assuming big bang theory is true and the universe actually has a beginning? In the case of an infinite universe with no beginning or end, the number of neutrinos swells to infinity, no matter how it is calculated.

The implications of the standard neutrino theory rather defy belief and leave one breathless. Of course, it is pure theory; nobody really knows for certain how many neutrinos there might actually be floating around in front of our noses. The more central question is, *are neutrinos real*, or merely a necessary theoretical illusion of c.1930 empty-space physics? Pauli's postulate of the neutrino came at the time when most of physics was content to ignore all the positive evidence for a cosmic ether, the same period when Miller was being isolated and erased.

Physicist Paul Dirac recognized the implications of neutrino theory to some extent, and proposed the existence of a *neutrino sea,* an ocean of countless neutrino particles, which sounds increasingly like the cosmic ether by a different name. This has led to a theoretical conundrum, as obviously an infinite or even limited "big bang" universe could not allow for infinity-squared numbers of neutrinos all crammed into every nook and cranny of cosmic and subatomic space, which anyhow is supposed to be empty. In more recent years, neutrino specialists started giving neutrinos a multitude of different qualities and properties, to yield up corrections to why their theory keeps running into problems. In fact, the whole theory of neutrinos – which is essential for classical nuclear decay theory – is today so overburdened with contradictions and complexities, and about which so little is factually known, that we can postulate that Dirac's "neutrino sea" **is** the *theoretical recognition of a contiguous ocean of fluid, non-particulate cosmic energy.*

By this view, a decaying radioactive atom does not discharge a "neutrino *particle*", but rather the excess cosmic energy frozen within the unstable atom is discharged to merge back into the cosmic ether ocean, from which it originally came. The cosmic ether is not merely a squeezed-together complication of discrete particles, but a thin gas-fluid ocean of exceedingly low density with great power. Out of it, luminating particles can emerge, flash a bit of UV-blue light energy, and either form into new matter, or recede back into the cosmic ocean, losing their specific identifiable properties. Considered fresh from this viewpoint, which is derived from the ideas of Reich on the formation of *visible orgone-units*, we can view the neutrino sea concept as being one and the same thing as the *cosmic ether/life-energy of space.* A look at neutrino-detection equipment adds much evidence to this.

Neutrino detectors as constructed in modern times are huge projects. They make use of massive arrays of photomultiplier tubes (PMTs) – a device which can detect and amplify very tiny and weak flashes of light. In one method of detection, numerous PMTs are placed around the interior walls of large metal tanks filled with water. The PMTs are oriented to peer into the center of the dark water-filled tank, which might be 20 or 40 meters in diameter. During the first periods of operation, those large neutrino detectors with multiple PMTs ran constantly for months or years, making very few detections, maybe one or two per month or year, of what were basically tiny flashes of light in the UV-blue frequencies. Those light flashes were then amplified by the PMT detector array. Such massive detector arrays must be operated deep underground, in old mines, or under mountains, to shield out ordinary light, as well as a chaos of other above-ground radiation particles. Or sometimes arrays of PMTs are lowered deep into the ocean, or into deep boreholes in the Antarctic and Greenland ice caps, where neutrino-reactive (blue-flashing) ice is used instead of water.

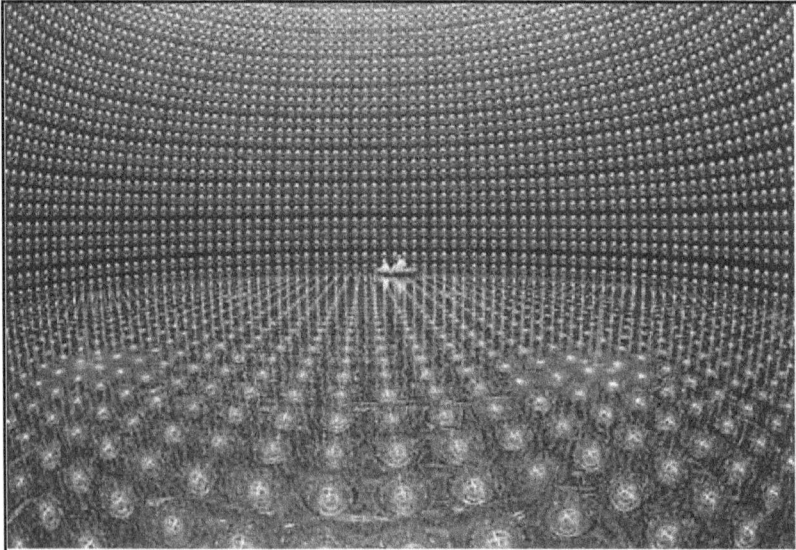

Figure 88. The Super-Kamiokande Neutrino Detector in Tokyo, Japan. 39 meters in diameter. The object in the center is a rubber boat with three men inside. Each of the globes in the walls carries a photomultiplier tube (PMT). When operating, the entire tank is filled with water and darkened, while the PMT array looks for flashes of ultraviolet and blue light. The experiment is based upon empty-space assumptions. Could the cosmic ether also produce such UV/blue flashes of light? The answer is probably *yes*, as discussed at the end of this chapter.

In more recent years after c.2010, neutrino detection has allowed for inclusion of "neutrino-like" particles, so that the physicists have more to study than merely a few light-flash events per month or year. By broadening the definition of "neutrino" to include "neutrino-like" and "neutrino candidate" particles, and therefore with more "neutrinos" to study, an actual "neutrino wind" has been identified, a motional component to the neutrinos (or light flashes). This "neutrino wind" has an azimuth similar to the cosmic ray anisotropy, but of a much broader region of space, covering azimuths of RA 0-12hrs sidereal (average at 6 hrs sidereal) and declinations of ±60°. That indicates the "neutrino wind" is "blowing" on average along the same vectors as the interstellar wind, the cosmic ray wind, and as we shall see, the dark matter wind. *They are moving in the same general direction as the cosmic ether wind,* determined by Dayton Miller in the 1920s. Coincidence?

Regarding the actual color frequencies of these claimed "neutrino light flashes" of predominantly UV-blue, I direct the reader to the discussion at the end of this chapter, on recent spectrographic research undertaken at my laboratory. This frequency spectrum of UV/blue emissions is associated with Reich's orgone energy charging of water and vacuum tubes. This association may also exist for not merely neutrino detectors employing PMTs in water tanks, oceans and glacial ice, but also for "dark matter" and other PMT-type detectors.

Dark Matter Wind

Another ubiquitous ether-like feature has been found, the so-called "dark matter" and "dark-matter wind". Both began as unseen and undocumented postulates formulated to better understand the dynamic behaviors of spiral galaxy rotation – but only within the theoretical straightjacket of the big bang and empty space theories, devoid of cosmic ether. Most of the mass of spiral galaxies is concentrated in a condensed, flattened ball at the center of their rotating disk. By conventional theory, the inflow velocities of spiral vortices should start slow in the outer reaches of their disk, with a gradual increase in velocity within the spiral arm bands, followed by a sharp velocity reduction as the inflowing stars are confronted by a gravitational pull from all different directions within the central ball. Instead, the observed rotational velocities at the outer reaches of the spiral galaxies start off at a maximum, which continues at the same general velocity as the stars spiral inwards, only gradually slowing down as the stars reach the more interior parts. By conventional thinking, this implies there must be a greater amount of "hidden matter" producing high gravitational forces within the outer parts of galaxies, than is identifiable from the various stars and nebulae which compose it. From this discrepancy came the concept of "dark matter", proposed to exist as a mysterious invisible halo of "something", necessary to create additional gravitational energy in the outer parts of spiral galaxies.

The two curves in Figure 89 show the theoretical versus the actual situation, in the example of M33, the Triangulum Galaxy. The outer galactic arm bands have a higher absolute velocity than the mid or interior sections, and maintains a high velocity of over 100-200 km/sec, slowing down only near to M33's core region. That velocity distribution is different from theoretical expectations as observed in other natural rotational motions, such as planetary orbits, whose velocities are significantly slower for outer planets than for interior ones.

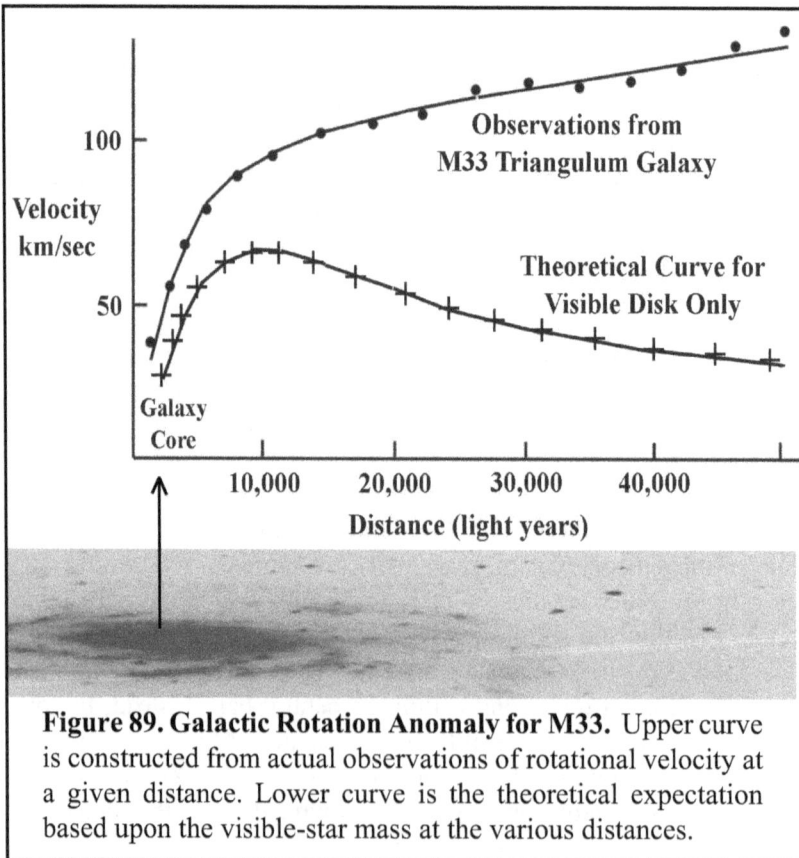

Figure 89. Galactic Rotation Anomaly for M33. Upper curve is constructed from actual observations of rotational velocity at a given distance. Lower curve is the theoretical expectation based upon the visible-star mass at the various distances.

Overall there is said to be more gravitational energy within the spiral galaxies than can be seen or identified. By the conventional maths, *as much as 90% of the mass of the universe is thereby said to be "dark" and unseen*, which is yet another mystical theoretical proposition, of invisible things and forces which cannot be seen having fantastic influences over all of existence. New classes of unseen particles or objects were therefore invented, such as the "MACHO" (Massive Compact Halo Object), and the "WIMP" (Weakly Interacting Massive Particle).

The MACHOs are described as a vast multitude of large planet-sized objects in the outer galactic halos, each perhaps as big as Jupiter, along with uncounted thousands or millions of "remnant" and "primordial" black holes, "dwarf dark" stars, and "various exotic stable configurations of quantum fields", whatever that means. This menagerie of proposed highly gravitating objects remains invisible, a list of *imaginaria mechanica,* as found in the entry on "Wimps and Machos", in the authoritative *Encyclopedia of Astronomy and Astrophysics* (Griest 2002).

By contrast, WIMPs are ultra-tiny particles said to exist everywhere, like a "dark matter sea", with an abundance which competes with that of neutrinos. Countless numbers are said to pass through every square centimeter of space and matter per second, with very few material interactions. WIMP particles are detectable only by use of PMTs similar to what is used for cosmic ray and neutrino detections. However, in the WIMP/dark-matter detection, the PMT is attached to a transparent sodium iodide crystal, doped with a slight bit of thallium, with a chemical formula of $NaI(Tl)$. The $NaI(Tl)$ crystals react to all kinds of cosmic and radioactive particles with flashes of UV/bluish light, which are then picked up by the PMTs as individual "counts", like "clicks" on a Geiger counter. When used in an above-ground environment, such devices will pick up thousands of counts or flashes of light per minute. In that case, those flashes are mostly identified as muons, as described in the cosmic ray section above. When taken deep underground, however, the $NaI(Tl)$ PMT detectors show a greatly reduced but nevertheless continued activity, as by convention the underground environment shields out all but the most penetrating particles. This residual radiation is considered to be the "WIMP" expression of "dark matter". The count rate for dark-matter PMT-$NaI(Tl)$ "WIMP" detectors was always much higher than the neutrino PMT detector counterparts immersed in water tanks or polar ice.

In one notable on-going experiment, an array of these special NaI(Tl) detectors was assembled by a team led by Rita Bernabei (2007, 2010, WebRef.10) at the DAMA (Dark Matter) Project, undertaken deep inside Gran Sasso mountain in northern Italy. A theoretical dilemma developed. The Gran Sasso WIMP residuals were unexpectedly variable over the course of the year, peaking in the months of June-July, with a minimum in December-January. This implies some kind of systematic effect upon "dark matter particles" as the Earth moved through "empty space". It was a result like Frank Brown's potatoes or Piccardi's chemical tests, they "shouldn't be responding" to sidereal cosmic factors, and yet, they were. And so, this mysterious new phenomenon, suggesting a literal *wind* in the already mysterious ill-defined "dark matter" (UV-blue flashes), continued to be detected.

Is dark matter wind a proxy for cosmic ether wind, and "dark matter" itself merely a confused mis-identification of a substantive gravitational cosmic ether? The rates of dark matter detection reveal an annual or seasonal variation, suggesting a literal stream or wind which mirrors the Miller ether wind. *Dark matter wind "blows" across the axis of 5-6 hrs sidereal towards 17-18 hrs sidereal.* Sound familiar?

Figure 90, from a DAMA project press release, shows 16 years of variable dark-matter wind detections. Bernabei acknowledged the logical understanding, that the DAMA residual variations were the consequence of the Earth's spiral-form motion through the cosmos, though without reference to the cosmic ether. When combined with the

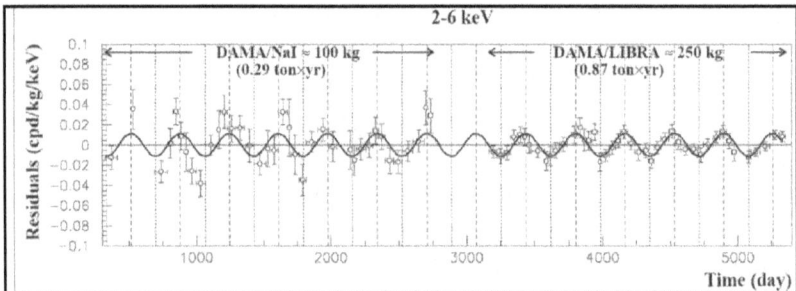

Figure 90. Seasonal Variations in Dark Matter Wind "WIMP" Residuals over approximately 14 years, from the Bernabei DAMA Group at Gran Sasso Mountain in northern Italy. The measured WIMP maximum for each yearly cycle occurs in June, at RA 17-18 hrs sidereal, with a WIMP minimum in mid-December, at RA 5-6 hrs sidereal. (2010 WebRef.10) These variations match the Earth's spiral-form velocities.

velocity of the Earth around the Sun, and the velocity of the solar system through space, Bernabi's team postulated a "dark matter wind" velocity maximum on 2 June and minimum on 2 December. Those dates are about a month off from standard perihelion and aphelion determinations, and with a reversal of the Keplerian velocity max-min expectations, as previously discussed. Miller's ether-wind velocities were also assumed to be slower in December and faster in June, though as noted he didn't make systematic ether-velocity measurements at those exact times. By pure Keplerian expectations, Miller's ether velocities and Bernabei's DAMA residuals make no sense. However, viewing the Earth-Sun motions as a spiral, as Reich argued, and by sidereal hour, as Miller argued, they make perfect sense. (see Figures 75-77)

The entire "dark matter" and "missing gravitation" mystery can be solved if we consider the subtle glowing dark blue halo surrounding the galaxies to be a visible expression of the negatively-entropic gravitational ether/life-energy phenomenon. This is identical in its basic features to Reich's blue-glowing orgone energy, previously documented as a bluish energy field around microbes and blood cells, in fluorescing water, in blue atmospheres and natural-forest phenomena, and in the halos of spiral galaxies such as Andromeda. (DeMeo 2011c, 2018) My work on the spectroscopy of orgone-charged water documented this UV-blue emission, and Reich demonstrated a similar blue-glowing reaction in ORAC-charged high-vacuum tubes. Without a change in theory, precise determination of what's going on in spiral galaxy dynamics may ultimately be as difficult for conventional astrophysics as it was for ornithologists to understand bird behavior by examining only stuffed dead birds in a museum. The comparison is not precise, of course, but astronomy needs to be more open to provocative thinking *outside the box*.

Higgs "God Particle" Field

The Higgs boson, or "god particle", appears as the most recent incarnation of the long-denied cosmic ether/life-energy. It is described not as merely another category of infinite particles but as *an entire contiguous field which encompasses all of matter and existence, filling all space, existing everywhere.* Ergo its "god-like" nature. And that, too, is functionally identical to the ubiquitous cosmic ether in its life-energetic expression. So far I do not know if a "wind" has been detected in this Higgs field, but if so, I will make a bet with Higgs' ghost, as to what its azimuthal direction of motion will be.

The Dynamic Ether of Cosmic Space

Summary of 17 Independent Vectors of Cosmic Motion

Figure 91 presents the final update of my star chart of the northern polar region, now marking out *17 different, independently derived cosmic vectors*, including the new ones just described in this chapter:

* The mystery "Planet 9" gravitational anomaly at 16 hrs sidereal.
* Interstellar Wind vector at 17.3 hrs sidereal.
* Cosmic Ray Anisotropy, blowing from 4-5 hrs to 16-17 hrs sidereal.
* Neutrino Wind, blowing across 0-12 hrs, centered on 6 hrs sidereal, aiming at 18 hrs sidereal.
* Dark Matter Wind, blowing from 5-6 hrs (Dec.-Jan.) toward 17-18 hrs (June-July) sidereal.

Those factors are approximately in the same cosmic directions as the other vectors previously discussed and/or plotted, which are:

* The northern pole of the solar system ecliptic at 18 hrs sidereal, with Dec. +66.5°.
* The modern Sun's Way towards Vega at 18.5 hrs Dec +39°.
* The Sun's rotational northern pole axis at 19.4 hrs Dec +64°.
* The center of the Milky Way Galaxy 17.5 hrs.
* Miller's northern axes of ether drift of 1928 at 17 hrs Dec.+68° and 1933 at 16.9 hrs Dec +70.5°.
* Galaev's ether vector at ~16 hrs.
* Munera's (dynamic) ether vector at ~17.5 hrs, Dec +79°N.
* Cahill's ether vector at ~17 hrs.
* Miller's 1931 average, Table of Earth's Motions at 18.2 hrs.
* Reich's implied maximum Earth velocity at 17 hrs.
* Piccardi's June-July ~17 hrs max velocity of Earth.
* Brown's August ~18-19 hrs max biological reactivity.

This final revised Figure 91 should remove any doubt regarding the special significance of this common axis of motion through cosmic space. What astonishes me is how this high degree of correlation of cosmic motional factors wasn't previously observed, being overlooked by the astronomy/astrophysics professionals, and never being noticed, acknowledged or discussed! How is it that so many researchers on

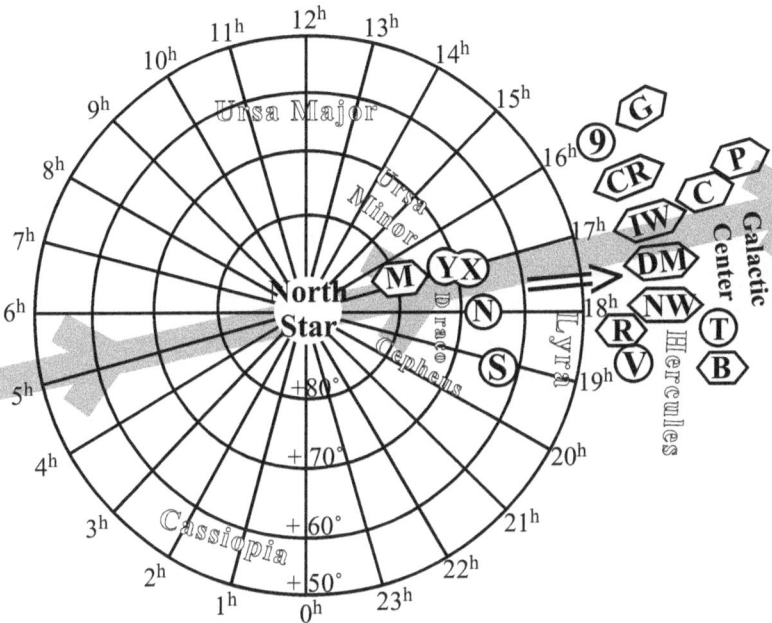

Figure 91.
17 Independent Vectors of Cosmic Motion

Looking North and Up to Polaris (Pole Star)
from Earth's Northern Hemisphere

Northern Pole of the Solar System Ecliptic = Ⓝ
Modern Sun's Way Towards Vega = Ⓥ
Sun's Rotational Axis Northern Pole = Ⓢ
Center of the Milky Way Galaxy = ⟹
Miller's Northern Axes of Ether Drift 1928: Ⓧ 1933: Ⓨ
Reich's "Point C" Cosmic Velocity Max = ⟨Ⓡ⟩
Galaev ⟨Ⓖ⟩ Munera ⟨Ⓜ⟩ and Cahill ⟨Ⓒ⟩ Ether Vectors
Piccardi ⟨Ⓟ⟩ & Brown ⟨Ⓑ⟩ Chemical & Biological Vectors
Miller's 1931 Average, Table of Earth's Motion = Ⓣ
Mystery Point "Planet 9" Gravitational Anomaly = ⑨
Interstellar Wind Vector = ⟨ⒾⓌ⟩
Cosmic Ray Wind Vector = ⟨ⒸⓇ⟩
Neutrino Wind Vector = ⟨ⓃⓌ⟩
Dark Matter Wind Vector = ⟨ⒹⓂ⟩

different continents, working around the same periods of time, in similar scientific disciplines and laboratories, who knew of each other's work, all made findings *pointing to a motional cosmic vector in the same general regions of cosmic sidereal space, and yet did not see it! How could they fail to recognize the profound cosmic pattern in their own works!?* And why, it must be asked, did it take a specialist in the Earth and atmospheric sciences, myself, from outside the disciplines of physics or astronomy, to spot these now-obvious patterns?

The answer to these question is a testament to the *self-blinded nature of academic specializations*, and to the fact that *one must firstly ask the right questions, to know where to look and hence to find the most telling and important answers.* The joke when I was a graduate student was as follows: The specialist learns *more and more about less and less, until they finally know everything about nothing.* In today's increasingly politically-correct world of scientism, I wonder if telling that joke in the universities would get one expelled or fired.

The UV-Blues of Cosmic Energy and Water

Around 2009, I began a study on the spectral changes in water induced by the Reich orgone accumulator (ORAC). It was discovered that a pronounced far-UV *absorbance*, with an equally strong near-UV and blue *fluorescence*, would appear in ORAC-charged water when excited by far-UV light. Identical control distilled water samples yielded no such spectral reactions (DeMeo 2018). The ORAC device used in these experiments is seen in the top portion of Figure 92. The upper graph, in the center of Figure 92, shows the increased absorption of far-UV light by ORAC-charged distilled water, while the lower graph depicts the near-UV and blue fluorescence reactions in a similar ORAC-charged water sample. In each case, the water was placed inside the ORAC for a few days, with a separate identical sample of control water kept under a cardboard box. These spectroscopic signatures indicate a new property was induced into water by the ORAC charging.

Similar spectral signatures were also found in vigorous Pacific storm rainwater and snow-melt samples, obtained at my high-altitude laboratory in rural southwest Oregon, just off the Pacific Ocean with no industrial pollution to speak of. ORAC-charged water has similarities to the phenomenon of Exclusion Zone (EZ) water, as discovered by Gerald Pollack (2009, 2013). Other investigators found corresponding spectral properties in aqueous solutions of DNA. This led me to a study

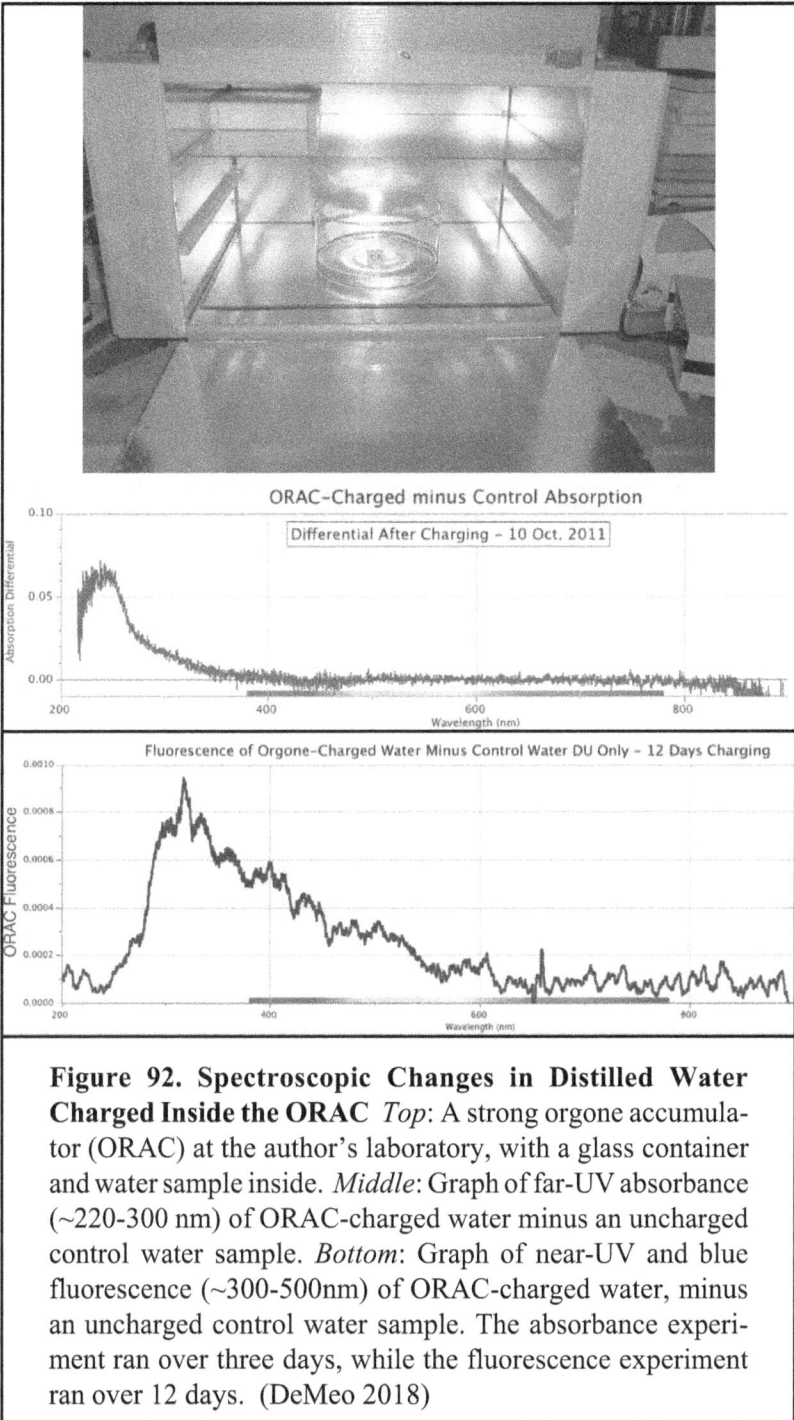

Figure 92. Spectroscopic Changes in Distilled Water Charged Inside the ORAC *Top*: A strong orgone accumulator (ORAC) at the author's laboratory, with a glass container and water sample inside. *Middle*: Graph of far-UV absorbance (~220-300 nm) of ORAC-charged water minus an uncharged control water sample. *Bottom*: Graph of near-UV and blue fluorescence (~300-500nm) of ORAC-charged water, minus an uncharged control water sample. The absorbance experiment ran over three days, while the fluorescence experiment ran over 12 days. (DeMeo 2018)

of blue glowing waters and glacier ice, and from there to similar blue-glowing phenomena in nature and the cosmos, where abundant similarities were found. Conventional explanations, such as Rayleigh light-scattering theory, or assumed preferential absorption of red colors, were explored in my study, but rejected notably due to the profound spectroscopical results from the ORAC-charged water experiments.

That such a spectral signature could be induced into distilled water by ORAC charging, without introduction of any physical or chemical influences, not only was yet another proof indicating Reich was on to something quite important, but that it had serious implications for chemistry, biology and cosmology. My findings added strong legs to Reich's overall theory of *orgonotic lumination,* which is a confirmation also of the luminiferous properties of the cosmic ether, as equated with Reich's orgone phenomenon. Similar UV-blue light flashes in the open sky or in water tanks, currently attributed to various "particles", may now be understood from a common ether/life-energy mechanism.

The blue-glowing galactic halos as seen around Andromeda Galaxy M31, appear to be a direct expression of the cosmic ether/life-energy. The halos have a dark blue radiance that can be seen and photographed, but which is unrelated to star fields. The blue glow extends from the galactic arm bands outwards to great distances, with a fairly sharp drop-off of intensity at a maximum distance. There also are blue-glowing regions connecting different clusters of stars or galactic objects to a "parent" galaxy, as if the galaxy got overcharged and then ejected part of its core, leaving an energetic trail between them. For example, Seyfert galaxies and their ejected quasars, as identified by Halton Arp, often have connecting energy bridges, as discussed in the next chapter. This is surely correct for Andromeda M31 and its two largest companions, M32 and M110. "Empty space" sometimes has a dark blue luminescence.

Astronomy explains the colors seen in cosmic space in part based upon the color variances identified in gas glow-tubes, which are filled with different gasses such as hydrogen or helium in a partial vacuum. When exposed to strong DC, pulsed or high frequency electrical charges, they glow with specific colors. Fluorescent light bulbs make use of this principle. However, these laboratory vacuum-tube conditions do not reflect the actual deep-vacuum, ultra-low particle density and weak electrical charge conditions out in open intergalactic space.

We do know from the glow tubes that nitrogen, argon and mercury vapor all give off a bluish glow when excited by pulsed DC or high frequency electricity, but there is no significant quantity of those gasses in cosmic interstellar or intergalactic space. Hydrogen is abundant in open space, but its color is a mix of predominating blue and red, appearing violet-red or pink to the eye. A hydrogen glow tube emits a light that in no way agrees with what is seen in the dark blue galactic halos, where there is no trace of green or red in photographs. Helium is also abundant, but it glows in the near-UV and yellow. One can see a rainbow of colors in star-rich nebulae, but neither stellar excitation, nor strong electrical fields, nor any high density of matter exists in the dark-blue regions of outer galactic halos. The most amazing example I have seen of this cosmic blue halo is in an astrophoto of Andromeda, made by Travis Rector at the Kitt Peak National Observatory. (WebRef.23)

The phenomenon of UV-blue light flashes, as seen in space, in water, in glacial ice, in certain crystals and dielectric plastics, and as detected in photomultipliers sensitive to those frequencies, is something that needs further investigation. Are the light flashes of a common origin, or even of an *identical nature?* PMTs sensitive to near-UV and blue frequencies, aimed into darkened water tanks or placed deep into polar ice sheets, detect UV-blue flashes of light. The same is true with PMTs attached to scintillator crystals composed of sodium-iodide, doped with tellurium (NaI-Tl) and used deep inside mountains to detect similar UV-blue light flashes from "dark matter". Or they are used above ground to detect cosmic or gamma ray atmospheric "Cerenkov" blue flashes. To the PMT device, they all cause the same reaction, yielding a flash and a "count". The only difference between them is the location and altitude where they are used, the scintillator material placed in front of them (water, ice, NaI crystals, transparent dielectric plastics, etc.), and the shielding material employed, if any. The same applies to gamma-ray telescopes, using PMT devices in multiple water containers, or aimed down into an upwards directed mirror, filtering out all but the UV components, and looking for direct light flashes out in the darkened sky. Likewise the UV-blue sensitive PMTs detecting biophotons in experimental biology.

Consider the UV emissions from helium, used to define the interstellar medium, or the dark blue glow surrounding many galaxies such as Andromeda M31, or the blue nature of quasars (discussed in the next chapter), or the UV emissions of the Earth's "plasmasphere", detected by the IMAGE satellite. Also the UV-blue flashes as picked up by neutrino, dark-matter, cosmic-ray and gamma-ray PMT detectors. *They all have arguably identical cosmic ether/life-energy origins*, as do other phenomena dependent upon PMTs for their detection. As a common theory, we have Reich's original finding on the orgone energy units, blinking in and out of visible existence, some believed to form small bits of matter, others sinking back into the cosmic ocean from whence they came. And as an interesting anecdote on this, we may ask how "background radiation" was counted before the invention of the GM tube or the spinthariscope? The counts were made by young scientists, usually physics graduate students with good vision, sitting in dark rooms with pencil and paper, marking down how many light flashes they saw with their eyes! Were these only "blood corpuscles in the eye" as some ignorant skeptics have laughingly brayed, like donkeys, to dismiss Reich's various findings?

The physical scientist and astronomer may object to my lumping all these blue-glowing and light-flashing phenomena into the same basket, and giving them a Reichian / cosmic ether explanation. However, are their own split-apart explanations any better or more reasonable? Can they see "WIMPS", "neutrinos", "cosmic rays", "gamma rays" or "Planet 9's" and "MACHOS" with their eyes, or put them on ordinary camera films? With cosmic and gamma rays, yes. But their own long search for a common understanding, as with the search for a "unifying field theory", has so far been for naught. By contrast, *the prematurely rejected research on cosmic ether, and the banned and burned texts of Wilhelm Reich on orgone energy, provide a real-worldly, unifying principle for the sciences, with a far better understanding of how nature and the universe function.* The academic world has largely slammed the door shut on such ideas, unfortunately, and has too often ignored or misapplied the cautions of Occam's Razor, thereby missing great opportunities for scientific discovery. But not always. Below is a short clip from my well-received 2018 lecture and subsequent paper in the *Water* journal, presenting provocative questions and new ideas that would have created an academic riot only a few years earlier.

"Do we live in an energetic water universe? Besides water-drenched quasars at the edge of known space, significant water is known to exist within our own solar system. Many of the moons of Jupiter and Saturn are known to be water-ice in surface composition, as are many comets. Mars and other planets show surface drainage or scouring features, along with direct evidence in ice-caps, suggesting they once possessed a much greater quantity of water than today. It appears, the more the scientific community looks for water in the cosmos, the more it is being found.

With growing evidence that water might be one of the most abundant molecules in the universe, an "outrageous hypothesis" forces itself: Is it possible, as with "water drenched" Quasar APM 08279+5255, that cosmic water exists in open space near to Andromeda, and thereby is excited into fluorescence by Andromeda's own intense stellar radiations? Is the blue glow surrounding M31 yet another expression of "blue-fluorescing cosmic water"? Or alternatively, is this blue color exactly as the heretic Wilhelm Reich described it: a cosmic energy which attracts and infuses into water within the Earth's atmosphere and crust, concentrating inside the ORAC, into cyclonic storms, into high dielectric materials and DNA...? And where under proper excitation, we may observe it with our eyes in the visible part of the spectrum, as well as detect it in spectrometers and by other means? Is the blue-glowing "energy-water stuff" the long-sought *self organizing principle, the "galactic glue" or cosmic-creation, negative-entropic force* which, by conventional empty-space dead universe theory is termed "dark matter"? And shouldn't exist? Except, it does exist, and isn't dark either. It is blue.

I am reminded of the example of a dozen blind scientists in a room with an elephant, each of whom is grasping a different part of the beast – the trunk, the tusks, the legs, the body, the tail – and each giving widely divergent descriptions, all of which are 100% accurate, but also suffer from great magnificent error. By focusing upon the differences, they miss the larger pattern of similarities that clarifies just what is being observed and measured. By Occam's Razor, this cosmic energy-water postulate – for which I credit Reich above all the very significant others – rather forces itself, and commands rational attention, open debate, and new lines of experimental investigation." (DeMeo 2018, p.72-73)

Now, in the last chapters, let us grapple with some additional "outrageous hypotheses" on the *implications* for modern physics, astronomy and astrophysics, from what has already been documented.

A Tale of Small Creatures on an Isolated Planet

A thousand near-sighted ants gather on a leaf floating along in a river, discussing the trees on the river-bank which they see unclearly at a distance. One ant proposes, *"The trees are in motion, passing by us rapidly, as our leaf is fixed in the center of the universe."* Nearly all the ants agree. One radical ant disagrees, saying *"Oh no, our leaf is floating on a river of water-space, and the trees are fixed and unmoving. We, however, are in motion."* The other ants get dizzy and upset at this postulate. A few of them bravely peer over the edge of the leaf, seeing only their reflections. Other ants extend their antennae in various directions, sensing no fast motion approaching the velocity of the trees. *"We observed nearly no movement of our leaf, and therefore, you are crazy and water-space does not exist!"* The radical ant is then repeatedly stung, and thrown overboard.

Implications and Consequences of a Material-Motional Cosmic Ether for Modern Astrophysics

The Cosmic Ether Changes Everything!

For more than 100 years, empirical experimental evidence identifying a real material and motional ether has been consistently ignored, overlooked and suppressed, while at the same time, ambiguous and speculative, mystical theories have been promulgated and hungrily devoured. And whenever evidence was asserted to support such mysticisms, it was never so unequivocal that opposing ether theory could not equally or better explain it. *Factually, proof for a motional and material cosmic ether changes everything!* It upsets the modern applecart, and forces us back to unfinished discussions of the early 1900s. To this we must add the considerable work of Reich, who independently and experimentally confirmed an ether-like life energy, and described how it moved in living tissues, in the atmosphere and in the cosmos, thereby adding additional detail to what is known about cosmic ether. The two objective discoveries, and their respective bodies of evidence – of cosmic ether and cosmic orgone – are at root, functionally identical. And not accidentally, some of the same players, notably Einstein and his followers, worked towards the erasure of both Reich and Miller.

In this closing chapter I will review modern cosmological concepts and experiments currently underway, and will challenge their basic foundational assumptions from the viewpoint of a dynamic cosmic energy in space. As a prelude, I would remind the scientific reader of a major fallacy in contemporary physics, where modern theories as from Einstein, the big bang and quantum entanglement, are stretched so thin in efforts to basically "explain everything", that in the process must resort to *increasingly fantastic and contradictory claims.* By contrast, the cosmic ether of space already has considerable independent evidence and equally valid explanatory and predictive power, resting firstly upon *the historical proofs of its own existence.*

The LIGO/ALIGO Giant Michelson Interferometers
Detecting Gravity Waves? Or Cosmic Ether Turbulence?

The Laser Interferometer Gravitational Observatory (LIGO) employs several giant Michelson-type dual beam interferometers, each with two cross-armed light paths of enormous proportions. However, with the cosmic ether neatly discarded, all significant interference effects observed within the LIGO apparatus are attributed to "gravitational waves". This latter concept was postulated by Einstein in 1916, in accordance with the larger scientific denial of published evidence of real ether-drift signals: Michelson-Morley, Morley-Miller, Sagnac, Miller, all were ignored. The consequences are that, today, Big Science experiments such as the LIGO and later ALIGO (Advanced LIGO) provide results which are *fully equivocal between Einstein's theory versus cosmic ether theory. Either could be true.* LIGO and ALIGO may be acquiring cosmic ether signals with changes in light speed, possibly at higher levels of turbulent velocity, notwithstanding the current interpretation according to Einstein's "gravity wave" postulate.

By the conventional gravity-wave theory, any significant variance in the LIGO/ALIGO interferometers is attributed to *changes in the lengths of the interferometer arms by sub-micrometer compressions or expansions,* rather than to *changes in velocity of the light beams moving through those arms.* And yet, there is no way to independently determine if the changes are due to variation in arm length or in light speed. The variable light speed argument is simply not allowed, as it would negate "gravity waves" as well as Einstein's relativity theory. The LIGO/ALIGO results are therefore interpreted *only* within theoretical terms agreeable with the empty-space assumptions of Einstein's theory. The experiment is described as such:

> " LIGO's interferometers are the largest ever built. With arms 4 km (2.5 mi.) long, they are 360 times larger than the one used in the Michelson-Morley experiment (which had arms 11 m (33 feet) long). This is particularly important in the search for gravitational waves because the longer the arms of an interferometer, the farther the laser travels, the more sensitive the instrument becomes. Attempting to measure a change in arm length 1,000 times smaller than a proton means that LIGO has to be more sensitive than any scientific instrument ever built,

**Figure 93. The Two LIGO Experiments: in Hanford,
Washington (top) and Livingston, Arkansas (bottom)**

so the longer the better. ... Each arm has ... mirrors near the beam splitter that continually reflect parts of each laser beam back and forth within the 4 km long arms *about 280 times* before they are merged together again. ... LIGO's interferometer arms are effectively 1120 km long, making them *144,000 times bigger than Michelson's original instrument!* ...the two beams are superimposed after their long journey through the interferometer. ...we expect to see particular interference patterns when a gravitational wave passes by..." (WebRef.11)

The LIGO interferometers, located in Hanford Washington and Livingston Arkansas, were upgraded in 2015 to form the Advanced LIGO, given the poor performance of the original LIGO devices. Today there is a similar VIRGO project located in Italy. So upgraded, ALIGO claims it can now detect 1/10,000th of the diameter of a proton, a sensitivity by which they hope to detect gravitational waves from "1 nanosecond after the big bang". ALIGO embraces the empty space not only demanded by Einstein's theory, but also by the big-bang creation myth, about which I have more to say shortly. This is Big Science, big-time, big-money research:

"More than 1,200 scientists from around the world participate in the effort through the LIGO Scientific Collaboration, which includes the GEO Collaboration.... The Virgo collaboration consists of more than 300 physicists and engineers belonging to 28 different European research groups: six from Centre National de la Recherche Scientifique (CNRS) in France; 11 from the Instituto Nazionale di Fisica Nucleare (INFN) in Italy; two in the Netherlands with Nikhef; the MTA Wigner RCP in Hungary; the POLGRAW group in Poland; Spain with IFAE and the Universities of Valencia and Barcelona; two in Belgium with the Universities of Liege and Louvain; Jena University in Germany; and the European Gravitational Observatory (EGO), [and] the laboratory hosting the Virgo detector near Pisa in Italy." (WebRef.12)

From that description, it is no wonder mainstream physical science today often boasts there is no longer any room for the independent-minded research scientist. Science publications now often have dozens or *hundreds* of co-authors, whose names alone take up several pages in their publications. One primary paper from the ALIGO experiment had over *1100 co-authors,* the first *six pages* of the paper taken up by their names and institutional associations. (Abbot 2018) Gone are the days of Galileo, Newton, Michelson and Miller, whose isolation from the pressures of major institutions was necessary for clear-minded focus.

In spite of LIGO's large set of claims and its significant army of workers and believers, we can nevertheless critique the claim that it detects "gravitational waves" from distant stars, or from still unproven "black holes". It employs a detector so basically similar to the Michelson interferometer that the variations in its measured data can just as easily be attributed to large transient waves or turbulent surges in the cosmic ether itself, creating more than the usual quantity of light-wave velocity variations. Again, there is nothing built into the LIGO devices so as to rule out the explanation of a dynamic ether turbulence or wind.

The LIGO light paths are oriented perpendicular, one arm along a north-south axis, the other on an east-west axis, laid out on relatively flat terrain. The light path of the first LIGO device in Washington is composed of a half-buried heavy concrete culvert-pipe enclosure, measuring about 2 meters in diameter. Inside this outer concrete shell is a metal tube of 1.2 meters diameter enclosing the light-beam paths, which is evacuated down to a hard vacuum. All optical component mirrors at the ends of these pipes are mounted in special cradles for dampening out mechanical vibrations, as from seismic, meteorologi-

Figure 94. Massive Mirror Ends of the LIGO Light-Paths

cal, tidal or vehicular sources. In size, it is somewhat reminiscent of the earlier large-scale Michelson-Pease-Pearson Irvine Ranch experiment, or that of Michelson-Gale. However, LIGO makes those prior efforts look quite small by comparison, and was constructed using modern materials and optical components, with space-age clean room procedures. It nevertheless follows a basic design nearly identical to the original 1887 Michelson-Morley experiment, reinterpreted along the lines of Einstein's theoretical assumptions and conclusions.

A powerful laser beam is split into two parts by a large half-silvered mirror, each beam separately travelling down and back through the two long perpendicular LIGO tubes, reflecting back and forth multiple times off mirrors at the tube-ends, prior to being recombined to form optical interference patterns. The interference patterns are monitored and recorded by the microsecond, using video cameras and computers, as the Earth turns to expose the LIGO interferometer arms to different cosmic orientations. Their efforts are not without significant problems.

Given how the entire LIGO light beams are contained inside heavy metal beam-tubes and concrete covers, it is not surprising that *for the "Initial LIGO" periods from 2002 through 2014, they observed no significant variations in the apparatus sufficient to claim detection of anything remarkable.* After upgrading the instrumentation, by the end of 2017 a total of 11 different "gravity wave" detections were listed in the *Gravitational Wave Transient Catalog 1* (WebRef.12), covering September 2015 through August 2017. A separate publication by LIGO scientist Crestwell in the same year boasted three new LIGO signals, but the "detected" signal was about 1/100th of the background noise, leading the *New Scientist* magazine to engage in rare public criticism of a mainstream effort. (WebRef.13) A new LIGO/ALIGO announcement was made during the preparation of this book in December 2018, retrospectively claiming a 2017 detection of "two colliding black holes". This could only add to the mystery, given that black holes remain hypothetical things which become unnecessary with a gravitational ether in space. I will speak to the issue of "black holes" momentarily, particularly with regard to the most recent claim of the "first image" of a black hole in galaxy M87.

The question remains, are they detecting "gravity waves" from imaginary "colliding black holes", or major surges or turbulence in the cosmic ether ocean of space? Such a turbulent ether phenomenon would be more intensive and chaotic than the regular variations in sidereal-day ether-drift measures. I recall, when reviewing Miller's

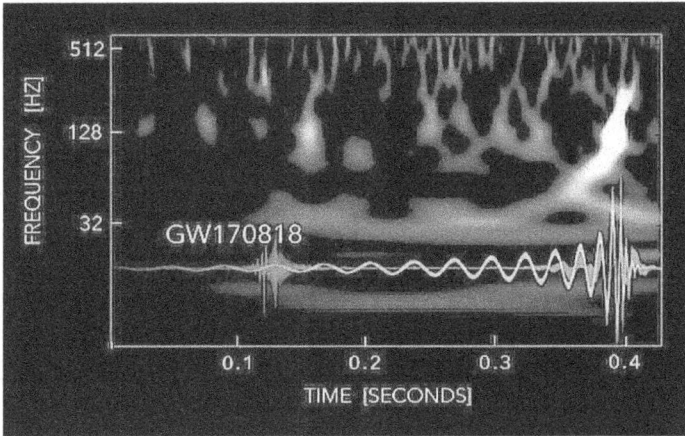

Figure 95. One LIGO Event, on 18 August 2017, lasting less than one second. It reveals "something", but is it a signature of uncommon ether turbulence creating a brief variation in light speed? Or a "gravity wave"? Which theory is most accurate?

original graphs, seeing several huge spikes in his ether-drift data. Could those also have been ether surges or turbulence? LIGO/ALIGO are providing an important set of measurements, but why does it only get interpreted according to Einstein's theory, and not by the theories for which the Michelson interferometer was originally invented?

From previous chapters we know the cosmic ether actually exists and has been measured. And that it also appears to play a fundamental role in gravitation. We must ask, are the LIGO/ALIGO detections really *ether waves or turbulence,* analogous to a rare large "sleeper wave" as occasionally observed along the Pacific Coast of North America? Or are they comparable to an unanticipated tidal wave due to a faraway earthquake in the years before such distant events were known, rendering their consequent waves unexpected and mysterious? Those examples are entirely different from the usual water waves or lunar tides affecting the surface of the oceans. Theoretically, ether-wave turbulence or surges affecting light speed appear as equally valid speculations for the LIGO/ALIGO empirical observations, without reference to Einstein, the big bang or "black holes". Nevertheless, the reference to invisible and unproven "colliding black holes" remains a rather popular, though mystical ad-hoc "explanation". These differences in theory could be sorted out by looking for a more *systematic sidereal hour* signal in the lower-intensity LIGO/ALIGO "background noise" data, which is currently filtered out. While the metal and concrete shielding of the LIGO and ALIGO light beams is such that it would block out most ether wind, the long length of the light paths suggests that even a lesser ether-wind signal might be detected. A sidereal pattern might already exist in the LIGO "noise", indicative of light-speed velocity variations, and not merely "changes in path-length" which then calculate over to "gravity waves". However, to find such an ether-wind signal they would have to first *take the idea seriously, and look for it!*

Atomic Clocks Warped by Space-Time? Or by Variable Inertia?

The chapter on *Einstein Rising* has already suggested how a dense ether layer close to a planet, star or our Sun, could bend light waves moving through it, by ordinary refraction effects similar to how light is bent when moving from one medium into another, as when passing from the air into water. But how, we can ask, would an increased ether density affect the *inertial properties of matter itself?* Inertia is as mysterious to us as is time and gravity, modern theory notwithstanding. What determines that a given "push" of a known force, will move an object of known mass along a given distance? Out in cosmic space, friction and gravity can be ignored. We know the equations by which to calculate the *magnitude* of such a force, as formalized by Newton, but not for *understanding* it. Calculations of inertia reference the mass of an object, as well as distance moved. It is therefore considered to be an innate property of matter, described by its mass and motion. However, this classical view is not complete.

The cosmic ether medium may, by its density or concentration, also be a determinant of inertial forces, altering the properties of the space in which matter rests and where motion take place. I propose something similar to Michael Faraday's argument that the magnetic field sur-rounding his rotating homopolar generator device did not itself rotate, and therefore was *a property of space itself*, and not of the wires and magnets used in building the apparatus. His experiment is worthwhile to consider, described for home-experimenters on internet (see WebRef.14). By convention, inertia is defined as a characteristic of mass, but *inertia may be variable, imparted to mass by the density and motility of the cosmic ether medium.*

The cosmic ether appears to have dielectric properties which could give rise to electrical charge, and may also impart motional influences upon matter, as Lorentz considered in his original ether-matter compression theory. This was never proven to cause an actual compression of matter, however. Reich developed experiments with electroscopes and static electricity inside his passive orgone accumulators (ORAC), showing the standard expectations of conventional theory did not hold. His work proved that the energetic properties of matter and living substance are altered inside the ORAC, within its higher density of the background medium. If this line of reasoning is correct, then inertia itself may be variable, through changes in properties of the local ether or orgone medium.

Atomic clocks – the large machines used in laboratories, not the commercial "atomic clock" you hang on your wall – function on the basis of an assumed constancy of inertia. The construction of such atomic clocks requires a signature atom, such as cesium, to be vaporized into a vacuum tube of exceedingly low pressure, a *hard vacuum*. The vacuum tube is then bombarded with microwaves, triggering emission of its specific peak frequency. That frequency is then precisely measured (for cesium it is 9,192,631,770 cycles per second), maintained and locked into place. And that gives a very precise determination of "one second" of time. But is that number merely an expression of clock-time, or of the actual flow of time itself? Can the cesium frequency be changed by its ordinary environmental factors?

When atomic clocks are moved to higher altitudes, or flown around the Earth, they have been experimentally documented to change their frequencies and record a slight gain or loss in timekeeping. An atomic clock moved from sea level to about 1 kilometer of altitude for a week, will run faster by around 40 nanoseconds (40 billionths of a second) by comparison to a master atomic clock kept at sea level over the same time period. Such effects have been measured (WebRef.15). While some have heralded this result as a "proof of Einstein's relativity", *claiming time itself had been altered*, Louis Essen, the scientist who invented the atomic clock in 1955, rejected Einstein's theory utterly, and not to the benefit of his professional standing. (Essen 1971, WebRef.16) Those experiments only proved that the atomic frequencies generated by their internal mechanisms and governing their timekeeping, speeded up by an exceedingly tiny amount over a few days. When they are returned to sea level, the atomic clocks resume their ordinary rate of internal timekeeping.

The Dynamic Ether of Cosmic Space

Since the cosmic ether penetrates into the vacuum, and also into all the pores of matter, *the atomic clock may run at a specific time-rate in accordance with the viscous density of the local ether medium, which is greater closer to the Earth's surface.* Higher altitudes with a higher ether-wind velocity, and lowered density and viscosity, might allow such a clock to race ahead. A comparison would be, to imagine a spring-wound mechanical clock immersed into a jar of fluid motor oil. It's "tick-tock" gearwork mechanism would thereby be slowed down. If the fluid oil is heated up to a lower viscosity, the clock speeds up.

An important atomic-clock experiment was undertaken by Hafele and Keating in 1971, who flew cesium-based atomic clocks in opposing directions around the world, close to the equator. The *eastward* flying Hafele-Keating atomic clock *slowed down by ~59 nanoseconds* in comparison to an identical master clock on the ground at the US Naval Observatory. The *westward* flying clock *raced ahead by ~273 nanoseconds, running faster* than the master clock.

In that experiment, we may imagine time was changed, in keeping with Einstein. However, it is far simpler to suppose that the atomic clocks temporarily changed their resonant frequencies – possibly the inertial motional properties of the cesium atoms as they bounced around inside their ether-saturated vacuum tube confinement – due to changes in the ether density and velocity . The trip east, with the Earth's rotation and also generally with the ether-wind's overall west to east and southwest to northeast component, would provide a slightly slower and more condensing ether condition within the cesium vacuum tube, making the atomic clock function more sluggishly. The trip west was against Earth's rotation, and also into the face of a faster flow of the ether wind, leading to a more "stirred up" condition of lower density and viscosity for that atomic clock, which ran faster. We can therefore postulate that a faster ether-wind velocity correlates with lower ether density and viscosity, allowing atomic processes to speed up. And in opposition, a slower ether-wind would lead to a more "calmed down" or sluggish, stagnant condition, slowing atomic processes.

These changes can be deduced from the results of the various ether-drift experiments, notably those of Miller and Galaev. Recall Miller's point on doing ether-drift experiments at higher altitudes to avoid entrainment effects, confirmed by Galaev in an ether velocity dependency upon altitude. Also recall Reich's observation that the orgone energy was "more active" at higher altitudes, a point confirmed at my high-altitude laboratory, and also echoed by Piccardi.

306

While the Einstein followers may feel comfortable explaining atomic clock behavior by their complicated and rather mystic space-time explanation, the effects of ether density and velocity appear to provide an explanation of equal merit. A jet aircraft flying through the ether at ~500 mph (0.22 km/sec) creates changes in an atomic clock's functions right down to the atomic level. The 10 km/sec ether-wind experienced by an eastward moving jet, moving *with* that ether wind, would be around 9.78 km/sec, while a similar westward moving jet, moving *against* or into an ether head-wind, would experience that wind at 10.22 km/sec, about a 4.5% difference. From the above, we may therefore postulate, *ether density and motility (viscosity) is a factor determining the inertial properties of matter, at the atomic level.*

Also very probably, the metal skin of the aircraft and housings around the atomic clocks, or the buildings in which they are placed during the altitude-only experiments, would inhibit their reactions to such ether influences, lowering the percent differences in ether-velocity depending upon placement and movement. Other experiments could be undertaken, such as placing an atomic clock mechanism in a thin glass cover up on a mountain, with a second identical atomic clock placed into a heavily-shielded enclosure at the same altitude. The unshielded atomic clock would then mirror the successful ether-drift experiments. Atomic clocks should be tested for these considerations, which if confirmed, might then become a new type of ether-detection device. All the factors known to affect the ether-drift experiments, or atomic clock experiments, would have to be taken into account for a new set of experiments.

The main points are, one needs not conclude Einstein's "changes in time" *is the only possible explanation. Inertial properties of atomic clock mechanisms might be altered in moving from one region of space to another, but not time itself!* This has profound implications for other unanswered questions, such as "what is gravity", and "can gravitational forces be changed or harnessed, by modifications to inertia"?

Open Questions About GPS and Ether Wind/Drift

The Global Positioning Satellite (GPS) network requires a coordinating timekeeping and adjustment signal from an Earth-based master atomic clock, partly due to variations in their on-board atomic clocks as they orbit around the Earth at around 4 km/sec. This is acknowledged, but given an Einstein relativistic explanation. However, GPS

signal corrections ought to be investigated *according to Miller's determinations of an axis of drift oriented mostly south-north, with a lesser but variable west-to-east component.* GPS satellites on a northward track move more decidedly *with the ether wind,* but later move on a southerly track, *against the ether wind.*

Furthermore, there could be a separate, additional Sagnac-type of effect at work, in that radio signals sent in the forward direction of their motion would have a slightly different velocity, relative to the ether, from signals sent in the opposite backward direction. The old "c+v is greater than c–v" issue would be at work. Prior arguments about atomic clocks being affected by an ether wind would also be operative in the GPS satellites. All such factors, taken together, may be responsible for the necessary regular correction adjustments to their master timekeeping signals.

While complicated, an ambitious electronic engineer with access to the actual GPS timekeeping data could evaluate for such ether-related

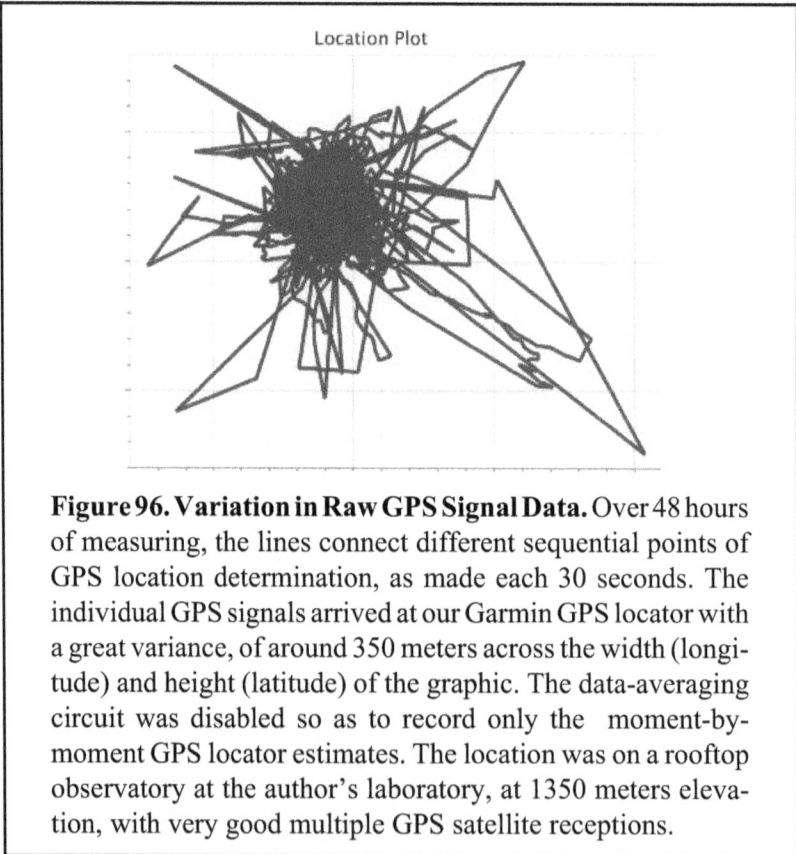

Location Plot

Figure 96. Variation in Raw GPS Signal Data. Over 48 hours of measuring, the lines connect different sequential points of GPS location determination, as made each 30 seconds. The individual GPS signals arrived at our Garmin GPS locator with a great variance, of around 350 meters across the width (longitude) and height (latitude) of the graphic. The data-averaging circuit was disabled so as to record only the moment-by-moment GPS locator estimates. The location was on a rooftop observatory at the author's laboratory, at 1350 meters elevation, with very good multiple GPS satellite receptions.

effects. To my knowledge, nobody is reviewing the GPS variances from the standpoint of possible cosmic ether velocities and azimuths. It is a problem similar to the LIGO/ALIGO data, where nobody is looking for such an ether-affected signal.

In a line of related evidence, I set up a stationary Garmin GPS receiver in the fiberglass telescope dome on the roof of my laboratory, to determine the variation in the GPS signals for our location. It was not moved over the period of 2007 through 2008, measured once per month over a 48 hour period. The averaging function of the GPS device was shut off, such that it acquired uncorrected satellite data, recorded digitally each 30 seconds in my lab. In the attached graphic, one can see the variation in our stationary GPS signal, at 1350 meters elevation in southern Oregon. Over a period of one or two minutes, the raw GPS data would dramatically shift from one location to another, even though the GPS detector was never moved. The magnitude of error was up to 350 meters across both latitude and longitude usually within a minute or two. A variation of ~100 meters was noted in elevation data.

This suggested to me, while the data provided by the civilian GPS system is very accurate when averaged out in normal mode, the data location point it yields for any one moment in time has a significant error. Why is this? I speculate, this may be due to the problem of ether wind creating radio-wave velocity variations at our 1350 meter altitude. Our reception during the experiment was good, with six or more of the satellites always within the range of the receiver.

Low Intensity Double-Slit Experiment: Light Particles or Waves?

One of the central examples always given for the particulate nature of light is the double-slit experiment, even though it reveals a clear wave-function property of light. A single light source is allowed to shine on an opaque surface with two narrow slits. The slits are a fraction of a millimeter in width, and separated by about one millimeter. As the light waves progress through those two slits, they spread out and the two light beams interfere with each other, revealing a pattern of alternating bands of light and dark regions, or interference bands. Large open slits will not show the effect, they must be sufficiently narrow before it can be seen. Also the light used in the experiment must be of the same general color for the best results. This effect was discovered by Young around 200 years ago, discussed on page 38.

Clearly the double-slit experiment reveals wave functions as shown in the prior chapter, quite similar to what is observed in sound waves and in water waves. Light waves move through the two slits, creating two wave fronts which then interfere and create dark and light bands. By the particle theory, isolated light photons move through the slit, creating similar interference patterns. However, each individual photon particle is said to somehow "know" where all the other particles have already landed, so that it can find a place to land in keeping with the laws of optical interference. Determination of where a given light particle actually lands on the display screen, but *only on the light-band regions*, is claimed due to mathematical probabilities.

What is more interesting, however, is how *apparently individual light particles* can be identified when the double-slit experiment is reproduced at very low levels of light, where one could not even see the light fringes on a projection screen or detector. In this case, the interference bands *slowly emerge over time, spot by spot,* on photographic film emulsions, or in modern charge coupled devices (CCDs) as in digital cameras. Interference fringes initially appear as individual points of light which accumulate apparently randomly over time, but *only on the light-band areas*, slowly building up to reveal the highly organized pattern of light and dark bands. This gives a superficial impression that individual light units, the photons, are obeying some hidden existential law of cosmic probability.

With the individual photons so guided by quantum magic, to land on specific *preferred locations* on the projection screen, but also to avoid landing on other specific *non-preferred locations*, the low-light double-slit experiment became a cornerstone of the revived particle theory of light. It also stimulated the development of quantum theory, that particles cannot be exactly identified as to velocity and location, only by reference to probability tables which the photons have somehow memorized. It substitutes yet another mystical view of unseen invisible forces at work, where cause and effect that govern ordinary life, nature and matter, are no longer operative. The implication in this case is, each light photon particle somehow "knows" where to land in obedience to probabilities, a process termed "quantum entanglement", where photons magically communicate with each other. By this explanation, the wave theory of light, and the necessity of a medium for transmission of waves (the ether) becomes irrelevant. But in fact there is a very simple explanation for how/why individual low-intensity light waves may generate apparent individual photon "events".

310

Figure 97. Individual Unit-Elements of Film Emulsions and CCDs. Each unit, with slightly variable sensitivity, when exposed to low intensity light waves, slowly accumulates a charge and reacts individually. Left: Electron micrograph of film emulsion grains, from around 0.1 to 1 micron in size. Right: Magnified CCD pixels of around 0.5 micron in size.

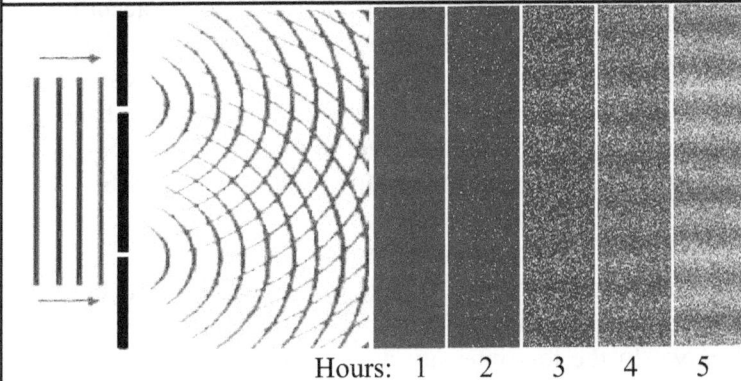

Hours: 1 2 3 4 5
Gradual low-intensity exposure of a photographic plate.

Figure 98. Interference Patterns Slowly Emerge from Low-Intensity Lightwaves through a Double Slit, but are misinterpreted as individual light photons moving to a given location based upon probabilities, as if each photon "knows" its proper place in the universe. Such magical solutions are unnecessary. The constant steady pressure of light waves at low intensity will incrementally trigger reactions in microscopic detector unit elements with variable thresholds of reactivity. Examples are: Individual photo-reactive silver nitrate domains of a photographic plate, individual pixels of a CCD imager, or individual photomultiplier tubes in an array. The so-called "quantum entanglement" is a magical, mythical beast.

311

The Dynamic Ether of Cosmic Space

The problem for the quantum-photon theory lies in the nature of the light detectors. For example, standard photographic film emulsions, when viewed microscopically, are composed of photosensitive chemicals deposited on cellulose material. In the deposition process, individual tiny crystal lattices of around 0.1 to 0.5 micron size are formed. At high light intensity, the entire sheet of film will react pretty much all at once, producing a final film image that depends upon which areas receive more light, and which areas receive less or no light. But at very low light intensity, with a time-exposure, the process of crystal lattice activation takes much longer, and thereby at first appears random in nature. A crystal lattice will first react here, then there, then someplace else, etc., without any uniform pattern being obvious. Each crystal lattice will slowly absorb light-wave energy until reaching a "triggering" threshold level, causing it to independently react. This is similar to how a smooth thin layer of popcorn in a flat warming pan slowly absorbs thermal energy, until each of the kernels "pops", according to its *own unique and variable low threshold level,* reacting at different times. Pop, pop... pop pop pop ... pop ... pop pop... pop, etc. In this analogy, the light-wave "heat" is distributed unequally in bands, in accordance with wave interference, even at low light intensities. Some of the photo-reactive lattices remain dormant, creating a dark band over time. Others receive the wave-energy only slowly, but eventually react in larger numbers to create parallel bands of light.

The detector units, be they film crystal lattices or CCD bits, will have a variable sensitivity and threshold of reaction, in the manner of a bell curve. In the low intensity double-slit experiment, each tiny crystal lattice or pixel will react slowly, as a tiny dot of light appearing here or there, and the dots build up over time, to reveal a coherent pattern of interference bands which appear similar to those created at higher light intensities. So what appears at first to be "random events" of individual light photon particles obedient to probabilistic targeting, are more easily understood as a function of *ordinary interfering light waves stimulating a slow triggering of the detector's variably sensitive components.*

From this viewpoint, a cornerstone of the particle theory of light actually is no cornerstone at all, but yet another example of light-wave functions being misinterpreted and replaced by a highly metaphysical speculation. I was led to this simple understanding by physicist Carolyn Thompson, a longtime critic of quantum theory's "entanglement" claims. Excommunicated from established universities, she neverthe-

312

less participated in internet discussions on such matters, and published papers in physics journals. Like many dissenting physical scientists, she accepted the cosmic ether of space, was a critic of Einstein's relativity, of the big-bang theory, and also generally of quantum theory and the particle theory of light. (WebRef.17) So am I, and for similar reasons.

The bottom line is, this cornerstone of quantum theory cannot be sustained. Its advocates have simply substituted a highly complex and metaphysical theory to explain the critical low-intensity double-slit experiment, ignoring the more simple and real-worldly explanation. Light waves in an ether medium provide a very straightforward understanding, without resorting to metaphysical multi-universes, space-time warps, creation-event singularities, or quantum magic.

The related issue of light passing through a transparent calcite crystal, for example, branching into two separate light-beams with opposing properties, also finds a reasonable explanation in wave theory. A rough analogy is how a farmer's plow divides the moist soil into two oppositely curling halves, a phenomenon also seen at the bow of a ship moving through water – but the two separated halves are still the same soil, the same water. The "magic" is in the crystal structure of the "plow", which acts to divide a light wave into two expressions which may or may not find exact oppositional components at any given interval of time. Thompson was highly critical of the claims from these experiments, which by her analysis rarely showed photon pairs of opposing properties in a statistically-significant manner. Her criticisms fell on mostly deaf ears. Referencing "particle-waves" or "wavicles" merely continues a needless contradiction. (WebRef.17)

The Heisenberg Uncertainty Principle? Or the Heisenberg Error?

According to the 1927 postulate of Werner Heisenberg, one cannot properly ascertain both the position and velocity of a given particle at the same time. He argued there are limitations which, by natural law, restrict one from making any such determination beyond a certain small region of uncertainty. He referenced only the particle theory of light, or the behavior of subatomic particles, which when understood as wave functions automatically lacks the sharp-boundary features of particles. The modern scientists working with subatomic particles, capturing particle tracks on film, have advanced sufficiently to objectively identify different species of particles beyond what was possible in

1927. Their photos of particle tracks very well document the mass, location, time and velocity of the particles they record, in ways Heisenberg could not imagine. The related idea that one disturbs a phenomenon merely by observing or measuring it may be true for flocks of birds, deer or humans, or for fluid or gaseous matter, but in general, inanimate matter cares not and responds not to the scientist's rulers and probes.

Consequently, as with the double-slit experiment, it is not scientifically accurate to proclaim that Heisenberg's uncertainty is some new law of nature. Heisenberg's speculation merely expressed a theoretical limitation and declared observational error within experimental science of 1927. *It is Heisenberg's Error.* Unfortunately, modern physics has built a cloud-castle in the sky around this error, which constitutes yet another blow to the metaphysical components of the increasingly fantastic "quantum universe", with poorly demonstrated "particle entanglements", and other illusory components. Likewise, regarding thought experiments about dead cats in boxes.

These ideas are expanded to their illogical limits in various science fiction films, and also by eBay hawkers selling "quantum vitamin pills", or gurus peddling "quantum meditation". Murray Gell-Mann protested this nonsense in his 2002 essay about "Quantum Flapdoodle". However, is their own understanding of nature and the universe so much better, given how they insist, wrongly, that the ether – which might provide an alternative mechanism for all what they claim – was "disproven" or "never detected"?

Big Bang Creationism?

The big bang theory of the origins of the universe, described as the theory of galactic recession by advocates, or as "big bang creationism" by critics, was founded upon the 1929 discovery by Edwin Hubble, of the red-shifting of light from distant galaxies. Most galaxies, when their light is passed through an optical spectrometer attached to a telescope, spreading it out into a rainbow of colors, will show a slight shifting of specific absorption lines and colors towards the red or longer wavelengths of the spectrum. That was never in question. However, the understanding of *how* such red-shifting could be produced, was constrained by the anti-ether thinking of the period.

In 1931, around the same time Miller was presenting his most substantial ether-drift evidence, the Roman Catholic priest and astronomer George Lemaitre proposed the first rendition of the big-bang theory. He argued, if the widespread red-shifting of galactic light was interpreted as an effect of Doppler recession, then it implied all of material existence was rushing away from the Earth, in an expanding universe. This further implied how, in the most distant past, all matter in the universe was located in the same lump, or pin-point of space. The universe Lemaitre envisioned was, not surprisingly, quite agreeable with Biblical theology. He also was a student of Arthur Eddington and advocate of Einstein, and so found much approval for his metaphysics within the astrophysical community. This stood in contrast to Miller, whose experimental findings continued to be ignored and erased by the same scientists rushing to embrace the Einstein/big bang metaphysics.

The *Doppler effect* is real, and was first documented as an auditory phenomenon. To a stationary observer, a moving car blowing its horn will sound at a higher frequency pitch when approaching, followed by a lower frequency after the car has passed by, when moving away. Applied to astronomy, Doppler shifting is used to explain galactic redshifts, as an effect of recessional velocity, and hence distance. And since the light from galaxies appears dominantly redshifted, towards the lower, longer wavelengths, galactic recession and hence the big-bang theory of creation was "logical", at least to those who deliberately ignored other possibilities to explain galactic redshifts.

Redshifts are today conventionally considered to be reliable astro-nomical distance indicators. However, the big-bang theory also in-cludes the rather astonishing proposition that the universe has a firm limited outer boundary, much like the geocentric crystal spheres were once considered to signify. Galactic recession also implies that all matter in the universe was previously bound up into one extremely dense object, or tiny point in space, the "singularity" out of which the big-bang explosion event somehow occurred, to scatter all matter and energy outwards, and create all that is known to exist. *These are all metaphysical speculations!*

Regarding an outer boundary to the universe, every year there are new astronomical observations which go against big-bang expecta-tions. Before the big-bang theory gained supremacy, astronomers mostly embraced the theory of an infinite universe with no beginning or end. After the big bang gained in popularity, the universe was declared to have a limited size, based upon what Earth telescopes could see or register on time-exposed photographs. Then came the Hubble telescope, with its ultra deep-field imaging camera, which in 1995 basically doubled the size of the observable universe overnight. In 2004, a tiny patch of sky was sequentially observed by Hubble, showing it to be filled with thousands of new and previously unseen galaxies. In 2012, the same patch of sky was revisited by the *Hubble Extreme Deep Field 09* team, who overlapped ten years of Hubble images of that same region, resolving even more galaxies, and thereby expanding the size of the universe even more. Figure 99 shows a negative image from that Hubble *extreme* deep-field image, from a small patch of night sky in a region about one-tenth the diameter of the Moon as seen from Earth. Around 5,500 galaxies are imaged within this poorly-reproduced negative, each dot representing a far away galaxy. A similar abundance of new galaxies can now be seen in just about every direction one could point. The orbiting James Webb Telescope, planned for launch in 2021, will predictably double the size of our universe once again. *Horreurs!*

Such findings as these have forced the advocates of the big-bang theory to further stretch their already over-stretched theory, now

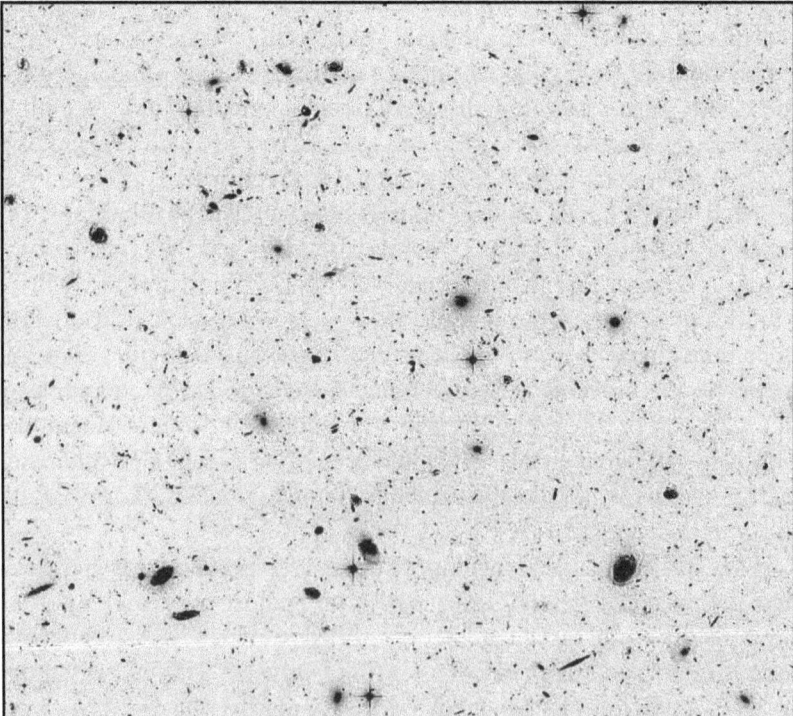

Figure 99. Distant Galaxies from the Hubble Extreme Deep Field Image. Each little point is a galaxy. (negative of image from NASA/ESA/HUDF09)

claiming that there is a difference between the size of the *observable* universe, by compared to the size of the "real" universe, whatever that means. The "real" universe is still claimed by the conventional thinkers to be finite in size, but still expanding outwards away from Earth in all directions. By some calculations, this expansion occurs at velocities greater than the light speed-limit. (*more Horreurs!*) So no matter what new evidence is found, *they've got it covered!* The ideas of an infinite universe filled with a cosmic medium that can explain redshifts by different arguments, is powerfully rejected with a dogmatic insistence that *redshifts **must be** cosmic distance indicators, no matter what!*

Beyond this difficulty, big-bang cosmology is also challenged by observations that known galaxies are not evenly distributed in space, but form a kind of "cosmic web", with an even higher degree of organization than previously considered. This constitutes yet another blow to the idea that cosmic space is an empty void, governed by entropy. Organizational energy is necessary to build such a complex structure, as seen in Figure 100. That image suggests to me a biological

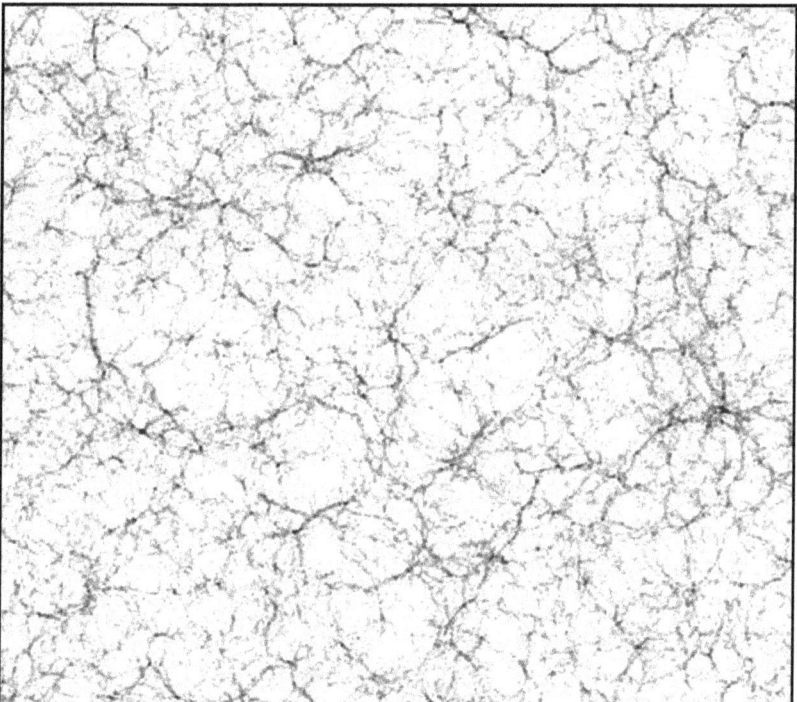

Figure 100. The Large Scale Dendritic Structure of Galaxy Distribution in the Universe. (after N. Hamaus, WebRef.18)

fabric, such as brain neurons and dendrites, a very high degree of organization and connectivity between galaxies. *This is further evidence of an ether/life-energetic universe, shaped and structured by its own self-organizing principles.* However, this is forbidden territory. The mantra is "redshifts are distance indicators" and "space is empty". *But, is it so?*

Firstly, the Doppler recession theory places the Earth into a special favored position at the central point from which all galactic matter is claimed to be moving away. This is in agreement with older ideas of geocentrism, the Earth being at the center of the universe. The Catholic Church tortured and burned heretics for defying geocentrism, as history shows, while students or heretic professors who today question big bang or relativity are silenced and centrifuged out of the universities. Such psychological overreaction to dissent, by itself, should cause great skepticism as to the big-bang's validity. At root, such impulses to punish dissent reveal an emotional insecurity for one's own ideas.

Secondly, the big-bang concept that all the matter and energy of the universe, *including space and time itself,* was compacted into a ball of infinitely dense matter of rather modest size, which then was squeezed down even further into literal nonexistence, can only be a mystic's mental imagery, no matter how much maths are applied and numbers crunched. It is theological at root, a new creation myth, as Nobel laureate Penzias stated, and described in the *Introduction*: Before the big bang, "as best as we can determine, space, time, matter and energy did not exist." In other words, *everything in the universe was spontaneously created out of nothing.* This is modern empty-space astrophysics, not some ancient belief of the Mesopotamian astrologers, or the geocentrists.

This fantastic creation story has not been lost on the theologians, who proclaim it to be one and the same as the creation myth presented in their various holy books, though they quibble about the time required for such a process. While many books directly criticize the errors of the big-bang theory on scientific grounds (Lerner 1992, Mitchell 2002), other books attempt to unite Biblical creation myths with modern astrophysics (Schroeder 1991). Articles have also appeared in newspapers, magazines and on internet with the same theocratic message, which few academics dare oppose, especially if big grant money is being put on the table. Certainly religious people are entitled to their opinions, but scientists are supposed to be duty-bound to experimental and empirical facts, no matter whose politics or religion (or atheism) might be offended. Unfortunately, modern big-bang advocates have a different set of heretical taboos, including the cosmic ether of space, non-Doppler explanations for galactic redshifting, and of course, Wilhelm Reich.

In 1982 Pope John-Paul II was invited by scientists to visit the big particle accelerator at CERN, Switzerland, where they chatted up the agreeability of their respective creation myths. The big bang takes us back to the elder Newton's theological preoccupations, which also declared the nonexistence of an unpermitted motional and material cosmic ether, favoring an immobilized static ether, or "absolute space". Einstein proposed a similar metaphysics, two centuries later. And from this, space is empty of any light-affecting medium. Ergo, big bang.

To say that all of known creation came from the explosion of a big bang singularity event, can be compared to a massive explosion and fire in a large book-printing factory, where after the fire is put out, one would observe, in the midst of the ash, smoke and rubble, the entire Science and Nature collections of the Library of Congress having self assembled in the midst of the rubble, where previously there was only stacks of paper, twine, glue and ink. Nobody has ever seen even one bit of graffiti text scratched on a wall after an industrial explosion and fire, much less a paragraph. How could all the planets, stars, solar systems, galaxies, etc., with their stone and mineral compositions, the oceans, atmospheres and even life in all its myriad varieties, arise from such a chaotic big-bang explosion, without reference to a cosmic self-organizing principle which would stand against mechanical entropy?

If the cosmos contains a self-organizing principle that sounds like a creative universal God-force, a Universal One that acts as gravitational prime mover and life-energy, providing our emotional and sensory bond with the universe, it surely isn't what Big Empty-Space Dead-Universe Science, or the various "holy men" of Big Religion (or the atheistic commissars) have in mind. However, it might be something similar to what the humble ether-drift scientists discovered, as merged with Reich's discovery of an ether-like cosmic ocean of creative, pulsating life-energy. (Reich 1934, 1938, 1949) Oliver Lodge (1925) also carried early ether theory into the realms of "life and mind", including biology and psychology, but only in a philosophical and not in an experimental context as Reich developed decades later. Unfortunately, with the doors slammed shut to such moderates, the extremists have dominated the debate.

The Dynamic Ether of Cosmic Space

Little of modern theory can coexist with a tangible, material and motional cosmological ether, much less one with water-reactive and life-energetic properties. The modern theories were borne from suppression of the evidence for cosmic ether and life-energy, utilizing all the tools of erasure and censorship, to include 20th Century book-burning. Cosmic space was declared empty and the universe dead. The big bang theory became the New Truth. Redshifts were pronounced, *ex-cathedra*, to be cosmological distance indicators, ignoring their obviously mystical and theological origins from a devout Catholic priest. However, *space is not "empty"*, but filled with a light-affecting cosmic medium, the ether. And it has life-energetic properties. Below are additional discussions on the redshifts, also from critics of the big bang. Other cosmological curiosities and outstanding questions are also addressed, pulling us back into the real world.

Despite being metaphysical in construct, big-bang creationism nevertheless provides *no creative principle or mechanism by which energy and matter can aggregate to form new and complex forms*, to oppose entropy and create and sustain the universe. Nor does it propose any solutions as to how life arose on our planet, something which was identified in Reich's microbiological work (1938), in his discovery of the *self-organizing bions*. Big-bang theory was founded upon empty-space concepts wholly governed by entropy, the dissolution and randomizing of matter which drives complex structures towards disintegration into lesser complexity. How, then, could the universe emerge from a cosmic explosion without the agency of a negatively entropic, self-organizing energetic force? The missing ingredient is fulfilled by a motional dynamic cosmic ether, with additional properties in accordance with the life-energetic sense of Reich's line of discovery. However, *such a creative force in nature, by its very existence, negates the big-bang theory, and takes us back to the older understanding of a timeless, infinite universe. Biblical creationism also gains no support from the cosmic ether and life-energy as described in this work, though the search for a creative force in nature surely does.*

Galactic Redshifts from Ether-Friction Light-Wave Attenuation

There are several promising ideas to explain cosmological redshifts and avoid the galactic-recession and big-bang consequences. There always is a loss of wave intensity of light (or sound waves, or water waves) by mere distance travelled, according to the inverse

320

square law. Double the distance (2x) and the light wave or sound wave intensity drops by a factor of one-fourth. Double that distance again (4x), and the light intensity drops by a factor of 16. The concept of "tired light" was proposed, that light from the most distant stars and galaxies would eventually spread so thinly as to be diminished down to near-zero intensity. Beyond this factor of intensity reduction, there is the issue of *attenuation, the resistance or absorption of a wave by the medium through which it is travelling,* to shift higher-frequency waves into lower-frequencies. This understanding offers easier explanations for both galactic redshifts and the "boundary of the universe" claims.

In water, waves lengthen as they move away from the point of excitation where the wave is formed. Throw a rock in a pond, and one immediately sees a high frequency wave form at the point of impact, with a gradual spreading outwards of the wave, whereupon it loses intensity and broadens the wavelength, eventually to nearly vanish. Sound waves are similar. High frequency sounds do not travel as fast or far as low frequencies, and even low frequencies do not continue forever. Regarding light, the cosmic ether would absorb a small portion of energy from the light waves, and shift their frequency and absorption lines slightly off to the longer or red wavelengths. None of this negates the existence of Doppler effects on light frequency, or that other mechanisms could be at work. However, it deprives the Doppler effect of its *exclusive* powers to define *all* redshifts as distance indicators.

Halton Arp's Intrinsic Redshifts and Luminous Galactic Bridges

Another blow was given to the big-bang theory by the astronomer Halton Arp. Once considered as the "Dean of American Astronomy" in part for his role in creating the large Palomar Observatory and its 200-inch telescope near San Diego, California. However, starting in the 1970s, Arp began making photos of adjacent astronomical objects which yielded discordant redshifts. He photographed pairs of galaxies which were connected by bridges of luminating substance, but which possessed vastly different redshifts. By conventional redshift determinations, one galaxy was relatively close, and the other very far away, the objects only appearing to be in close proximity due to their coincidental positions along the same line-of-sight. However, the visible luminous bridges connecting the objects were not easily dismissed, especially when Arp began making numerous astrophotos of them, and published his findings. Arp also noted how Seyfert galaxies, known as a highly energetic type of spiral galaxy with a bright central core, were frequently photographed with quasars or quasar-like objects just above or below the plane of their central axes. Luminous bridges connected the Seyferts to the quasars, indicating as Arp argued, that the quasars had been ejected from the Seyferts. The problem is that quasars have extremely high redshifts and are by convention believed to exist very far off, near the edge of the known universe. By contrast, the Seyfert galaxies have relatively modest redshifts, indicating they are much closer.

As Arp's findings were published and became better known, especially from his 1987 book, *Quasars, Redshifts and Controversies*, it created a furious storm in the astrophysics community. He suffered academic punishments, censorship and loss of access to the large telescopes, being forbidden to make any more disturbing photographs. He was told, literally, to "get with the big bang program" or suffer the consequences. Facing repression and future unemployment, he moved to Germany and took a job at the Max Planck Institute, where he continued his work at a modest level. His later book, *Catalogue of Discordant Redshift Associations* documented the findings, while his book *Seeing Red: Redshifts, Cosmology and Academic Science*, exposed the unethical abuses and roadblocks which attended his work.

While Arp appeared ambiguous about the cosmic ether, his findings agree with it nonetheless, as ether provides a more common sense mechanism for redshifting, and allows for luminous bridges of energy connecting discordant redshift objects. Arp's theory of young, hot astronomical objects as having intrinsically high redshifts, such as ejected quasars, offered a more functional understanding of redshifting by real-world gravitationally-related mechanisms, quite different from Einstein's relativistic theory. A variable density ether might also be at work, building up to higher levels of concentration, intensity and matter-creation within the cores of Seyfert galaxies, increasing their brightness and gravitation along the plane of their rotating disks. This suggests a build-up of internal tension with subsequent "birthing" of quasar and/or globular star cluster "eggs" which are then ejected into open space above and below rotating galactic disks. This proposal comes from Reich's "life-formula" of tension-charge-discharge-relaxation, which governs cell-division in biology, but is also seen at work in thunderstorms, earthquakes, and perhaps also, over very long periods of time, with the accumulated energetic tension inside galaxies, which then creates and slowly ejects new stellar conglomerations.

The 3-Degree Cosmic Background Radiation: Ether Friction?

Detection of a super cold but nevertheless very slight thermal energy in cosmic "empty space" was puzzling for an astronomy which had prematurely discarded the motional and substantive ether. This slight temperature of –270° Celcius (or 3° Kelvin) was termed the Cosmic Microwave Background Radiation, or CMBR. In truth it isn't exactly 3° Kelvin, but measures in at 2.72548 ± 0.00057°K. Discovered firstly by Earth-based radio telescopes, it was eventually mapped to reveal a predominant thermal pattern most intensive along the plane of the Milky Way Galaxy. By subtracting the assumed contribution of the Milky Way, a more smooth background of thermal energy was mapped, which nevertheless contained a residual, small amount of unequal distribution, termed the *CMBR anisotropy.*

Big-bang theory offered an explanation for this anisotropy, or variation, in keeping with its theology, identifying this small temperature as the "residual smoke and heat left over from the big bang"– even though "empty space" has few molecules, and no "smoke" per se. Cosmic ether offers another and even simpler explanation, without resorting to metaphysical speculations on "instant creation" of everything from nothing. As light and other electromagnetic waves move through the cosmic ether of space, a slight friction is generated, just as when sound waves or water waves move through their respective mediums of air and water. The theory of the CMBR as due to light-wave motions through the ether was taboo, however, and hence ignored.

The big bang theory predicted a more isotropic, or equally distributed amount of thermal energy, so the anisotropy at first created a puzzlement. This puzzle was eventually "resolved" by assuming the CMBR indicated a real motion of our solar system and the entire galaxy through space. This motion, it was claimed, was along an axis which was aligned between the "hot spot" of constellation Leo, as compared to the "cold spot" near to Aquarius, on the other side of the universe. This Leo-Aquarius CMBR axis is substantially off from the identified vectors for ether drift. Leo is located at RA ~11 hrs sidereal, and ~0° Declination. If the CMBR anisotropy were more substantial, such as one degree (or a tenth of a degree), the postulate of the CMBR anisotropy reflecting a motion through the universe might have some credibility. However, the thermal variation mis-attributed to cosmic motion is tiny, only ± 0.00057°K, suggesting the CMBR anisotropy is an *intrinsic thermal variation, without motion, marking actual slightly warmer and cooler regions of the cosmos.* It is yet another example of astrophysics giving *massive* importance to a very tiny quantity, upon which an entire upside-down pyramid of gargantuan complicated theory is teetering. Meanwhile, more substantial quantities – as with the repeatedly detected cosmic ether drift at ~10 km/sec along a very different cosmic axis of motion – are ignored.

Black Holes or Ether-Gravitational Vortices?

According to current astrophysical theory, when a particularly large star reaches the end of its life, its mass is too great to explode into a supernova. Instead, in a kind of "reversed big bang", it supposedly collapses into an ever-tighter ball, until all substance compresses into a gigantic "gravitational well" (in accordance with Einstein's relativity), from which not even light can escape. Ergo, a "black hole". By conventional theory, they are intrinsically invisible, only being apparent through the behavior of visible matter in their surroundings. As pure postulates, the mythical "black holes" nevertheless are conveniently declared into existence wherever gravitational anomalies are observed that cannot otherwise be explained within an empty-space universe.

In reality, "black holes" are yet another bit of modern astrophysical metaphysics, a product of empty-space theory. One reads about "black holes" powering nearly every kind of astronomical process. Thousands or millions of them, at variable intensities, are supposed to exist in the Milky Way alone, with innumerable more throughout the universe. But the truth is, *not one has ever been proven to exist.* They remain fully hypothetical, like unicorns, in spite of all the big claims, including the 2019 claim of "the first photo of a black hole", to be discussed shortly.

Large black holes are supposed to inhabit the central cores of spiral galaxies, thus providing the internal gravitational attraction necessary to create the spiraling of matter towards the center. They are also said to "fuel" other very high energy astrophysical phenomena, with the claim that matter is being "torn apart" as it falls into the black hole, to create massive bursts of energy along the "event horizon" in the process. However, no *direct evidence* for black holes or "event horizons" has ever been provided. They remain pure speculations.

A black hole is supposed to exist in the core of our Milky Way Galaxy, at a point called Sagittarius-A. There, a cluster of stars has been observed to be swirling around in vortex motions at relatively high velocities. This star cluster has been photographed over a period of 20 years, with a graphic representation of those motions given in Figure 101. The swirling of stars is real, as shown in astro-images arranged sequentially, revealing a clear circulating vortex motion. One of the swirling stars, identified as "S2", has made a full orbit around this core region, with a velocity determined to be around 5000 km/sec at its closest approach within that *ether vortex tornado*, as I call it. There are many articles and videos on internet, showing the motions of this interesting group of stars, detected in the infrared frequencies by radio telescope. The actual images show only individual stars moving against a background of open space with a few other stars scattered about in

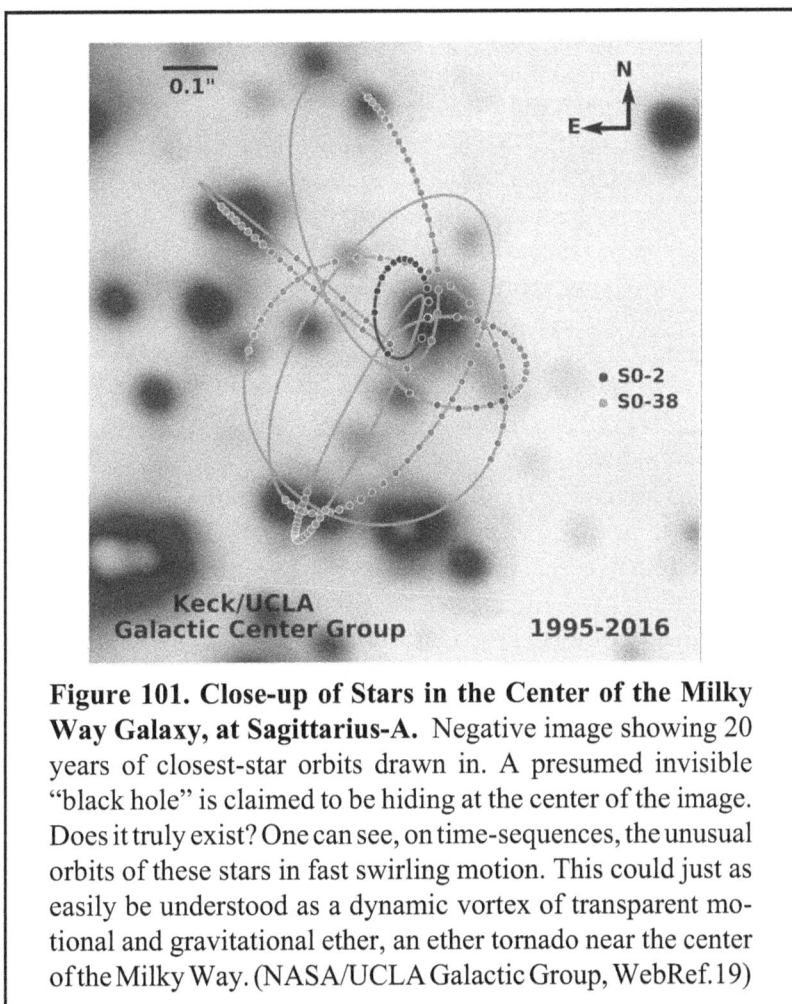

Figure 101. Close-up of Stars in the Center of the Milky Way Galaxy, at Sagittarius-A. Negative image showing 20 years of closest-star orbits drawn in. A presumed invisible "black hole" is claimed to be hiding at the center of the image. Does it truly exist? One can see, on time-sequences, the unusual orbits of these stars in fast swirling motion. This could just as easily be understood as a dynamic vortex of transparent motional and gravitational ether, an ether tornado near the center of the Milky Way. (NASA/UCLA Galactic Group, WebRef.19)

slower motions. However, there is no "black hole" apparent in these photo sequences.

The conventional argument on the unproven nature of the black hole, is that no star has yet fallen into or has been pulled close enough to an event horizon where it could be swallowed up or torn apart by severe gravitational stress. Only then could a strong burst of energy be observed, it is asserted. Star S2 made such a close passage to the hypothetical black hole in late 2018, but without being torn apart as predicted. Insignificant infrared radiation bursts were detected, and while this is not so unusual, black hole advocates claimed it as "proof".

Then, in early 2019, a global media blitz announced that, for the first time, a "black hole" had finally been photographed. (WebRef.20) Or so it was claimed. This new "black hole" was located in the core region of M87, a supergiant elliptical galaxy of nearly spherical shape, in the constellation Virgo. It was imaged by a team using the Event Horizon Telescope (EHT), which is a network of eight different radio telescopes around the world, linked together by computers and satellite communications. Images so obtained from the EHT network are combined and subjected to computer processing to yield an output similar to a single radio dish nearly as large as the diameter of the Earth itself. It is quite a technological accomplishment, but burdened with many unstated assumptions and problematic data processing.

M87 is an unusual galaxy to begin with, considered to be the most massive galaxy in the known universe, having a very high redshift, with over 12,000 separate globular clusters surrounding it (the Milky Way Galaxy has around 175). Viewed optically through telescopes, M87 appears as a massive ball of gas, with two enormous jets of plasma emerging from its core. The jets extend outwards in opposing directions with a spiraling twist, to a great distance beyond the galactic boundary. Hubble images of this plasma jet were studied in the late 1990s, and determined to be *moving away from M87 at from four to six times the "speed of light"*. That itself is a rather obvious violation of Einstein's theory. Einstein supporters of course claim that the observed and empirically-calculated superluminal speed was an "apparent velocity", a "space-time illusion", to preserve their favored theory. More on that in a moment.

Figure 102 presents the M87 "black hole", in black and white, but is otherwise exactly as circulated by the media blitz in April 2019. The original images were scanned over four days in early April of 2017, requiring two years to process and figure out. Upon closer examination, serious problems are apparent in the image made available to the public in 2019. For example, by ordinary logic and non-relativistic reasoning, if a "black hole" is formed from the collapse of a spherical star, pulling in matter, energy and light from all directions, then why does it appear so perfectly flat and face-on to planet Earth, in such a circular manner? Would not an event horizon incident at a spherical black hole create a dispersal pattern of energy more like a skullcap or an umbrella, fanning out around the surface of the sphere, but not creating a "ring" effect, like a donut or an automobile tire (see next page)? And since a black hole is a spherical phenomenon, why don't we see the illuminated lower parts extending up the Earth-facing side to partly illuminate the lower half of its "donut hole"? The nearly perfectly round nature of the dark center of the image and the smooth dispersal pattern over the top half are also suspect. Such perfection and smoothness is rarely seen in nature or astronomy. And what about the sheer coincidence of fleeting probabilities, that M87's supposed "black hole", the first ever imaged, just happens to lie flat-faced towards Earth, shaped like a donut, and is not a spherical object. Further, this image is from radio telescopes, at the 1.3 mm microwave/RF band, in the far infrared. That being the case, *where are the other stars in the dense M87 galactic core region near to this "black hole", which also emit energy in the same general frequencies?*

A recent paper on this specific M87 black hole was published in *Astrophysical Journal* (Bouman, et al 2019 - with around 350 different co-authors). There it was stated: "We constructed images from the calibrated EHT visibilities, which provide results that are independent of models". In the next paragraph, however, they admitted "...imaging algorithms incorporate additional assumptions and constraints that are designed to produce images that are physically plausible (e.g., positive and compact) or conservative (e.g., smooth), while remaining consistent with the data." This suggested, they processed the M87 data in accordance with an expected "plausible" final image. My criticism here is validated in another step used by the EHT team, in that "...four known geometric models (ring, crescent, filled disk, and asymmetric double source)" were fed into their "imaging algorithms", which then produced identical image outcomes. In short, their algorithms (computer programs) massaged an unknown original image to produce the circular ring image seen in Figure 103. Beyond that, a slight "circular blurring" was added, albeit of a very small angular quantity. While that paper appeared to be written by a committee, and lacked clarity on many

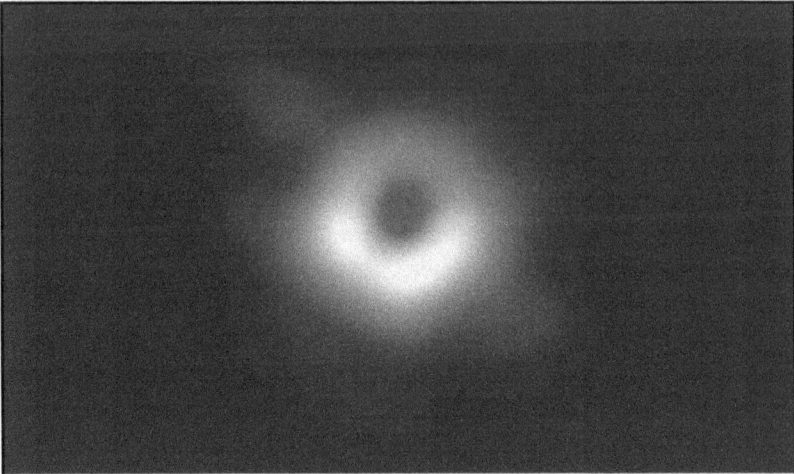

Figure 102. The M87 "Black Hole" Image
Black and white rendition of the EHT publicity photo. Where are the outlying stars, which emit similar energy spectra? Why does it appear like a donut or auto tire, and why is it so exactly "face on" to Earth? Why is the center "donut" hole not illuminated over the bottom half, given the brilliant region that dominates the lower half of the ring – that should illuminate the full lower facing side of a spherical black hole. What does the unprocessed original raw image look like?

Figure 103. Is This the Original Unprocessed "Black Hole" Image?
A *ring of stars*? It appeared on the computer screen of a second publicity photo, without comment, at the same time as the April 2019 announcement. (See text and WebRef.21)

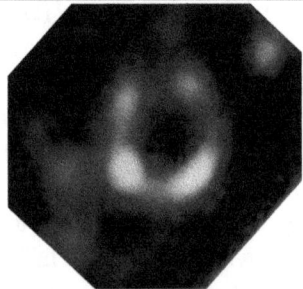

points, it sounded as if the EHT data was processed to shape up "something" out there to make it look like what they expected a "black hole" to look like, beyond reasonable corrections as for atmospheric distortions and coherence between the different radio telescopes.

Another possibility exists, perhaps the most troubling, that the original unprocessed raw image was merely a *ring of stars,* from which their computer algorithms produced such a nice-neat outcome. For example, such a ring of stars was posted to the Facebook page of a leading member of the EHT team who personally oversaw development of the complex computer algorithm, at the time of the April 2019 press announcement. (WebRef.21) The Facebook photo included a laptop computer upon which a nice ring of stars was displayed, in the same orange color with brighter stars at the bottom, uncannily similar to the "final version" M87 black hole image. Was that the original unprocessed M87 image? *And if so, is the M87 black hole image merely a "ring of stars" highly processed by computer "enhancements" into what they are calling a "black hole"?* I captured the image from the computer screen photo seen in WebRef.21, flipped it 90° horizontal and rotated it a bit, as seen in Figure 103. I also experimented with that ring-of-stars image, using the primitive adjustment features on my ancient PhotoDeluxe program, and came up with renditions that looked even closer to the "black hole" image.

For the record, twice I emailed the EHT team's press contact asking about the original unprocessed images, expressing my concerns, but never got a reply, only silent treatment. Hence, I put the information out to the public. Maybe an answer will be forthcoming? Or maybe, we will see dozens of more "black hole" photos that look very similar, leaving the issue of computer enhancements of "rings of stars" unanswered?

A related issue is, how to define the "centers" of far away galaxies, by which to start a search for such "black holes" (or rings of stars)? M87 occupies an angular diameter of 7 arc-minutes in the sky, within the constellation of Virgo. Earth's Moon is 31 arc-minutes in diameter, so M87 in its full extent is about two-ninths the diameter of the Moon, as seen from Earth. It is too dim to be seen without a powerful telescope. The smaller core of M87 extends over 46 arc-*seconds*, which is about 1/40th the diameter of the Moon.

M87's "black hole", by conventional redshift-distance assumptions, is even smaller, only 1/2500th the diameter of the Moon. Details of its core features required use of the huge EHT radio telescope array, which produces images quite different and more fuzzy than a conventional optical telescope. At that distance, four of our solar systems are said to fit inside the M87 black hole center, indicating it is far more gigantic than the claimed black hole at Saggitarius-A in the center of the Milky Way Galaxy. However, *if redshift conventions are wrong, M87 could be much closer, and hence smaller in overall size, allowing blurry rings of stars to be misidentified as gigantic black holes, especially if the stars had luminous bridges connecting them, as is the case in Figure 103*. For full clarifications, we must wait for publication of the unprocessed M87 images, as it appeared before being sent through their data massaging "corrections". Meanwhile, this M87 "black hole photo" seems very likely to be a misinterpreted and more ordinary *ring of stars*.

As noted earlier, my postulate of *ether tornados*, or extreme gravitational ether vortices, remains as a viable explanation for "black hole" galactic and gravitational anomalies. Such vortices could animate stellar and planetary matter into rotational spiral forms or whirlpool motions. Nature is filled with whirlpools and vortexing matter, which can be seen with our own eyes. There are water whirlpools, large gyres in the oceans, and vortexing spirals in atmospheric cyclonic storms and hurricanes as seen by satellites from space. Then we have the swirling solar system, spiral galaxies, and even spiral clusters of galaxies. The swirl of stars at Sagittarius-A might well be a kind of ether tornado of exceptionally high velocity, only giving off the occasional burst of light when it would shear off a bit of stellar surface material.

Consider a huge star, racing along at 5000 km/sec and making a sharp perihelion turn as it swirls in a tight vortex near to a galactic core. It might lose some part of its mass when making such a sharp turn, merely from centrifugal forces. As to the mystery cause of these stellar motions, by returning to a motional and substantive gravitational ether, all such features currently attributed to "black holes" can more easily be understood. Neither Einstein's "space-time warps" nor invisible black holes with "gravity wells" are necessary.

If one can muse about black holes "tearing stars apart" by gravitational shear forces, one can just as easily muse about a fast vortex of cosmic ether energy swirling around like a giant cosmic tornado, producing similarly strong gravitational shear forces. There might be dozens or hundreds of ether-vortex "tornados" swirling around the regions of galactic cores, creating star circles or clusters of orbiting stars, the ether being basically transparent, as in Sagittarius-A.

One last point on "black holes" and Einstein's space-time warps. Today the science journalists are prone to parroting things overheard at conferences from "scientists gone wild", speculating about "white holes" and "time reversals" in their Magical Mystery Tour of the universe. "Time is an illusion" says more than one scientific sage, as if they had inhaled volcanic vapors, like the Oracles of Delphi. Is everyone supposed to uncritically and quietly swallow such unreal ad-hoc speculations? In most universities, unfortunately, this is demanded.

Super-Luminal Objects, Moving Faster Than Light

Many objects in the universe, observed by deep-space telescopes, reveal *superluminal* light-speed velocities greater than the "official" speed limit of ~300,000 km/sec. By Einstein's relativity theory, no object can go faster than the declared "Speed of Light". Only a few years ago, one could speak about superluminal objects as a theoretical challenge to Einstein. Today, however, his theory is used to smack down such forbidden velocities, and few dare to question it. Whenever a superluminal or other theoretically "impossible" moving object is found (impossible only by empty-space thinking), an invisible "black hole" is usually claimed to be lurking nearby, which then *corrects everyone's maths and theoretically slows down time and velocities* at the location of the superluminal object, even if standard methods of measurement show it to be breaking the purported "cosmic speed limit". Keep in mind, we are speaking about material objects moving faster than light, not necessarily the light from the speeding objects (although that can also happen).

Galaxy M87, previously discussed regarding the hypothetical "black hole" in its center, is a good example of this. A series of Hubble telescope photographs were made of a small segment of the M87 jet over the years 1994 through 1998, by astronomer John Biretta of the Space Telescope Science Institute. His series of yearly images documented the M87 jet clouds moving at 5.5 to 6.1 times light speed, as shown in Figure 104. However, the Einstein theory does not permit such superluminal velocities, no matter how exacting the evidence.

The press release for his study sounded like it was written by two different people, the first half by the empirical astronomer Biretta, which began with the bold and true statement that "Astronomers reported today discovering clouds which appear to move many times faster than the speed of light." (WebRef.24) After presenting the basics of the discovery, the entire second half of the press release had a completely different tone, trying to reconcile those documented superluminal observations with the Einstein theory. That section argued the superluminal motion was "an illusion", proclaiming only a sub-light speed of the M87 jet. Nevertheless, the actual measured light speed, determined by standard geometry from the Hubble images, showed the M87 jet had moved at 5.5 to 6 times the velocity of light. Whatever else, those are facts.

The theory of relativity rejects such superluminal velocities, claiming they are impossible, even if determined by objective photographic evidence and straightforward maths and logic. Such velocities are simply "forbidden". End of discussion. We observe and measure a moving object, and by its estimated distance from us, distance travelled, and elapsed time, we can calculate its velocity. Such straightforward simplicity is today no longer allowed.

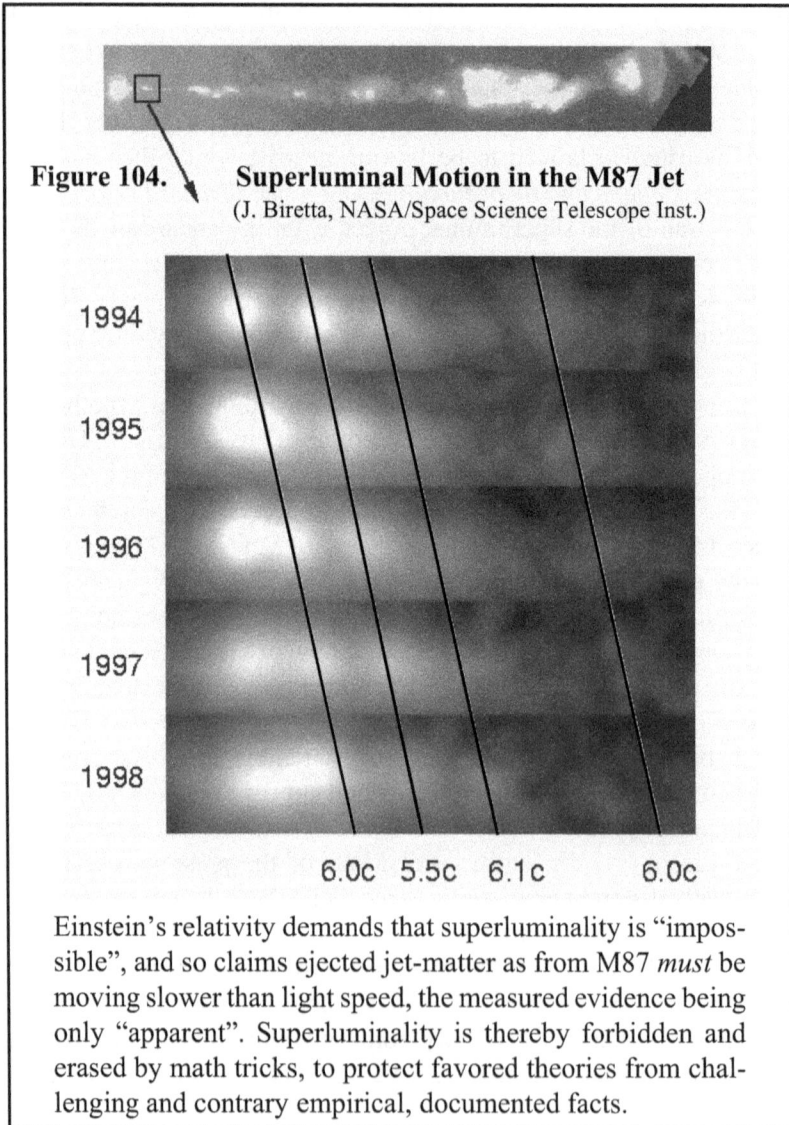

Figure 104. **Superluminal Motion in the M87 Jet**
(J. Biretta, NASA/Space Science Telescope Inst.)

1994

1995

1996

1997

1998

6.0c 5.5c 6.1c   6.0c

Einstein's relativity demands that superluminality is "impossible", and so claims ejected jet-matter as from M87 *must* be moving slower than light speed, the measured evidence being only "apparent". Superluminality is thereby forbidden and erased by math tricks, to protect favored theories from challenging and contrary empirical, documented facts.

By math tricks, conventional astrophysics can, like Harry Potter with his magic wand, slow down all observed and documented super-luminal speeds into something merely "apparent". Ordinary logic, empirical observational astronomy and photographic proof must be subordinated to inferior status, with a sub-light conclusion.

Unlike the Copernican Heliocentric model, which did away with the complexity of geocentric epicycles to explain planetary motions more simply, allowing for further human advancement, the Einsteinian astrophysics adds additional layers of metaphysical complexity. Documented empirical observations and facts are claimed to be "illusions", while true illusions propped up by fuzzy maths are proclaimed to be "reality". It is not too different from how Galileo's telescope observations were declared to be impossible heresy, even while his tormentors laughingly boasted how they refused to look through it! Reich said it best: "Perfectly exact physics isn't so very exact, just as holy men are not so very holy". (Reich 1953, p.2) (In reply, the astrophysicist says: "Who you gonna believe? My funny maths, or your lyin' eyes?")

Superluminal M87 jets are not alone in this battle between observation and theory. Most superluminal objects are high-speed jets of glowing matter, as found in quasars, or as ejecta from exploding supernovae. They include quasar 3C273, whose ejecta races at an estimated 9.6 times the speed of light, and quasar 3C279, whose jet moves at twice light speed. Cohen (1979, WebRef.25) surveyed a series of 33 different superluminal objects, with velocities from 2 to 12 times light speed. Abundant examples are found, notably for objects with high redshifts, a point I will return to momentarily. But note the date of Cohen's paper, of 1979. That was before black holes had gone from a speculative few dozen to "absolutely must be" *thousands or millions*, as it was necessary to seed the empty-space universe with invisible gravitational forces of time-space warping capabilities. Today, with the observed velocities of superluminal objects being subjected to "relativistic slowing of time", their velocities are systematically *revised downwards*. Consequently one often finds the term "superluminal" prefaced by "supposed", "conjectured" or "apparent", so as not to cause headaches for the Einstein followers.

However, there are rational arguments by which some of these very distant quasar jets might be slowed in velocity, if they were indeed closer to us, and not so far away. The distances of these quasar and quasar-like objects is calculated by standard Hubble redshift-distance theory. While I have no interest to rescue Einstein's relativity theory in this regard, if redshifts are not truly distance indicators, then the observed objects may actually be closer to the observer than the redshifts would demand. And if closer, then the superluminal characteristics become significantly slower, possibly below the "official" cosmic speed limit. This would probably not erase all superluminal objects down to sub-light speeds, but it might cut their velocities in half, or more.

I've already discussed the work of Halton Arp on this matter. The repression he experienced at the hands of the astrophysicists was substantial, so it seems unlikely the relativists would turn around and attack the Hubble Constant (of redshift-distance indications) to salvage their own favored theory. As noted, a cosmic ether is incompatible with both the Einstein relativity and redshift/big-bang theories, allowing us to formulate a more steady-state and rational universe, something infinite in time and space, without beginning or end, and absent the other-worldly fantasies of empty-space astrophysics.

A few additional points on superluminality:

Cosmic ether could be at work in these cases, to supply the needed energy to cause an "overcharge" of sorts, whereby the jets vent their energy and matter in a discharge which might last over millions of years. The cosmic ether surrounding the galaxy or quasar would be providing a substantial influx of energy along the plane of its vortex, which is then discharged along the pathway of least resistance, *along its axis of rotation, and perpendicular to its inward-moving vortex.* An energetic ether moving from open space towards such a galaxy would provide it with the necessary "fuel" and make unnecessary the postulate of "black holes", either at the core of M87 or within the Milky Way Galaxy at Sagittarius-A. Likewise, no "millions of invisible black holes" needs be postulated to seed the universe with equally invisible "space-time gravity wells", to overcome the contradictions of the "missing gravitational mass", which in fact is the ignored and taboo ubiquitous and gravitational dynamic ether.

Too much intellectual and emotional energy, money and reputations have been invested in Einstein's relativity and the black-hole theory to allow any serious challenge to even one of their unproven foundational assumptions. We must not forget, all this talk about Einstein's relativity got its start from the falsification of science history more than 100 years ago, with the erasure of the positive results of the original ether-drift experiments. With empty space, lacking a cosmic medium, everyone was led into the cul-de-sac of the Einstein cult, the big-bang theory, the imaginary black holes, dark matter, wimps, machos, and quantum magic. Additional layers of complex fantasy-illusions have continually been piled on. By comparison to 100 years ago, the universe today is described as a complicated and other-worldly mess, a mystical conglomeration filled with hypothetical objects and other-dimensional unrealities, governed by invisible, unproven and *unprovable* forces that lay beyond the realm of direct perception or instrumental measurement – *beyond the realm of reality!* How different is this from the various other-worldly heavens and hells of formal theology? Modern astrophysical theory largely remains opaque to ordinary conception and approach, as well as to authentic criticism.

Meanwhile, the public hysteria and media hoopla attending each new mystic claim of modern astrophysics grows in proportion to the unreality of those claims. Black holes remain unproven, in spite of one seriously questionable photograph. Curved space-time cannot be directly observed or perceived, no more than various "multiple universes", "cosmic strings" and so forth. And yet, today such fantasy talk dominates scientific discussions and publications, and mainstream media presentations, no less than did geocentrism and epicycles dominate "leading expert" conversations in the times of Copernicus and Galileo. True progress in space science by the engineers and space program, meanwhile, plods along as always, using old-fashioned wave theory and conventional Galilean-Newtonian physics to achieve their magnificent objectives.

When will modern scientists take off the blinders, recall their duty to facts and truth, and defend the *freedom of inquiry and freedom of speech* of their fellows, if only to preserve it for themselves? When will they throw the money-changers from their sadly crumbling temples, which today often seem more like political indoctrination centers, complete with student rallies preaching *against* freedom of speech and inquiry. If the situation does not change, soon enough the radicalized faculty and students will be burning books in front of university libraries. Neither institutionalized theology nor scientism offer an escape from the current situation. A reformed and genuine empirically-driven science might do so, if we are fortunate.

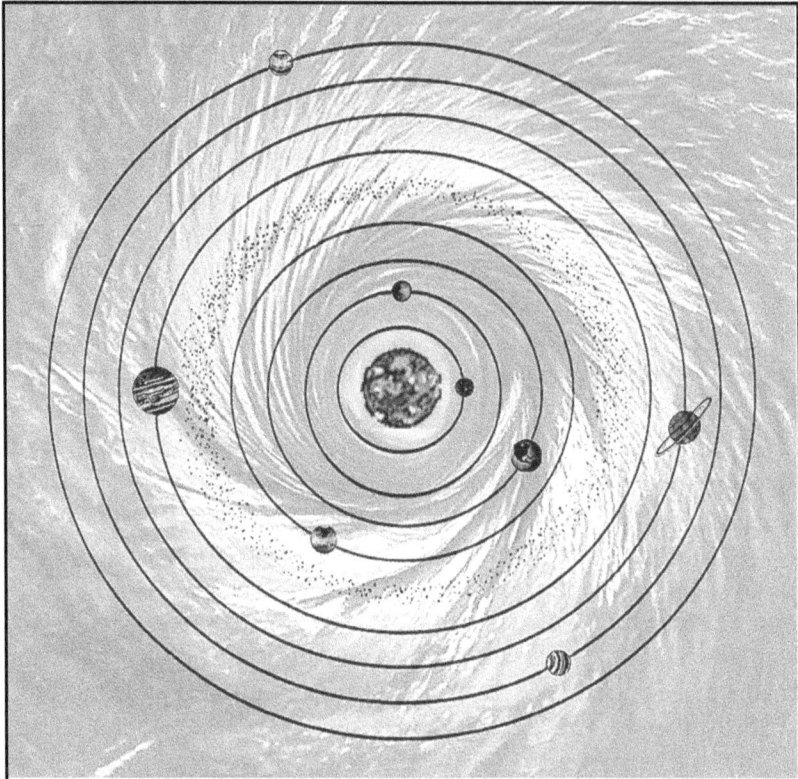

**Figure 105. Curved Cosmic Ether/Life-Energy
Superimposition Produces Gravitation (not to scale)**

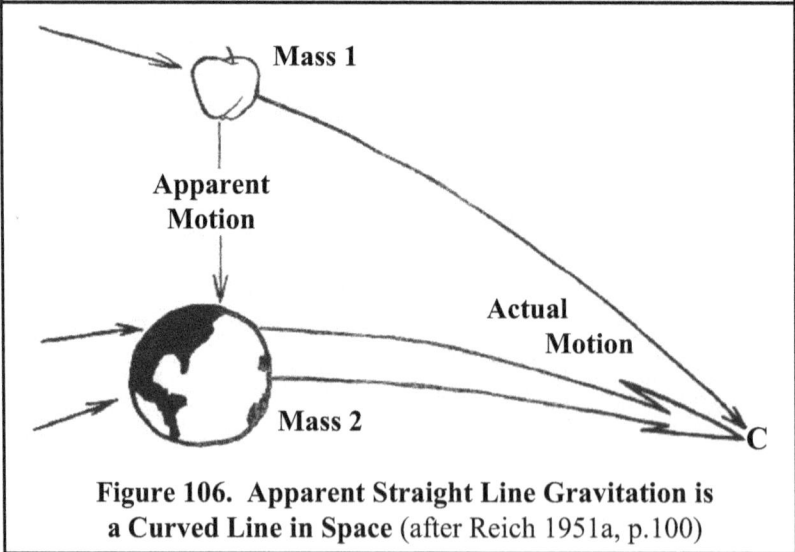

**Figure 106. Apparent Straight Line Gravitation is
a Curved Line in Space** (after Reich 1951a, p.100)

Conclusion

Figure 105 on the facing page was first presented in the *Introduction*. The lower part is Figure 106, also reproduced for emphasis from the chapter on *Ether and Cosmic Life Energy*. Together with all the other figures in Part III, they present a concluding, though generalized and non-mathematical, *ether/life-energetic* understanding of cosmic forces ruling celestial motions, gravitation and other aspects of matter and life.

Figure 106 depicts Mass 1 moving towards Mass 2 in a straight line, but only if one is standing upon the rotating Earth. It is only an *apparent* straight-line motion. Standing as an observer out in space, what we call "gravitation" is seen as a curve of motion, with both objects captured in a sweep of merging, negatively entropic and self-attracting cosmic energy, carrying matter with it. Ether/life-energy, orgone energy as Reich described it, superimposes in a curve of energetic attraction which brings the two objects together. This is standard old-fashioned *Galilean relativity,* which often gets lost in the modern discussions about Einstein's relativity and imaginary "space-time gravity wells".

The old master Galileo had, in the 1600s, already proven basic properties of gravitation in his famous experiments at the Tower of Pisa, where balls of unequal weight were dropped from a height of around 55 meters, arriving at the ground at the same time. This refuted the older view of Aristotle that different weight objects fell at different velocities. Galileo also wrote several logical premises, today called "thought experiments". He imagined (and possibly confirmed by experiment) a man on horseback riding in a straight line, who holds a ball off to the side of his direction of motion, and then drops the ball which falls downwards. From the perspective of the horseback rider, the ball moves downwards in a straight line, with a forward motion the same as the horse, until landing directly next to where the horse is galloping. From the perspective of someone standing on the ground, watching the horseman ride by and drop the ball, however, the ball falls downwards in a long curve, not a straight line. This observation led Galileo to go beyond his initial weight-drop experiments at Pisa, to develop his own *Galilean general relativity*, centuries before Einstein.

The Dynamic Ether of Cosmic Space

Galileo formulated similar logical and accurate outcomes in his famous thought experiment of "Salvaitius' Ship". Salvaitius was the wise narrator in Galileo's *Dialogues Concerning the Two Chief World Systems*, in reference to heliocentrism versus geocentrism. He wrote:

> "Shut yourself up with some friend in the main cabin below decks on some large ship, and have with you there some flies, butterflies, and other small flying animals. Have a large bowl of water with some fish in it; hang up a bottle that empties drop by drop into a wide vessel beneath it. With the ship standing still, observe carefully how the little animals fly with equal speed to all sides of the cabin... When you have observed all these things carefully... have the ship proceed with any speed you like, so long as the motion is uniform and not fluctuating this way and that. You will discover not the least change in all the effects named, nor could you tell from any of them whether the ship was moving or standing still." (Galileo 1629)

By such common sense reasonings, Galileo's relativity presumed all events take place within the same time and space, in the real world 3-dimensional frame of reference, as simple differences in location and perspective. Galileo's universe is also compatible with a physical and material cosmic ether in motion. Einstein, by contrast, demanded that space be an empty void, deprived of all properties, and fully separated the two observers into different "frame of reference" unrealities, while wrongly insisting that *light speed must be held as a constant.*

Galileo also reasoned how gravitation appeared differently to one person standing on the deck of a moving ship, and to another standing on the nearby shore, watching that same ship pass by. On the ship, a ball dropped from the upper mast to the deck appears as a straight line motion. Meanwhile, to the person standing on-shore, a *slightly curved pathway* of greater length and velocity is seen. Velocities and lengths are thereby relative to the observer's position and perspective. That is *Galilean relativity*.

A *Reichian relativity* is similar, invoking a superimposing cosmic energy which constitutes the force called "gravity". Both Galileo's and Reich's relativities take place in the here and now 3-D reality, differing only in that Galileo had few clear ideas about the ether's motions or properties, and none about its life-energetic aspects. From this simple but powerful viewpoint, *all gravitational motions can be similarly described and understood.* There is no need to invoke hidden properties of matter, that cause it to be attracted to other matter through a vague "action at a distance" mechanism (which not even Newton agreed with), nor to reference mystic invisible "space time gravity wells".

The Earth is moved through the heavens, along with the Sun and other planets, by a gravitational ether/orgone which is firstly attracted to all forms of matter, suffusing into and binding with it, and secondly is also attracted to itself, in a negatively entropic manner. These motions are contained within superimposing energetic vortices, which have no straight-line or purely "downward" motions. All objects move, apple from tree to ground, the Moon in endless free fall orbit around the Earth, and Earth in endless orbit around the Sun. It is the curved superimposing energetic motions that brings material objects together, although large-mass moons and planets in orbital trajectories of rotation can create an equally powerful outward centrifugal force which balances against the inward superimposing centripetal forces. So neither the Moon crashes into Earth, nor the Earth into the Sun.

Summary of Cosmic Ether/Life-Energy Properties and Behavior

1. Ether/life-energy is a moving ubiquitous ocean, filling all space and encompassing every living and non-living thing on Earth, extending out to the most distant observable galaxies, and beyond. It is the primordial energy of an *infinite universe*, without beginning or end. By its motions and material properties, it floats, pushes and carries matter with it as it moves. It is the prime mover and cause of gravitation, as well as the carrier of light and electromagnetic waves, also able to luminate and create its own radiant energy when sufficiently excited. Cosmic energy in open space moves from a mass-free towards a mass-bound and mass creating condensation. It is attracted to all matter, and to itself. As it condenses, it gains a denser, more viscous property.

2. The absolute and angular velocities of planets around the Sun follow Keplerian maths when viewed in a flat 2-D plane, but Kepler's equations, however useful, are insufficient when orbital motions are considered in their real-world 3-D spiral-form trajectories. One may visualize this most readily by viewing the solar system diagram in Figure 105, moving up and out of the page towards the reader, in a counter-clockwise rotation, along with a side-slipping off to the reader's right side, towards their right ear, at a 60° angle up from the flat page.

3. Each planet and star has a similar moving energy vortex surrounding it, which condenses into a more substantive quality or viscosity, slowing in velocity as it approaches planetary or stellar surfaces. This viscous layer was described by Lorentz as *condensed ether,* by Miller and others as an *entrained ether*, and by Reich as Earth's *orgone energy envelope*. Such a condensed but transparent ether-orgone layer could account for refractive effects such as stellar aberration, as well as for starlight bending within the Sun's *extended corona*, which itself is a condensed layer of the same cosmic substrate.

4. Such a motional, entrained and condensed ether layer, in spiral-vortex rotational motion, is the force which sets the planets into axial rotation, puts moons into orbit around planets, propels planets into spiral-vortex orbits around the Sun, and swirls the stars with their planets towards the galactic center. Objects moving close to the Sun get caught in its faster-rotating extended ether-field, adding to or wholly creating Mercury's perihelion shift. This may also influence the orbital dynamics and velocities of close-passing comets and their tails.

5. The laws of cosmic ether and cosmic orgone energy functions govern all natural spiral and vortex features, such as whirlpools, tornados, hurricanes, stellar-planetary systems and galaxies. Spirals within spirals within spirals, as with fractal patterns which change but nevertheless appear identical no matter at what scale they are viewed. The energetic and creative superimposing energy vortices of the universe repeat at different scales. The vortical motions of the smallest wiggling of microbes, the rotational spiral forms seen in living DNA and in embryology, in snail or seashell formations, are reflected in progressively larger scales of universal functions, such as ocean gyres, tornadic storms, hurricanes, and out to spiraling moons, solar systems and galaxies. *These spiralling motions are apparent only when viewed from the exterior. When viewed from inside the vortexing motions, the vortex itself may not be apparent, and gravitating objects would appear to move in purely "downward", straight lines.*

6. The distances of planets from the Sun are a function of the balance between vortical cosmic energy inflow versus outwards pushing centrifugal forces. The balance point of the two motions is a likely causal factor for Bode's Law of planetary spacings, and similar astronomical patterns of current mystery. The Fibonacci and "golden section" math ratios may be an expression of superimposing spiral ether dynamics, much as fractal patterns are shaped by an underlying mathematics.

7. All the above points flow from a central causal simplicity, of a cosmic ether/life-energy permeating all of existence, in pulsating and superimposing attraction, functioning as the prime mover and primordial self-organizing principle of the natural world, working in opposition to the chaotic, dissipating and mechanical forces of entropy.

8. Beyond the above points, we may also postulate that, as cosmic ether/life-energy flows across and down into the ground, it may be the source of internal heating and transmutation of new mineral matter at subterranean depths. (Kervran 1971) A buildup of internal energetic tension at depth might play a fundamental role in deep-Earth volcanic processes, earthquakes and the motion of continental plates.

The above points are generalized, and don't address the fine details, but lay down a set of basic principles that appear to be universal in nature. Our understandings are thereby simplified and improved by invoking *only one singular mechanism where dozens are currently postulated to explain the same cosmic, atmospheric and biological factors.* New patterns and similarities have also been identified in this work, which moves the dead universe, empty-space celestial mechanics towards a new understanding of cosmic ether and life energy. Both have substantial empirical and experimental proofs, as presented. This is an advancement beyond those theories which reference metaphysical other-worldly mechanisms, such as "space-time gravity wells", "big bangs", invisible "black holes", or magical "quantum entanglements". The ether/life-energy also holds significance for medicine, atmospheric science, chemistry and biology, as already summarized. While the modern physics and astronomy speak about a long sought "unified field theory" within a mechanical "billiard-ball" universe, the research documenting the commonality of a dynamic ether and life-energy offers *a unification of theory with reality across multiple disciplines.* It is a new understanding which literally *breathes life into the cosmos,* raising new sets of important questions previously unformulated.

Towards a Dynamic Life- and Cosmic-Energetic Future

The clash of ideas outlined in this book is basically that of *experimentalists versus theoreticians.* Michelson, Morley, Miller, Sagnac, Galaev and Munera represent the original ether experimenters who created marvelously sensitive apparatus, and used them to examine the nature of light and the universe. Reich, Piccardi and Brown figure harmoniously into this same line of universal cosmic functions, as experimental scientists ruled by the results of their experiments. By contrast, the empty-space advocates who prize themselves for their arm-chair mathematical wizardry represent the opposition theoreticians. They offer little hope for scientific breakthroughs, but they always come running behind each new scientific discovery, *a-posteriori,* claiming their complex other-worldly theories predicted the new findings, or can explain them (or explain them away). The experimental empirical scientist is then viewed as somehow inferior to the theory-math wizards. Biology and other disciplines are simply irrelevant to their line of thinking, as living empiricism cannot be so easily reduced to mathematical formalism. *The Einstein theories, by virtue of their dominance, thereby apply a powerful brake against any new concept or invention which might take humanity out of the empty space, dead universe doldrums.* While Einstein once correctly stated "experimentum summus judex" (experiment is the final judge), he rarely acted with such graciousness towards experimental challenges. His true persona and beliefs were presented most clearly in a 1929 article in the *New York Times*, "Field Theories, Old and New", exposing an alarming bit of hubris, in the claimed superiority of mental gymnastics over empirical experimentally determined facts:

> "The ether was invented, penetrating everything, filling the whole of space, and was admitted as a new kind of matter. Thus it was overlooked that *by this procedure space itself had been brought to life.* It is clear that this had really happened, since the ether was considered to be a sort of matter which could nowhere be removed. It was thus to some degree identical with space itself...
>
> The characteristics which especially distinguish the general theory of relativity and even more the new third stage of the theory, the unitary field theory, from other physical theories, are *the degree of formal speculation, the slender empirical basis, the boldness in theoretical construction, and finally the fundamental reliance on the uniformity of the secrets of natural law and their accessibility to the speculative intellect. It is this feature which appears as a weakness to physicists who incline towards realism or positivism, but is especially attractive, nay, fascinating to the speculative mathematical mind.*" (Einstein 1929. Emphasis added).

The above words come direct from Einstein. He presents a glowing boast of his mental powers and mathematical prowess, with a clear bias *against* empirically determined facts, and *against any concept that would breathe life and motion into cosmic space.* His form of speculative arrogance and pseudo-intellectualism, with its explanatory monopoly over experimental findings, flooded into science of the post-ether period. Einstein's relativity, the big-bang theory, quantum entanglement, and similar other-worldly concepts collectively smothered out the flames of critical scientific curiosity and rational dissent, placing theoretical shackles upon just what was possible, and what was not.

Consider how the negative attitudes that "humans cannot fly" or "going to the Moon is folly" affected generations of young inventors and scientists. And when airplanes were finally successful (invented by two bicycle mechanics!) the mantra of "cannot" shifted to the "sound barrier". Aircraft were predicted to crash into a thick wall of air which could not possibly move out of their path at velocities above the speed of sound. By clever engineering, humans flew, broke the sound barrier, and went to the Moon. But firstly the imaginative genius of inventors and engineers had to believe it was possible, and then devote themselves to all-consuming tasks with real, practical results that frequently were opposed *and feared*, by most of the Academy and the citizenry. Now consider how the Einstein theory, invoking a cosmic speed limit, that "empty space" has no creative energy or power, became immensely popular in a quick and easy manner. It had a stifling effect upon modern thinking. And where defeatist thinking was insufficient to dissuade the inventive genius or dissenter, there always was censorship, public slander, book-burning and prison. All were used by 20th Century science, medicine and media in the post-ether period, to squelch competing ideas. That same ugly toolkit is still put to work, even today.

Aside from the research of Wilhelm Reich in his discovery of a biologically-active cosmic energy, I am reminded of the neglected work of Townsend Brown on *electrogravitics*, which is still a potential method of significant propulsion. He privately embraced the cosmic ether but later adopted the language of the Einstein relativists, hoping for funding. They rejected him anyhow. Also the heretic Immanuel Velikovsky who proposed an *electric universe* theory, was unethically attacked and battered by the astrophysicists, including by the skeptic-clubbers Carl Sagan and Harlow Shapely, who denounced his ideas as worthless. Shapley got Velikovsky's books censored, and later confessed he had *never read those books, and had no intention of doing so!* And he was hardly alone in such shabby conduct.

Tesla also embraced the cosmic ether concept, gave humanity our current system of electrical motors and generators, and spoke of extracting electrical power from the "wheelwork of nature". He made immense fortunes for Westinghouse and J.P. Morgan, after which they abandoned him to poverty in old age. Thomas Edison affirmed an "etheric force" as early as 1875, a guiding concept for virtually all his inventive life, but later in 1923, at 75 years old, he declared against the ether. Marconi and similar inventors of the early 1900s gave us fantastic new inventions, at a time when ether theory was used as a working fundament. Reich felt the rejection of the ether theory to be a "catastrophe" for science, but went on to discover a similar cosmic life-energy. He applied it to heal the sick, make rains in deserts, and even developed a small motor that ran on the cosmic energy. His reward was public slander, death in prison, his books burned. It is still the case that physicians go to prison for using healing methods "not approved" by the same book-burning FDA. The mystery of cosmic energy motors remains with us, but it seems, however they might function, the solution to their riddles is linked to the gravitational and motional cosmic medium, the dynamic ether/life-energy. *Einstein's relativity, the big-bang and quantum theory provide no help to the inventor or engineer, who continues to use older empirical Newtonian and wave theory, and materials science in the design of new electronics and products.*

By contrast, the post-ether world appears rather bleak in the absence of such major breakthroughs as made in the era of cosmic ether: the automobile, telephone, railroads, airplanes, radio, TV, vacuum tubes, electromagnetic discovery, the liquid-fuel rocket, electrical generators and motors, the battery, early computers, etc. They all got going during the era of widespread use of cosmic ether concepts. *Most of modern inventiveness has been limited to improvements upon those older discoveries when cosmic ether was our guiding light.* Medicine has also been a disaster in the post-ether era, when the medical/pharmaceutical cartel gained control over government institutions. Revitalizing and economical natural treatments for degenerative illness such as cancer were developed by poorly-funded medical pioneers of the early 20th Century, such as Reich and Max Gerson, or the herbalist Harry Hoxsey. They were all attacked and shut down by force. Life-Energy Medicine is today confined to small private clinics, or practiced underground, banished from the hospitals where they are needed most.

The average citizen is equally to blame, as they embrace death-medicine just as surely as astrophysic embraces empty space and a dead universe. At his trial, one of Reich's detractors, FDA agent Maguire said it best, *"They talk about pre-atomic, orgone energy! What's that? ...we are getting the H-bomb!"* (Sharaf 1983, p.451)

Where To From Here?

The ether was detected, its general velocities and basic properties identified, with other aspects generally inferred. With the ether described as a material thing, one can now work with it. It is my hope that this book will stimulate new experimental investigations along these lines. Reich points the way as well, his orgone accumulator for example, the high orgone absorption by water, and the differential attractions of metals versus organic or dielectric materials, all provide suggestions on how the cosmic medium could be engineered, and not merely theoretically discussed. *This book is but a starting point, an unfinished but significant footpath into a new continent of discovery, following the early trailblazing of Miller and Reich.*

In some future age, when humanity chooses Life over Death and finally reaches out to the stars, certainly it could only happen with new technical breakthroughs rooted in cosmic energy concepts. Space travel, clean environments and stable economies can only develop meaningfully when energy is abundant and cheap. The next step for humanity is to extract such energy directly from the ether/life-energy medium, from the so-called "vacuum of empty space". New methods will be necessary, to excite and directionally induce propulsive force within matter, to liberate inertia, and thereby to *generate both massive electrical charge, and motional velocities beyond the speed of light, by thousands or millions of times over.* Mars in a few minutes, Alpha Centauri in a few days. The passage of time inside such a superluminal space ship would not be affected in the slightest. Time would pass at the same rate inside the spaceship as at the point of origin, and at the destination. Such a new kind of space engine would possibly have already been developed were it not for the widespread *failure of imagination created by mystic theories of inherent limitations.*

Humanity can do better, if human energy and inventiveness is liberated from the "empty-space, dead universe" world view. Our desires to solve Earthly problems and explore other planets and stars will never happen in any serious manner without new breakthroughs.

Gravitation will eventually be modified, and electric power extracted from the cosmic medium, in ways that "space time gravity wells" could never provide. Einstein's cosmic "speed limit"and the dictatorship of mystical theories in science must be rejected. Imaginative ideas must be given a chance to breathe, and not be smothered in the cradle.

Michael Faraday, who discovered the electromotive force was once dismissively criticized, being asked of what value was a simple wire jumping when exposed to a moving magnet. "Of what value is a newborn baby!" he replied, and how correct he was! Max Planck also summarized the problem succinctly, in his statement that "science progresses, funeral by funeral". From lofty perches, the Royalty of scientific institutions rigidly hangs on to power and blocks progress as a practiced art, no less than the bloated politicians, until finally dying.

Once Einstein is sent out to pasture, new motors extracting energy from the cosmic medium might finally be given support, to develop new power systems for our homes and industry. A spacecraft might then be developed that would move at superluminal velocities, as in the *Star Trek* or *Star Wars* films. This will be technically achieved, once gravitation is understood as a superimposing flow of dielectric and excitable orgone-ether substance. Modern inventors already investigate such ideas, below the radar of the establishment, just as in the years of early airplane and radio development. Empty-space concepts give us nothing to technically work with or develop, literally. They doom humanity to a depressing future of limited horizons.

While I remain optimistic about the future of humankind, I have no illusions about the dangers that lie ahead for our small planet, or that what I write in this book will be taken seriously by any more than a handful of those in the universities. It is therefore dedicated to the *future young scientists and innovators*, who will resolve the major problems facing those new generations, and then spring outwards to colonize Mars and Titan, out to the stars, leapfrogging over a sea of Kuiper planetoids, rich with every element imaginable, including vast quantities of life-sustaining water-ice. I therefore salute the open-minded thinker, inside or outside of existing institutions, those with the important combination of focused imaginations, with studied scientific and technical skills, and persistence. I encourage them to ignore the naysayers, steer around the institutional road-blocks, avoid the madness of crowds, cherish their isolation, to work and study with intensity, courage and optimism, and above all to follow their heart and inner compass.

Appendix 1

A Simple Model for Visualization
of Cosmic Ether Motions

A three-dimensional model and exercise can assist our understanding of how the Earth obtains variations in the velocity and direction of ether drift as described in this book, in a manner that most anyone can follow. Review the methods described below alongside the graphic, to gain a practical understanding of the ether-wind vectors.

First, obtain an Earth globe on a stand, oriented with its 23.5° axial tilt. Then place a "+" mark composed of two thin black tape pieces over Southern California, oriented N-S and W-E. The "+" mark represents the cross-arms of the interferometer at Mount Wilson. The plane of the solar system ecliptic is now defined by the flat table upon which the tilted Earth globe is standing, and also by the flat surfaces of the floor and ceiling of the room.

Second, position the globe such that the Earth's north pole points off to your left side, towards the left-top ceiling. That is where it would point to Polaris. The northern pole of the ecliptic is then almost directly above the globe, overhead on the ceiling. Note this is very different from the north pole of the Earth, which points to the star Polaris, at an angle of 23.5° towards the left-side ceiling or wall, and away from the north ecliptic pole, directly above the Earth globe.

With that orientation, we can now view the left side wall as the sidereal 6th hour. The right side of the table, or right-side wall, is then the sidereal 18th hour. That 18th hour is also where one finds the approximate center of the Milky Way Galaxy.

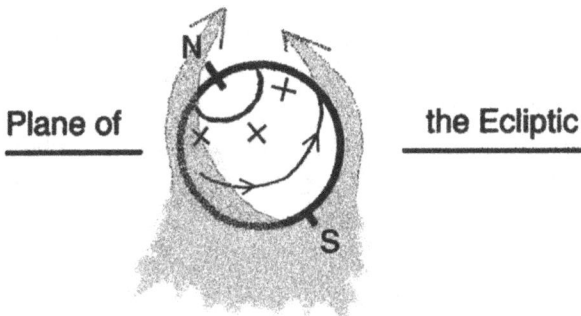

Plane of   the Ecliptic

The Dynamic Ether of Cosmic Space

Now we must imagine the flow of the cosmic ether wind, moving up from the floor, upwards through the table, sweeping in spiral form around the Earth from south to north, with a lesser west to east motion, aiming towards the ceiling. All these variables remain the same irrespective of the Earth's motion around the Sun, or the time of day.

We can now turn the globe on its axis, and see how, when the + mark of tape representing the interferometer is turned over the 24 hours, it is exposed to varying intensities and compass directions of the moving ether wind, just as observed by Miller. When the + mark is placed directly to the left side of the globe, pointing to the left wall of the room (or 6 hrs sidereal), it is exposed to the maximum velocity of ether drift. Then, when rotated by 180° to the 18 hrs sidereal position, the "+" mark is now aiming about halfway between the right wall of the room (galactic center) and the ceiling, or northern pole of the ecliptic. In this position, the Earth itself blocks the flow of ether wind, exposing the interferometer to a reduction in ether-wind velocity. When your globe is rotated over the full 24 hours, as Miller noted, the vectors of ether wind swing back and forth across the SW-NE and SE-NW axes, irrespective of season or time of year. It is a set of motions fixed into the sidereal hour cosmic vectors. As Miller noted: "When the observed azimuth of motion is charted, the resulting curve of directions crosses its own axis twice in each day."

This "globe on a table" is a reasonably precise and simple model to visualize Miller's ether-drift findings, though the important seasonal variations are not revealed using this model. For the location of Mount Wilson in the Northern Hemisphere, the highest velocity of ether drift is at 5 hrs sidereal when the Earth is more directly exposed to a near horizontal flow of ether wind, moving from the south to the north. The slowest ether velocity for Mount Wilson is at 17 hrs sidereal, when the Earth blocks the ether wind for that location. And yet, it is the 5 hr sidereal maximum ether wind which pushes the Earth along its general 17 hr sidereal direction, towards the center of the Milky Way Galaxy.

An interferometer device placed at other locations on the Earth, such as in the southern hemisphere in Argentina, South Africa or Australia, or at the north or south poles, would yield a somewhat different ether wind maximum and minimum velocity, and azimuth. An examination of the accompanying figure, along with the globe on the table, allows one to explore such possibilities.

Appendix 2

Isaac Newton's 1679
Letter to Robert Boyle, on the
Cosmic Ether of Space

Newton the Younger (1689)   Newton the Elder (1712)

Prefacing Comments

Below is a letter on the question of the cosmic ether of space, written by Isaac Newton in 1679 to Robert Boyle, a fellow scientist about 15 years older than Newton at the time, and who is remembered with a fame nearly equal to that of Newton. This letter first came to my attention when it was reprinted in a relatively-unknown journal edited by the heretic-scientist Wilhelm Reich, his *International Journal of Sex-Economy and Orgone Research* (vol.3 1944, p.191-194). The original reference from Reich's journal is found in the 1938 volume *Isaac Newton: 1642-1727*, by J.W.N. Sullivan (Macmillan, NY, p.118-124). However, a longer and more complete version of the letter was thereafter found in an 1846 publication of lengthy title by William Vernon Harcourt, cited (as Newton 1679) in the Reference section of this book, containing pertinent information not previously available.

The letter below is significant firstly because it is not well-known outside of a few historians. Where it is quoted, significant parts as I have now restored, are often left out.

The Dynamic Ether of Cosmic Space

The letter is secondly significant because of its contents. Newton's early embrace of a tangible ether creating a gravitational pressure, and able to penetrate into solids, was heresy not just to the Vatican in his time, but also for the modern departments of physics in nearly every university, where concepts of empty space, devoid of any tangible qualities are embraced.

The letter clearly shows the young Newton, who wrote this in 1679 when he was 37 years old, had a firm belief and working grasp of the ether, as a thing of *substance and ponderability*. For him, it was something that participated in the movement and ordering of the planets and universe, and was a working force in gravitation, chemistry and optics. In this, Newton was continuing the conceptual ideas of Galileo, which had been such an irritant to the Vatican Bishops, who tolerated no possibility of a motional force in nature other than God. The idea that natural ether and God (or Holy Ghost) might be identical descriptions for the "prime mover" was equally intolerable to the Church, as while one could scientifically know and measure the ether, one could not measure or know "the divine". The young Newton was not bothered by such conceptual difficulties, but the older Newton increasingly became preoccupied with theological matters, to the point that nearly all his biographers would agree he had become as much of a theologian as a scientist in his last decades. Even only 20 years after penning this Letter to Boyle, he writes the following in the last query of his *Optics*:

"Now by the help of these principles, all material things seem to have been composed of the hard and solid particles, above-mentioned, variously associated in the first creation by the counsel of an intelligent agent. For it became him who created them to set them in order. And if he did so, it's unphilosophical to seek for any other origin of the world, or to pretend that it might arise out of a chaos by the mere laws of nature; though being once formed, it may continue by those laws for many ages..." (quoted in Sullivan, p.125-126)

During those later periods, Newton would drop ideas such as a ponderable and moving cosmic ether in favor of more abstract concepts, such as the divine "prime mover" or deified "absolute space", which was foundational for most later astrophysical investigations into the nature of the cosmos. The most obvious result of this shift was, that in the original Michelson-Morley experiment for testing of ether-drift, everyone anticipated a very large ether drift effect, based upon the assumption that the Earth was racing at very high speeds through an intangible, substance-less static and immobile cosmic ether. No such intangible static ether has ever been demonstrated, nor could it be. But a material and substantive entrained ether, moving more slowly at lower altitudes and close to the speed of the Earth itself, something similar to that proposed by the young Isaac Newton, was detected repeatedly, as I have already summarized in this book, as well as in a few published papers on the subject.

I also have prepared a separate webpage which offers PDF downloads of most of the historic ether-drift research papers which obtained a positive result, by scientist-authors such as Michelson, Morley, Miller and Sagnac, plus more recent positive replications as by Galaev, Múnera and others. (WebRef.26)

Thirdly, this letter from Newton is significant for its insights into how the ether "adheres" to matter, and may work to bind matter together, to create optical, chemical and gravitational effects. He would later abandon all such ideas, and the world would basically forget about them until the first half of the 20th Century, when scientists such as Dayton Miller, Wilhelm Reich, Giorgio Piccardi and others detected exactly that kind and form of a cosmic energy force in nature, expressing itself in their experimental results.

Newer studies undertaken today, attempting to better understand the nature of the substance and structure of space, can benefit from this older work. There is a prejudice to be overcome in the sciences, that one must not drink too deeply from these old refreshing wells, in spite of their clear and deep waters, as if the plastic-bottled sugary fizz-water of modern vending-machine scientism will make us wiser or healthier. It is not so.

The young Newton was on the right track and had it right, and would have made the same basic mathematical and experimental proofs in optics and gravitation had he stayed with those original thoughts about the ether. Nothing would be different, except for a greater appreciation for the substance of space. The older Newton lost that track, descending into theological labyrinths, even while a few heretic scientists of later centuries found the "red thread of Ariadne", and continued onwards.

James DeMeo, PhD
Ashland, Oregon, October 2009
(WebRef.27)

PS: As a final point, one can reflect upon the difference between the "two Newtons" as revealed in their portraits. The younger Newton as painted by Godfrey Kellner in 1689 on the left side above, even while it is ten years after his "letter to Boyle", shows a man who still carries a vitality and spark of life. The elder Newton in the right side portrait, painted by Sir James Thornhill in 1712, reveals Newton the theologian, preoccupied with the hereafter.

Note: The symbols in Newton's letter appear to mean as follows:

$\mathbb{D}$ = Ag Argentum or Silver, symbolized by the color of the Moon.

$\odot$ = Au Aurum or Gold, symbolized by the color of the Sun.

ISAAC NEWTON to ROBERT BOYLE, 1679

Honoured Sir,

I have so long deferred to send you my thoughts about the physical qualities we spoke of, that did I not esteem myself obliged by promise, I think I should be ashamed to send them at all. The truth is, my notions about things of this kind are so indigested, that I am not well satisfied myself in them; and what I am not satisfied in, I can scarce esteem to fit to be communicated to others; especially in natural philosophy, where there is no end of fancying. But because I am indebted to you, and yesterday met with a friend, Mr. Maulyverer, who told me he was going to London, and intended to give you the trouble of a visit, I could not forbear to take the opportunity of conveying this to you by him.

It being only an explication of qualities which you desire of me, I shall set down my apprehensions in the form of suppositions as follows.

And first, I suppose, that there is diffused through all places an aetherial substance, capable of contraction and dilatation, strongly elastic, and, in a word, much like air in all respects, but far more subtile.

2. I suppose this aether pervades all gross bodies, but yet so as to stand rarer in their pores than in free spaces, and so much the rarer, as their pores are less; and this I suppose (with others) to be the cause why light incident on those bodies is refracted towards the perpendicular; why two well-polished metals cohere in a receiver exhausted of air; why mercury stands sometimes up to the top of a glass pipe, though much higher than thirty inches; and one of the main causes why the parts of all bodies cohere; also the cause of filtration, and of the rising of

water in small glass pipes above the surface of the stagnating water they are dipped into; for I suspect the aether may stand rarer, not only in the insensible pores of bodies; but even in the very sensible cavities of those pipes; and the same principle may cause menstruums to pervade with violence the pores of the bodies they dissolve, the surrounding aether, as well as the atmosphere, pressing them together.

3. I suppose the rarer aether within bodies, and the denser without them, not to be terminated in a mathematical superfices, but to grow gradually into one another; the external aether beginning to grow rarer, and the internal to grow denser, at some little distance from the superfices of the body, and running through all intermediate degrees of density in the intermediate spaces; and this may be the cause why light, in Grimaldo's experiment, passing by the edge of a knife, or other opaque body, is turned aside, and as it were refracted, and by that refraction makes several colours. Let ABCD be a dense body whether opake or transparent, EFGH the outside of the uniform aether, which is within it, IKLM the inside of the uniform aether, which is without it; and conceive the aether, which is between EFGH and IKLM, to run through all intermediate degrees of density between that of the two uniform aethers on either side. This being supposed, the rays of the sun SB, SK, which pass by the edge of this body between B and K, ought in their passage through the unequally dense aether there, to receive a ply from the denser eether, which is on that side towards K, and that the more by how much they pass nearer to the body, and thereby to be scattered through the space PQRST, as by experience they are found to be. Now the space between the limits EFGH and IKLM I shall call the space of the aether's graduated rarity.

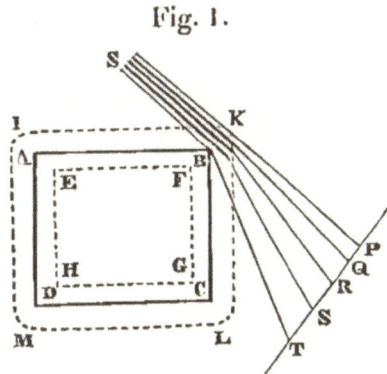

Fig. 1.

4. When two bodies moving towards one another come near together, I suppose the aether between them to grow rarer than before, and the spaces of its graduated rarity to extend further from the superficies of the bodies towards one another; and this, by reason that the aether cannot move and play up and down so freely in the strait passage between the bodies, as it could before they came so near

together: thus if the space of the aether's graduated rarity reach from the body ABCDFE only to the distance GHLMRS, when no other body is near it, yet may it reach further, as to IK, when another body NOPQ approaches And as the other body approaches more and more, I suppose the aether between them will grow rarer and rarer. These suppositions I have so described, as if I

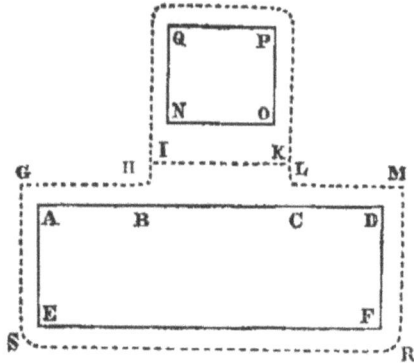

Fig. 2.

thought the spaces of graduated aether had precise limits, as is expressed at IKLM in the first figure, and GMRS in the second ; for thus I thought I could better express myself. But really I do not think they have such precise limits, but rather decay insensibly, and, in so decaying, extend to a much greater distance than can easily be believed or need be supposed.

5. Now, from the fourth supposition it follows, that when two bodies approaching one another come so near together as to make the aether between them begin to rarefy, they will begin to have a reluctance from being brought nearer together, and an endeavour to recede from one another; which reluctance and endeavour will increase as they come nearer together, because thereby they cause the interjacent aether to rarefy more and more. But at length, when they come so near together that the excess of pressure of the external aether which surrounds the bodies, above that of the rarefied aether, which is between them, is so great as to overcome the reluctance which the bodies have from being brought together; then will that excess of pressure drive them with violence together, and make them adhere strongly to one another, as was said in the second supposition. For instance, in the second figure, when the bodies ED and N P are so near together that the spaces of the aether's graduated rarity begin to reach to one another, and meet in the line I K, the aether between them will have suffered much rarefaction, which rarefaction requires much force, that is, much pressing of the bodies together; and the endeavour which the aether between them has to return to its former natural state of condensation, will cause the bodies to have an endeavour of receding from one another. But, on the other hand, to counterpoise this endeavour, there will not yet be any

excess of density of the aether which surrounds the bodies, above that of the aether which is between them at the line I K. But if the bodies come nearer together, so as to make the aether in the mid-way line I K grow rarer than the surrounding aether, there will arise from the excess of density of the surrounding aether a compressure of the bodies towards one another, which, when by the nearer approach of the bodies it becomes so great as to overcome the aforesaid endeavour the bodies have to recede from one another, they will then go towards one another and adhere together. And, on the contrary, if any power force them asunder to that distance, where the endeavour to recede begins to overcome the endeavour to accede, they will again leap from one another. Now hence I conceive it is chiefly that a fly walks on water without wetting her feet, and consequently without touching the water; that two polished pieces of glass are not without pressure brought to contact, no, not though the one be plain, the other a little convex, that the particles of dust cannot by pressing be made to cohere, as they would do, if they did but fully touch; that the particles of tingeing substances and salts dissolved in water do not of their own accord concrete and fall to the bottom, but diffuse themselves all over the liquor, and expand still more if you add more liquor to them. Also, that the particles of vapours, exhalations, and air do stand at a distance from one another, and endeavour to recede as far from one another as the pressure of the incumbent atmosphere will let them; for I conceive the confused mass of vapours, ar, and exhalations which we call the atmosphere, to be nothing else but the particles of all sorts of bodies, of which the earth consists, separated from one another, and kept at a distance by the said principle.

From these principles the actions of menstruums upon bodies may be thus explained: suppose any tinging body, as cochineal or logwood be put into water; so soon as the water sinks into its pores and wets on all sides any particle which adheres to the body only by the principle in the second supposition, it takes off, or at least much diminishes, the efficacy of that principle to hold the particle to the body, because it makes the aether on all sides the particle to be of a more uniform density than before. And then the particle being shaken off by any little motion, floats in the water, and with many such others makes a tincture; which tincture will be of some lively colour, if the particles be all of the same size and density; otherwise of a dirty one. For the colours of all natural bodies whatever seem to depend on nothing but the various sizes and densities of their particles, as I think you have seen described by me more at large in another paper. If the particles be very small (as are those of salts, vitriols, and gums), they are transparent ; and as they are supposed bigger and bigger, they pat on these colours in order, black, white, yellow, red ; violet, blue, pale green, yellow, orange, red ; purple, blue, green, yellow, orange, red, &c., as it is discerned by the colours, which appear at the several thicknesses of very thin plates of transparent bodies. Whence, to know the causes of the changes of colours, which are often made by the mixtures of several liquors, it is to be considered how the particles of any tincture may have their size or density altered by the infusion of another liquor. When any metal is put into common water, the water cannot enter into its pores, to act on it and dissolve it. Not that water consists of too gross parts for this purpose, but because it is unsociable to metal. For there is a certain secret principle in nature, by which liquors are sociable to some things and unsociable to others; thus water will not mix with oil, but readily with spirit of wine, or with salts; it sinks also into wood, which quicksilver will not; but quicksilver sinks into metals, which, as I said, water will not. So aquafortis dissolves ☽, not ☉; aqua regis ☉, not ☽ &c. But a liquor, which is of itself unsociable to a body, may, by the mixture of a convenient mediator, be made sociable; so molten lead, which alone will not mix with copper, or with regulus of Mars, by the addition of tin is made to mix with either. And water, by the mediation of saline spirits, will mix with metal. Now when any metal is put in water impregnated with such spirits, as into aquafortis, aqua regis, spirit of vitriol, or the like, the particles of the spirits, as they, in floating in the water, strike on the metal, will by their sociableness enter into its pores and gather round its outside particles, and by advantage of the continual tremor the particles of the metal are in, hitch them-selves in by degrees between those particles and the body, and loosen them from it; and the water enter-ing into the pores together with the saline spirits, the particles of the metal will be thereby still more loosed, so as by that motion the solution puts them into, to be easily shaken off, and made to float in the water: the saline particles still encompassing the metallic ones as a coat or shell does a kernel, after the manner expressed in the annexed figure, in which figure I have made the particles round, though they may be cubical, or of any other shape. If into a

Fig. 3.

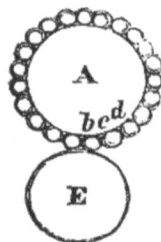

Fig. 4.

356

solution of metal thus made be poured a liquor abounding with particles, to which the former saline particles are more sociable than to the particles of the metal (suppose with particles of salt of tartar), then so soon as they strike on one another in the liquor, the saline particles will adhere to those more firmly than to the metalline ones, and by degrees be wrought off from those to enclose these. Suppose A a metalline particle, inclosed with saline ones of spirit of nitre, E a particle of salt of tartar, contiguous to two of the particles of spirit of nitre, b and c, and suppose the particle E is impelled by any motion towards d, so as to roll about the particle c till it touch the particle d, the particle b adhering more firmly to E than to A, will be forced off from A ; and by the same means the particle E, as it rolls about A, will tear off the rest of the saline particles from A one after another, till it has got them all, or almost all, about itself. And when the metallic particles are thus divested of the nitrous ones, which, as a mediator between them and the water, held them floating in it, the alcalizate ones, crowding for the room the metallic ones took up before, will press these towards one another, and make them come more easily together : so that by the motion they continually have in the water, they shall be made to strike on one another; and then, by means of the principle in the second supposition, they will cohere and grow into clusters, and fall down by their weight to the bottoln, which is called precipitation. In the solution of metals, when a particle is loosing from the body, so soon as it gets to that distance from it, where the principle of receding described in the fourth and fifth supposition begins to overcome the principle of acceding, described in the second supposition, the receding of the particle will be thereby accelerated ; so that the particle shall as it were with violence leap from the body, and putting the liquor into a brisk agitation, beget and promote that heat we often find to be caused in solutions of metals. And if any particle happen to leap off thus from the body, before it is surrounded with water, or to leap off with that smartness as to get loose from the water, the water, by the principle in the fourth and fifth suppositions, will be kept off from the particle, and stand round about it, like a spherically hollow arch, not being able to come to a full contact with it any more ; and several of these particles afterwards gathering into a cluster, so as by the same principle to stand at a distance from one another, without any water between them, will compose a bubble. Whence I suppose it is, that in brisk solutions there usually happens an ebullition. This is one way of transmuting gross compact substance into aerial ones. Another way is by heat ; for as fast as the motion of heat can shake off the particles of water from the surface of it, those particles, by the said principle, will float up and down in the air, at a distance both from one another, and from the particles of air, and make that substance we call vapour. Thus I suppose it is, when the particles of a body are very small (as I suppose those of water are), so that the action of heat alone may be sufficient to shake them asunder. But if the particles be much larger, they then require the greater force of dissolving menstrua ms to separate them, unless by any means the particles can be first broken into smaller ones. For the most fixed bodies, even gold itself, some have said will become volatile, only by breaking their parts smaller. Thus may the volatility and fixedness of bodies depend on the different sizes of their parts. And on the same difference of size may depend the more or less permanency of aerial substances, in their state of rarefaction.

To understand this, let us suppose A B C D to be a large piece of any metal, E F G H the limit of the interior uniform aether, and K a part of the metal at the superficies AB. If this part or particle K be so little that it reaches not to the limit EF, it is plain that the aether at its centre must be less rare than if the particle were greater; for were it greater, its centre would be further from the superficies AB, that is, in a place where the aether (by supposition) is rarer ; the less the particle K therefore, the denser the aether at its centre ; because its centre comes nearer to the edge AB, where the aether is denser than within the limit E F G H. And if the particle were divided from the body, and removed to a distance from it, where the aether is still denser, the aether within it must proportionally grow denser. If you consider this, you may apprehend how, by diminishing the particle, the rarity of the aether within it will be diminished, till between the density of the aether without, and the density of the aether within it, there be little difference ; that is, till the cause be almost taken away, which should keep this and other such particles at a distance from one another. For that cause explained in the fourth and fifth suppositions, was the excess of density of the external aether above that of the internal. This may be the reason then why the small particles of vapours easily come together, and are reduced back into water, unless the heat, which keeps them in agitation,

Fig. 5.

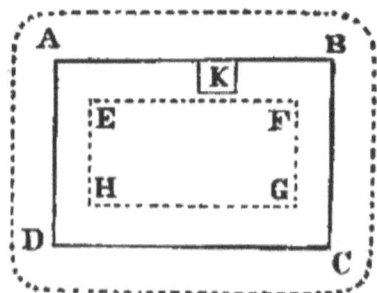

358

be so great as to dissipate them as fast as they come together; but the grosser particles of exhalations raised by fermentation keep their aerial form more obstinately, because the aether within them is rarer. Nor does the size only, but the density of the particles also, conduce to the permanency of the aerial substances; for the excess of density of the aether without such particles above that of the aether within them is still greater; which has made me sometimes think that the true permanent air may be of a metallic original; the particles of no substance being more dense than those of metals. This, I think, is also favoured by experience, for I remember I once read in the Philosophical Transactions, how M. Huygens at Paris, found that the air made by dissolving salt of tartar would in two or three days time condense and fall down again, but the air made by dissolving a metal continued without condensing or relenting in the least. If you consider then, how by the continual fermentations made in the bowels of the earth there are aerial substances raised out of all kinds of bodies, all which together make the atmosphere, and that of all these the metallic are the most permanent, you will not perhaps think it absurd, that the most permanent part of the atmosphere, which is the true air, should be constituted of these, especially since they are the heaviest of all other, and so much subside to the lower parts of the atmosphere and float upon the surface of the earth, and buoy up the lighter exhalations and vapours to float in greatest plenty above them. Thus, I say, it ought to be with the metallic exhalations raised in the bowels of the earth by the action of acid menstruums, and thus it is with the true permanent air; for this, as in reason it ought to be esteemed the most ponderous part of the atmosphere, because the lowest, so it betrays its ponderosity by making vapours ascend readily in it, by sustaining mists and clouds of snow, and by buoying up gross and ponderous smoke. The air also is the most gross unactive part of the atmosphere, affording living things no nourishment, if deprived of the more tender exhalations and spirits that float in it; and what more unactive and remote from nourishment than metallic bodies?

I shall set down one conjecture more, which came into my mind now as I was writing this letter; it is about the cause of gravity. For this end I will suppose aether to consist of parts differing from one another in subtilty by indefinite degrees; that in the pores of bodies there is less of the grosser aether, in proportion to the finer, than in open spaces; and consequently, that in the great body of the earth there is much less of the grosser aether, in proportion to the finer, than in the regions of the air;

and that yet the grosser aether in the air affects the upper regions of the earth, and the finer aether in the earth the lower regions of the air, in such a manner, that from the top of the air to the surface of the earth, and again from the surface of the earth to the centre thereof, the aether is insensibly finer and finer. Imagine now any body suspended in the air, or lying on the earth, and the aether being by the hypothesis grosser in the pores, which are in the upper parts of the body, than in those which are in its lower parts, and that grosser aether being less apt to be lodged in those pores than the finer aether below, it will endeavour to get out and give way to the finer aether below, which cannot be, without the bodies descending to make room above for it to go out into.

From this supposed gradual subtilty of the parts of aether some things above might be further illustrated and made more intelligible; but by what has been said, you will easily discern whether in these conjectures there be any degree of probability, which is all I aim at. For my own part, I have so little fancy to things of this nature, that had not your encouragement moved me to it, I should never, I think, have thus far set pen to paper about them. What is amiss, therefore, I hope you will the more easily pardon in

Your most humble servant and honourer,

Isaac Newton.
Cambridge, Feb. 28, 1678-9.

Robert Boyle, recipient of Newton's letter.

References

A chronologically-ordered list of the historical ether-drift citations, with download links, is found at: www.orgonelab.org/energyinspace.htm

Abbot BP (2018) GWTC-1: A Gravitational-Wave Transient Catalog of Compact Binary Mergers Observed by LIGO and Virgo during the First and Second Observing Runs. arXiv:1811.12907v2

Alfven H (1981) *Cosmic Plasma*, Springer.

Allais M (1997) *L'Anisotropie de L'Espace:La nécessaire révision de certains postulats des théories contemporaines*, Clément Juglar, Paris.

Allais M (2002) Experiments of Dayton C. Miller (1925-1926) and the Theory of Relativity. *Pulse of the Planet #5*, p.132-137.

Arp H (1987) *Quasars, Redshifts and Controversy*, Interstellar Media.

Arp H (1997) *Seeing Red: Redshifts, Cosmology and Academic Science*, Aperion. Canada.

Arp H (2003) *Catalogue of Discordant Redshift Associations*. Aperion, Montreal.

Asimov I (1966) *The Neutrino, Ghost particle*, Doubleday, NY.

Baker CF (pseud. Rosenblum CF) (1972) An Analysis of the United States Food and Drug Administration's Scientific Evidence Against Wilhelm Reich, Part II, the Physical Concepts. *J. Orgonomy*, 6(2) 222-231.

Baker CF (pseud. Rosenblum CF) (1973) An Analysis of the Food and Drug Administration's Scientific Evidence Against Wilhelm Reich, Part III, Physical Evidence. *J. Orgonomy*, 7(2) 234-245.

Baker CF (1980) The Orgone Energy Continuum. *J. Orgonomy*, 14(1) 37-60.

Baker, CF (1982) The Orgone Energy Continuum: the Ether and Relativity. *J. Orgonomy*, 16(1) 41-67.

Becker RO (1998) The Body Electric: Electromagnetism and the Foundation of Life, Wm. Morrow, NY.

Bauer I (1987) Erethrocyte Sedimentation: A New parameter for the measurement of Energetic Vitality. *Annals, Inst. Orgonomic Sci.* 4:49-65.

Baumer H (1987) *Sferics: Die Entdeckung der Wetterstrahlung*. Rowohlt Verlag, Hamburg.

Becker RO, Selden G (1985) *The Body Electric*. Wm. Morrow, NY.

Bernabei R (2007) DAMA sheds light on dark-matter particles. *Nature* 449, 24, 6 September.

Bernabei R (2010). Particle Dark Matter in the Galactic Halo: Recent Results from DAMA//LIBRA. *CRIS 2010* lecture, September, Catania Italy; see WebRef.10

Beloussov L, Popp FA, Voeikov V, Wijk RV, eds. (2000) *Biophotons and Coherent Systems: Proc. 2nd Alexander Gurwitsch Conference.* Moscow U. Press; republished by Springer Press 2007.

Benveniste J et al (1988) Human basophil degranulation triggered by very dilute antiserum against IgE. *Nature* 333:816-818, 30 June.

Blasband RA (1972) An Analysis of the United States Food and Drug Administration's Scientific Evidence Against Wilhelm Reich, Part I, the Biomedical Evidence. J. Orgonomy, 6(2) 207-222.

Bortels VH (1956) Die hypothetische Wetterstrahlung als vermutliches Agens kosmo-meteoro-biologischer Reaktonen. *Wissen-schaftliche Seltschrift der Humboldt-Universitat zu Berlin* VI:115-124.

Bortels, VH (1965) Das Gefrieren kleiner Wassermengen als solar-meteoro-biologische Modellreak-tion. *Naturwissenschaften* 52:118.

Brace DB (1904) On Double Refraction in Matter moving through the Aether, *Phil. Mag.* 6(7) 317.

Bradford CM, et al. (2011) The Water Vapor Spectrum of APM 08279+5255, Astrophys. J. Let. arXiv:1106.4301.

Brown FA, Hastings JW, Palmer JD (1970) *Biological Clock: Two Views.* Academic Press, NY.

Brown FA (1975) Evidence for External Timing of Biological Clocks. in *An Introduction to Biological Rhythms*, J. Palmer ed., Academic Press, NY, p.209-279.

Burns JT (1997) *Cosmic Influences on Humans, Animals and Plants: An Annotated Bibliography.* Magill Bibliographies, Scarecrow Press, London and Salem Press, Pasadena.

Burr HS (1971) *Blueprint For Immortality.* Neville Spearman, London 1971; republished as *The Fields of Life.* Ballantine Books, NY, 1972.

Cahill RT (2004) Absolute Motion and Gravitational Effects. *Apeiron,* 11(1)53-111, January.

Cahill RT (2006a) The Roland DeWitte 1991 Experiment (to the Memory of Roland DeWitte) *Progress in Physics,* July 2006, p.60-65.

Cahill RT (2006b) A New Light-Speed Anisotropy Experiment: Absolute Motion and Gravitational Waves Detected. *Progress in Physics,* Oct.2006, p.73-92.

Cahill RT (2007) Optical-Fiber Gravitational Wave Detector: Dynamical 3-Space Turbulence Detected. *Progress in Physics,* Oct. 2007, p.63-68.

Cahill RT (2008) Correlated Detection of sub-mHz Gravitational Waves by Two Optical-Fiber Interferometers. *Prog. in Physics,* April, p.103-110.

Campbell WW, Trumper R (1923) *Lick Obs. Bull.* 11, 41.

Campney DC et al (1963) An Ether Drift Experiment Based on the Maussbauer Effect. *Phys. Letters,* T:241-243.

Cedarholm JP et al (1958) New Experimental Test of Special Relativity. *Phys. Rev. Letters,* 1(9) 342-349.

Chai B-H, Zheng J-M, Zhao Q, Pollack GH (2008) Spectroscopic Studies of Solutes in Aqueous Solution. *J Phys Chem A* 112, 2242-2247.

Chappell J (1965) Georges Sagnac and the Discovery of the Ether. *Arch. Internat. d'Histoire des Sciences*, 18:175-190.

Church C (1993) Chapter on Sferics Studies, in *The Tornado: Its Structure, Dynamics, Prediction, and Hazards,* American Geophysical Union, 174-175.

Clark RW (1971) *Einstein: The Life and Times,* World Publishing Co., NY.

Cohen MH et al. (1978) Superluminal Variations in 3C 120, 3C 273 and 3C 345, Astrohysical Journal, 231:293-298. Also see WebRef.25.

Collins H, Pinch T (2012) *The Golem: What You Should Know About Science,* Cambridge Univ. Press.

Consoli M., Costanzo M. (2003) The motion of the Solar System and the Michelson-Morley Experiment

Cowen R (1997) Does the Cosmos Have a Direction? *Science News,* 26 April, p.252.

Creswell J, et al (2017) On the Time Lags of the LIGO Signals. arXiv:1706.04191.

DeMeo J (1979a) *Preliminary Analysis of Changes in Kansas Weather Coincidental to Experimental Operations with a Reich Cloudbuster.* Geography-Meteorology Department, University of Kansas, Thesis; republished by Orgone Biophysical Research Lab, Ashland, Oregon, 2010.

DeMeo J (1979b) Evidence for the Existence of a Principle of Atmospheric Continuity. monograph in DeMeo 1979a.

DeMeo J (1980) Water Evaporation Inside the Orgone Accumulator. J. Orgonomy 14(2) 171-175.

DeMeo J (1989) Postscript on the Food and Drug Administration's Scientific Evidence Against Wilhelm Reich. *Pulse of the Planet* 1:18-23 Spring.

DeMeo J (1990) The Orgone Energy Continuum: Some Old and New Evidence. *Pulse of the Planet,* 1(2) 3-8.

DeMeo J (1991) OROP Arizona 1989: A Cloudbusting Experiment to Bring Rains in the Desert South-west. *Pulse of the Planet,* 3:82-92.

DeMeo J (1993a) Anti-Constitutional Activities and Abuse of Police Power by the US Food and Drug Administration and Other Governmental Agencies. *Pulse of the Planet* 4:106-113.

DeMeo J (1993b) OROP Israel 1991-1992: A Cloudbusting Experiment to Restore Wintertime Rains to Israel and the Eastern Mediterranean During an Extended Period of Drought. *Pulse of the Planet* 4:92-98.

DeMeo J (1993c) Research Reports and Observations: OROP Namibia 1992-1993. *Pulse of the Planet* 4:114-116.

DeMeo J (1994) The Desert-Drought Map and its Implications. *Abstracts, 90th Annual Meeting, Association of American Geographers,* San Francisco, California, 29 March - 2 April, 81.

DeMeo J (1996) Cloudbusting: Growing Evidence for a New Method of Ending Drought and Greening Deserts. American Institute of Biomedical Climatology, *AIBC Newsletter,* September, 20:1-4.

DeMeo J (2000a) Critical Review of the Shankland, et al, Analysis of Dayton Miller's Ether-Drift Experiments. Conference, *Natural Philosophy Alliance,* Berkeley, Calif., May.

DeMeo J, ed. (2002a) *Heretic's Notebook: Emotions, Protocells, Ether-Drift and Cosmic Life Energy: with New Research Supporting Wilhelm Reich,* Natural Energy, Oregon.

DeMeo J (2002b) Dayton Miller's Ether Drift Experiments: A Fresh Look. *Pulse of the Planet,* 5:114-130.

DeMeo J (2002c) Green Sea Eritrea: A 5-Year Desert Greening CORE Project in the SE African Sahel. *Pulse of the Planet* 5:183-211.

DeMeo J (2002d) Reconciling Miller's Ether Drift With Reich's Dynamic Orgone. *Pulse of the Planet,* 5:137-146.

DeMeo J (2002e) Orgone Accumulator Stimulation of Sprouting Mung Beans. *Pulse of the Planet* 5:168-176.

DeMeo J (2002f) Origins of the Tropical Easterlies: An Orgone-Energetic Perspective. *Pulse of the Planet* 5:212-218.

DeMeo J (2004) A Dynamic and Substantive Cosmological Ether. *Proceedings of the Natural Philosophy Alliance,* 1(1) 15-20.

DeMeo J (2009) Experimental Confirmation of the Reich Orgone Accumulator Thermal Anomaly. *Subtle Energies & Energy Medicine.* 20(3) 17-32.

DeMeo J (2010a) *In Defense of Wilhelm Reich: Opposing the 80-Years' War of Mainstream Defamatory Slander Against One of the 20th Century's Most Brilliant Physicians and Natural Scientists,* Natural Energy, Oregon.

DeMeo J (2010b) *The Orgone Accumulator Hand-book: Wilhelm Reich's Life-Energy Science and Healing Tools for the 21st Century, With Construc-tion Plans.* 3rd Edition, Natural Energy Works, Ash-land, Oregon.

DeMeo J (2010c) Report on Orgone Accumulator Stimulation of Sprouting Mung Beans. *Subtle Energy and Energy Medicine.* 21(2) 51-62, 2010.

DeMeo J (2010d) *The Orgone Accumulator Handbook: Wilhelm Reich's Life Energy Discoveries and Healing Tools for the 21st Century, with Construction Plans,* Natural Energy, Oregon 2010.

DeMeo J (2011a) Dayton C. Miller Revisited, in *Should the Laws of Gravitation be Reconsidered?* Munera, ed 2011, Apherion, p.289-319.

DeMeo J, Ed. (2011b) Heretics Notebook: Emotions, Protocells, Ether-Drift and Cosmic Life-Energy, with New Research Supporting Wilhelm Reich, Natural Energy Works, Oregon.

DeMeo J (2011c) Water as a Resonant Medium for Unusual External Environmental Factors, *Water,* 3:1-47.

DeMeo J et al. (2012) In Defense of Wilhelm Reich: An Open Response to Nature and the Scientific/Medical Community. *Water,* 4:72-81.

DeMeo J (2013) *In Defense of Wilhelm Reich,* Natural Energy Works, Ashland, Oregon.

DeMeo J (2014) Does a Cosmic Ether Exist? Evidence from Dayton Miller and Others. *J. Scientific Exploration* 28(4) 647-682, 2014.

DeMeo J, Ed. (2015) *On Wilhelm Reich and Orgonomy*, Natural Energy Works, Oregon.

DeMeo J (2018) Anomalous "Living" Spectrographic Changes in Water Structures: Explorations in New Territory. *Water*, 10:41-81.

DeMeo J, Senf B. Eds (1997) *Nach Reich: Neue Forschungen zur Orgonomie: Sexual konomie , Die Entdeckung Der Orgonenergie*. Zweitausendeins Verlag, Frankfurt.

Díaz-Vélez, JC, et al. (2017) Combined Analysis of Cosmic-Ray Anisotropy with IceCube and HAWC. *Proceed. of Science, 35th Int. Cosmic Ray Conference* - ICRC217 -10-20 July 2017. Busan, Korea.

Dirac P (1951) Is there an Aether? Letter to the Editor, *Nature* 168:906.

Dudley HC (1975) Michelson's Hunch Was Right. *Bull Atom Sci,* January 47-48.

Dudley HC (1959) *New Principles in Quantum Mechanics: Their Application to Space Exploration and Relativity*. University Books.

Dufour A, Prunier F (1942) On a Fringe Movement Registered on a Platform in Uniform Motion. *J. de Physique. Radium* 3(9) 153-162.

Einstein A (1905) Concerning an Heuristic Point of View Toward the Emission and Transformation of Light. Original German in *Annalen der Physik.* 17:132, 18 March.

Einstein A 1905) Investigations on the Theory of the Brownian Movement. Original German in *Annalen der Physik.* 17:549, 11 May.

Einstein A (1905) On the Electrodynamics of Moving Bodies. *Annalen der Physik.* 17:891-921, 30 June.

Einstein A (1911) On the Influence of Gravitation on the Propagation of Light. Original German in *Annalen der Physik* 340(10) 898-908.

Einstein A (1915) Explanation of the Perihelion of Mercury from General Relativity Theory, Original German in *Königlich Preußische Akademie der Wissenschaften.*Berlin. Sitzungsberichte.

Einstein A (1916) Foundation of the General Theory of Relativity. Original German in *Annalen der Physik* 354, 354 (7), p.769-822.

Einstein A (1905) Does the Inertia of a Body Depend Upon its Energy-Content? Original German in *Annalen der Physik.* 8:639, 27 September.

Einstein A (1920) Ether and the Theory of Relativity. Lecture at the University of Leyden, 5 May 1920. Reproduced in many texts and on-line, for example in German in *Meine Weltbilt* 1933, and in English in *Essays In Science*, Wisdom/Philosophical Library, NY, 1934, p.98-111.

Einstein A (1926) Interview: Meine Theorie und Millers Versuche (My Theory and Miller's Experiment) *Vossische Zeitung,* 19 Jan.; Contained in Hentschel (1992).

Einstein A (1927) Interview: Einstein und das Berliner kilogramm (Einstein and the Berlin Kilogram) *Deutsche Allgemeine Zeitung,* 27 Nov.; Contained in Hentschel (1992).

Einstein A (1929) "Field Theories, Old and New". *NY Times,* 3 Feb.

Emery GT (1972) Perturbation of Nuclear Decay Rates. *Ann Rev Nucl Sci,* Annual Reviews.

Erylkin et al. (2019) Puzzles of the Cosmic Ray Anisotropy. arXiv:1901.00160v4

Essen L (1955) A New Ether Drift Experiment. *Nature.* 175:793-794.

Essen L (1971) *The Special Theory of Relativity: A Critical Analysis,* Oxford Univ. Press.

Faigl P (1990) The Scale Buoy Effect – An Unsolved Problem in Physical Chemistry. *Out of the Crucible Conference Proceedings,* 12-14th December 1990, Centre for Human Aspects of Science and Technology, University of Sydney NSW, Australia.

Field G, Bahcall J, Arp H. (1974) *The Redshift Controversy* (Frontiers in Physics) Addison Wesley.

Fitzgerald GC (1889) The Ether and the Earth's Atmosphere, *Science,* Letter to the Editor, 13(328) 390.

Fizeau H (1851) Sur les hypotheses relatives a l'ether lumineux, *Comptes Rendus* 33:349-355.

Fizeau H (1859) Sur les hypotheses relatives a l'ether lumineux, *Ann. De Chim. Et de Phys.* 57: 385-404.

Fizeau H (1860) On the Effect of the Motion of a Body upon the Velocity with which it is Traversed by Light, *Philosophical Magazine/Journal of Science,* Fourth Series, 19:245-260, April.

Fletcher H (1943) Biographical Memoir of Dayton Clarence Miller, 1866-1941. *National Academy of Sciences,* Vol.XXIII, Third Memoir, p.60-74, National Academy Meeting.

Fresnel A (1818) Lettre d'Augustin Fresnel a Francois Arago sur l'influence du mouvement terrestre dans quelques phenomenes d'optique, *Annales de chimie et de physique,* 9: 57-66.

Frisch PC, et al (2013) Decades-Long Changes of the Interstellar Wind Through Our Solar System. *Science* 341(6150) 1080-1082. WebRef.8.

Galaev YM (2001) Ethereal Wind in Experience of Millimetric Radiowaves Propagation. *Spacetime and Substance,* 3, 5(10)211-225.

Galaev YM (2002) The Measuring of Ether-Drift Velocity and Kinematic Ether Viscosity Within Optical Waves Band. *Spacetime and Substance,* 3, 5(15)207-224.

Galileo (1610) *Sidereal Messenger,* Venice.

Galileo (1629) *Dialogue Concerning the Two Chief World Systems.* trns. Stillman Drake, unabridged edition, U Calif. Press, 1967, p.216-218.

Gell-Mann M (2002) Quantum Mechanics and Flapdoodle, chapter in *The Quark and the Jaguar,* Owl Books.

References

Greenfield J (1974) *Wilhelm Reich Versus the USA. W.W.* Norton, NY.

Griest KÒ (2002) "Wimps and Machos" in the *Encyclopedia of Astronomy and Astrophysics*, Nature Publishing, UK.

Gurwitsch A, et al (1932) *Die Mitogenetische Strahlung: Zugleich Sweiter Band der "Problem der Zelteilung".* Monograph. See Beloussov 2000.

Hebenstreit G (1995) *Der Orgonakkumulator nach Wilhelm Reich. Eine experimentelle Untersuchung zur Spannungs- Ladungs- Formel.* Diplomarbeit, Universität Wien.

Hentschel K (1992) Einstein's Attitude Towards Experiments: Testing Relativity Theory 1907-1927. *Stud. Hist. Phil. Sci.* 23(4) 593-624.

Hicks WM (1902) The FitzGerald-Lorentz Effect. *Nature*, 65(1685) 343, 13 Feb., also in *Philosophical Magazine.*

Hicks WM (1902) *On the Michelson-Morley Experiment Relating to the Drift of the Aether.*

Ho MW, Popp FA, Warnke U (1974) *Bioelectro-dynamics and Biocommunication.* World Scientific Publishing.

Ho MW, (2012) Living Rainbow H2O, World Scientific Publishing, Singapore.

Hoffmann B (1972) *Einstein Creator and Rebel,* Penguin Books.

Jaseja TS et al (1964) Test of Special Relativity or Space Isotropy by Use of Infared Masers. *Phys. Rev.* 133a:1221-1225.

Joos G (1930) Die Jenaer Wiederholung des Michelsonversuchs, *Annalen der physik.* 5.7(4) 385-407.

Joos G., Miller DC (1934) Letters to the Editor. *Physical Review,* 45:114, 15 Jan.

Kavouras J (2005) *Heilung mit Orgonenergie: Die medizinische Orgonomie.* Turm Verlag, Bietighem, Germany.

Kehr RW (2002) *The Detection of Ether*, manuscript.

Kennedy R. (1926) A Refinement of the Michelson–Morley Experiment. *Proceedings, Nat. Acad. Sci.* 12:621-629.

Kennedy RJ, Thorndike EM (1932) Experimental Establishment of the Relativity of Time. *Phys. Rev.* 42:400-418.

Kervran CL (1971) *Biological Transmutations*, Crosby Lockwood, UK. Also same title with slightly different content by Swan House, NY 1972.

Kimball M (1981) An Interview with Dr. Robert S. Shankland, Subject: Dayton Miller. Transcript of audio tape, 15 Dec. 1981, Case Western Reserve University Archive, Cleveland, Ohio.

Kodera K, et al (1990) Downward Propagation of Upper Stratospheric Mean Zonal Wind Perturbation to the Troposphere. *Geophys. Res. Letters.* 17(9) 1263-1266.

Labitzke K, van Loon H (1997) The Signal of the 11-Year Sunspot Cycle in the Upper Troposphere-Lower Stratosphere. *Space Science Reviews* 80(3-4), 393-410.

Labitzke K (2001) The Global Signal of the 11-Year Sunspot Cycle in the Stratosphere: Differences Between Solar Maxima and Minima. *Meteorologische Zeitschrift* 10(2) 83-90, April.

Lerner E (1992) *The Big Bang Never Happened*. Vintage.

Livingston DM (1973) *The Master of Light: A Biography of Albert A. Michelson*, U. Chicago Press.

Lodge O (1892) The Motion of the Ether Near the Earth, *Proceedings, Royal Inst. of Great Britain, Weekly Meetings*, 1 April. Vol.XIII:565-580.

Lodge O (1894) Aberration Problems New Ether Experiments, *Philosophical Magazine*.

Lodge O (1909) *The Ether of Space*, Harper & Brothers.

Lodge O (1912) The Ether of Space and the Principle of Relativity. *Science Progress*, 6(337-334)

Lodge O (1919) Aether and Matter: Being Remarks On Inertia and On Radiation and On the Possible Structure of Atoms. *Nature* 104:16, 4 Sept.

Lodge O (1921) Relativity: The Geometrisation of Physics and its Supposed Basis on the Michelson-Morley Experiment. *Nature* 106:795-800 17 Feb.

Lodge O (1925) *Ether & Reality: A Series of Discourses on the Many Functoins of the Ether of Space*. Hodder & Stoughton, London.

Lorentz HA (1899) Stoke's Theory of Aberration in the Supposition of a Variable Density of the Aether, *Proc. Roy. Soc.* 1: 443-448.

Lorentz HA (1904) Electromagnetic Phenomena in a System Moving With Any Velocity Smaller than that of Light. *Proceedings of the Royal Netherlands Academy of Arts and Sciences* 6: 809-831.

Luckey TD (1991) *Radiation Hormesis*. CRC Press.

Maglione R (2007) *Wilhelm Reich and the Healing of Atmospheres*. Natural Energy Works, Ashland, OR.

Marett D (2002) West-East Asymmetry and Diurnal Effect of Cosmic Radiation. *Pulse of the Planet* 5:177-182. (*Heretic's Notebook*)

Marinov S. (1980) Measurement of the Laboratory's Absolute Velocity. *General Relativity and Gravitation*, 12(1) 57-66.

Martin JE (1999) *Wilhelm Reich and the Cold War*. Flatland Press, Mendocino CA. Reprinted 2014, Natural Energy Works, Oregon.

Michelson AA (1881) The Relative Motion of the Earth and the Luminiferous Ether. *American Journ. Science*, Vol.XXX(377) 120-129.

Michelson AA (1882) *Experimental Determination of the Velocity of Light, Made at the U.S. Naval Academy*, Annapolis, Naval Observatory.

Michelson AA, Morley EW (1887) On the Relative Motion of the Earth and the Luminiferous Ether, *American Journal of Science*, Third Series, Vol.XXXIV(203) 333-345 Nov.

Michelson AA (1888) A Plea for Light Waves, *Proceedings, AAAS,* Section B, 37:67-78.

Michelson AA (1903) *Light Waves and Their Uses*, Univ. Chicago Press.

Michelson AA (1904) Relative Motion of Earth and Aether. *Philosophical Magazine*, 6(8) 716-719.

Michelson AA (1913) Effect of Reflection from a Moving Mirror on the Velocity of Light, *Astrophysical Journal,* 37:190, April 1913.

Michelson AA (1924) Preliminary Experiments on the Velocity of Light. *Astrophysical Journal,* 60:256, November.

Michelson AA, Gale H, Pearson F (1925) The Effect of the Earth's Rotation on the Velocity of Light. (Parts I and II), *Astrophysical Journal*, 61:137-145, April.

Michelson AA, Gale H (1925) Letter to the Editor: The Effect of the Earth's Rotation on the Velocity of Light", *Nature*, 115:566, April.

Michelson AA (1927a) *Studies in Optics.* U. Chicago Press, 1927.

Michelson AA (1927b) Measurement of the Velocity of Light Between Mount Wilson and Mount San Antonio. *Astrophysical Journal,* 65: 1, January.

Michelson AA, Pease FG, Pearson F (1929a) Repetition of the Michelson-Morley Experiment, *Nature,* 123:88, 19 Jan.;

Michelson AA, Pease FG, Pearson F (1929b) Repetition of the Michelson-Morley Experiment, *J. Optical Society of America*, 18:181, 1929.

Michelson AA, Pease FG, Pearson F (1935) Measurement of the Velocity of Light in a Partial Vacuum. *Astrophysical J.,* 82:26-61.

Milian V (2002) Confirmation of an Oranur Anomaly, *Pulse of the Planet* 5:182.

Milian-Sanchez V, Mochol-Salcedo A, Milian C, Kolombet VA, Verdu G (2016) Anomalous Effects on Radiation Detectors and Capacitance Measurements Inside a Modified Faraday Cage. *Nuclear Instruments and Methods in Physics,* Research A, 828:210-228.

Miller DC (1922) The Ether-Drift Experiments at Mount Wilson Solar Observatory. *Physical Review*, 19:407-408, April.

Miller DC (1925a) Ether-Drift Experiments at Mount Wilson. *Proceedings, Nat. Acad. Sciences*, 11(6) 306-314, June.

Miller DC (1925b) Ether-Drift Experiments at Mount Wilson. *Science,* 61(1590) 617-621, 19 June.

Miller DC (1925c) Ether-Drift Experiments at Mount Wilson. *Nature,* 116(2906) 49-50, 11 July.

Miller DC (1926a) Ether-Drift Experiments at Mount Wilson in February 1926. *Physical Review A*, 27:812, April.

Miller DC (1926b) Significance of the Ether-Drift Experiments of 1925 at Mount Wilson", *Science*, 63:433-443, 30 April 1926.

Miller DC (1926c) Radio Talks on Science: The Ether-Drift Experiments of 1925 at Mount Wilson. *Scientific Monthly*, 22:352-355, April.

Miller DC (1926d) Interview: The Measurement of Ether Drift, Professor Dayton Miller's Experiments. *Modern Science,* July.

Miller DC (1928) Untitled Lecture, Conference on the Michelson-Morley Experiment, Held at the Mount Wilson Observatory, Pasadena Calif. 4-5 February 1927, *Astrophysical Journal*, LXVIII:341-402, Dec.; also in *Contributions From the Mount Wilson Observatory*, No.373, Carnegie Institution of Washington, 1928.

Miller DC (1929) Ether Drift Experiments in 1929 and Other Evidences of Solar Motion. *Science*, 70(1832) 560-561.

Miller DC (1930) Ether Drift Experiments in 1929 and Other Evidences of Solar Motion, with Table. *J. Royal Ast. Soc. Canada*, 24:82-84.

Miller DC (1931) Ether Drift Experiments in Cleveland in 1930. *Report of the Centenary Meetings, London 1931, Sept.23-30.* British Assn. for Advancement of Science, p.337-338.

Miller DC (1933a) The Ether-Drift Experiment and the Determination of the Absolute Motion of the Earth. *Reviews of Modern Physics*, 5(2) 203-242, July.

Miller DC (1933b) The Absolute Motion of the Solar System and the Orbital Motion of the Earth Determined by the Ether-Drift Experiment. *Science*, 77(2007) 587-588, 16 June. Shorter note of same title published in *Physical Review,* 43:1054, 1933.

Miller DC (1934) The Ether-Drift Experiment and the Determination of the Absolute Motion of the Earth. *Nature,* 133:162-164, 3 Feb.

Millikan RA (1938) Biographical Memoir of Albert Abraham Michelson, 1852-1931. *National Academy of Sciences,* XIX, Fourth Memoir, 120-140, National Academy Meeting.

Mineur H (1927) The Experiment of Miller and the Hypothesis of the Dragging Along of the Ether. *J. Royal Ast. Soc. Canada,* 21:206-214.

Mitchell W (2002) *Bye-Bye Big Bang: Hello Reality*. Cosmic Sense Books.

Morley EW, Miller DC, Eddy H (1898) The Velocity of Light in the Magnetic Field, *Proceedings AAAS*, 38:283-295.

Morley EW, Miller DC (1904) Extract from a Letter dated Cleveland, Ohio, August 5th, 1904, to Lord Kelvin from Profs. Edward W. Morley and Dayton C. Miller. *Philosophical Magazine,* S.6, 8(48) 753-754, Dec.

Morley EW, Miller DC (1905a) Report of an Experiment to Detect Change of Dimension of Matter Produced by its Drift through the Ether. *Science*, Vol.XXI. No.531, p.339, March.

Morley EW, Miller DC (1905b) On the Theory of Experiments to Detect Aberration of the Second Degree. *Science*, Vol.XXI. No.531, p.339, March.

Morley EW, Miller DC (1905c) On the Theory of Experiments to Detect Aberration of the Second Degree. *Philosophical Mag.*, May, p.669-680.

Morley EW, Miller DC (1905d) Report of an Experiment to Detect the FitzGerald-Lorentz Effect. *Philosophical Mag.*, May, p.680-685 (with separate plates and figures) .

Morley EW, Miller DC (1905e) Report of an Experiment to Detect the FitzGerald-Lorentz Effect. *Proceedings, Am. Acad. Arts & Sciences*, 41:321-328, August.

Morley EW, Miller DC (1907) Final Report on Ether-Drift Experiments. *Science*, 25:525, 5 April.

Mueller GO, Kneckebrodt K (2006) 5 Years of Criticism of the Special Theory of Relativity (1908-2003).

Munera HA (2002) The effect of solar motion upon the fringe-shifts in a Michelson-Morley interferometer a la Miller. Annales de la Fondation Louis de Broglie, 27 (3) 463-484.

Munera HA ed (2011) *Should the Laws of Gravitation be Reconsidered?: The Scientific Legacy of Maurice Allais*, Aperion Press.

Múnera HA, Hernandez-Deckers D, Arenas G, Alfonso E (2006) Observation During 2004 of Periodic Fringe-Shifts in an Adialeiptometric Stationary Michelson-Morley Experiment. *Electromagnetic Phenomena* (Inst Electromag Res, Kharkov, Ukraine), 6(1) 70-92.

Munera HA et al (2009) Observation of a Non-conventional Influence of Earth's Motion on the Velocity of Photons, and Calculation of the Velocity of Our Galaxy. Conference in Beijing.

Müschenich S (1995) *Der Gesundheitsbegriff im Werk des Arztes Wilhelm Reich*. Doktorarbeit am Fachbereich Humanmedizin der Philipps-Universitat Marburg, Verlag Gorich & Weiershauser, Mar-burg.

Müschenich S, Gebauer R (1986) *Die (Psycho-) Physiologischen Wirkungen des Reich'schen Orgonakkumulators auf den Menschlichen Organismus*. Universitat Marburg Dept. Psychology Dissertation; published as *Der Reichsche Orgonakkumulator*. Nexus Verlag, 1987.

Newton I (1679) Letter to Robert Boyle, in Letter to Henry Lord Brougham, FRS &c., Containing Remarks on Certain Statements in his 'Lives of Black, Watt and Cavendish'. by Rev. William Vernon Harcourt, With an Appendix Containing Newton's Letters on Air and Aether. London: Richard & John Edward Taylor, 1846, p.131-141.

Newton I (1686) Trans. Motte A. *Newton's Principia: The Mathematical Principles of Natural Philosophy*. Daniel Adee 1846, p.72.

Nassau JJ, Morse PM (1927) A Study of Solar Motion by Harmonic Analysis. *Astrophysical Journal*, v.65, March.

Nordenstrom B (1983) *Biologically Closed Electric Circuits: Clinical, Experimental and Theoretical Evidence for an Additional Circulatory System*. Nordic Medical Publications, Stockholm.

Petersen WF (1934) *The Patient and the Weather,* Vols I – IV. Edwards Brothers, Ann Arbor.

Petersen WF (1947) *Man, Weather, Sun*. Charles C. Thomas, Springfield.

Piccard A, Stahel E (1926) L'exp rience de Michelson, realise en ballon libre. *Comptes rendus*, 183:420-421, 1926.

Piccard A, Stahel E (1927) Nouveaux resultats obtenus par l'exprience de Michelson" *Comptes rendus,* 184:152.

Piccard A, Stahel E (1927) L'absence du vent d'ether au Rigi. *Comptes rendus,* 184:1198-1200.

Piccard A, Stahel E (1928) Realization of the Experiment of Michelson in Balloon and on Dry Land. *Le Journal de Physique,* Series IV, 9(2) 49-60, Feb.

Piccardi G (1962) *Chemical Basis of Medical Climatology,* Charles Thomas press, Springfield.

Piccardi G (1965) Intensity of Solar Corona, Wolf Number, Biological and Chemical Tests. *Geofisica e Meteorologia* 14(3-4) 77-78.

Piccardi G (1966) Causality and Astrogeophysical Phenomena. *Geofisica e Meteorologia* 15(3-4) 75-77.

Piccardi G (1968) Two considerations on Fluctuating Phenomena. *Geofisica e Meteorologia* 17(3-4) 95-97.

Piccardi G, Capel-Boute C (1972) The 22-year Solar Cycle and Chemical Tests. *J Interdiscip Cycle Res* 3(3-4) 413-417.

Pollack GH (2001) *Cells, Gels and the Engines of Life.* Ebner and Sons, Seattle.

Pollack GJ (2013) *The Fourth Phase of Water: Beyond Solid, Liquid and Vapor.* Ebner & Sons, Washington.

Pollack GH, Figueroa X, Zhao Q (2009) Molecules, Water, and Radiant Energy: New Clues for the Origin of Life. *Int. J. Mol. Sci.* 10:1419-1429.

Popp FA, Beloussov LV (2010) *Integrative Biophysics: Biophotonics.* Springer.

Presman AS (1970) *Electromagnetic Fields and Life.* Plenum Press, NY.

Rayleigh L (1892) Aberration? *Nature,* 24 March 1892, p.499-502.

Rayleigh L (1902) Does Motion thorough the Aether cause Double Refraction? *Phil. Mag.* 6(4) 678.

Reich W (1934) *The Bioelectrical Investigation of Sexuality and Anxiety.* reprinted Farrar, Straus & Giroux, NY 1982.

Reich W (1938) *The Bion Experiments: On the Origin of Life.* reprinted Farrar, Straus & Giroux, NY 1979

Reich W (1939) Three Experiments with the Static Electroscope (1939) *Orgone Energy Bull* III(3):144-145, 1951.

Reich W (1942) *Discovery of the Orgone, Vol.1: Function of the Orgasm,* reprint: Farrar, Straus & Giroux, NY, 1973.

Reich W (1944) Orgonotic Pulsation: The Differentiation of the Orgone Energy from Electromagnetism. *Int. J. Sex-Economy and Orgone Research.* 3:97-150.

Reich W (1945) Experimental Demonstration of the Physical Orgone Energy, *Int. J. Sex-Economy and Orgone Research.* 4:133-146.

Reich W (1948) *Discovery of the Orgone, Vol.2: The Cancer Biopathy,* reprint: Farrar, Straus & Giroux, NY, 1973.

Reich W (1949) *Ether, God and Devil. Orgone* Inst. Press. reprint: Farrar, Straus & Giroux, NY, 1973.

Reich W (1951a) *Cosmic Superimposition.* Wilhelm Reich Foundation. reprint: Farrar, Straus & Giroux, NY, 1973.

Reich W (1951b) *The Orgone Energy Accumulator, Its Scientific and Medical Use,* Wilhelm Reich Foundation, Rangeley.

Reich W (1951c) *The Oranur Experiment,* Wilhelm Reich Foundation, Rangeley, Maine.

Reich W (1953) *The Einstein Affair,* Orgone Institute Press.

Reich W (1953b) *The Einstein Affair,* Wilhelm Reich Biographical Material. Orgone Institute Press, Rangeley, Maine.

Reich W (1953c) The Murder of Christ.Orgone Inst. Press. reprint Farrar, Strauss & Giroux 1966.

Reich W (1957) *Contact With Space.* Core Pilot Press, NY.

Reich W (1960) *Selected Writings: An Introduction to Orgonomy.* Farrar, Straus & Giroux, NY.

Ruderfer M (1972) Power spectrum of the neutrino sea. *Lettere al Nuovo Cimento* 5(1) 86-88, 2 Sept.; Also see NASA Probe Finds Sea of Cosmic Neutrinos, New Evidence Of Early Universe. *Science Daily* NASA/ Goddard Space Flight Center (2008, March 7)

Sagnac MG (1913) L'Ether lumineux Demonstre par l'effet du vent relatif d'aether dan interferometre en rotation uniforme. Note de G. Sagnac, presentee par E. Bouty. *Comptes Rendus,* 157:708-710, 1913.

Sagnac MG (1913) Sur la preuve de la realite de l'ether lumineux par l'experience de l'interferographe tournant. *Comptes Rendus,* 157:1410-1413, 22 Dec.

Sagnac MG (1914) Effet tourbillonnaire optique. La circulation de l'ether lumineux dans un interferographe tournant. *Journale de physique* 5(4) 177-195.

Sanders CL (2009) *Radiation Hormesis and the Linear-No-Threshold Assumption.* Springer.

Schauberger V (1998) *The Water Wizard: The Extraordinary Properties of Natural Water.* Gateway Books, Bath, UK.

Schauberger V (1998) *Living Energies. Gateway* Books, Bath, UK.

Schmidt KB et al (2011) The Color Variability of Quasars, *Astrophys. J.* 744(2) 1-16.

Schroeder G (1990) *Genesis and the Big Bang: The Discovery of Harmony Between Modern Science and the Bible.* Bantam.

Seifriz W (1936) *Protoplasm.* McGraw-Hill, NY.

Shankland R (1936) An Apparent Failure of the Photon Theory of Scattering. *Physical Review,* 1 Jan.

Shankland R (1941) Dayton Clarence Miller: Physics Across Fifty Years. *Am. J. Physics,* 9(5) 273-283, Oct.

The Dynamic Ether of Cosmic Space

Shankland R (1949) Michelson at Case. *Am. J. Physics*, 17(8) 487-490, Nov.

Shankland R, McCuskey SW, Leone FC, Kuerti G (1955) New Analysis of the Interferometer Observations of Dayton C. Miller. *Reviews of Modern Physics*, 27(2) 167-178, April.

Shankland R (1963) Conversations with Albert Einstein. *Am. J. Physics*, 31:47-57, Jan.

Shankland R (1964) Michelson Morley Experiment. *Am. J. Physics*, 32:16-35, 1964; also in *Scientific American*, 211:107-114.

Shankland R (1973) Conversations with Albert Einstein. II. *Am. J. Physics*, 41:895-901, July.

Shankland R (1973) Michelson's Role in the Development of Relativity. *Applied Optics,* 12(10) 2280-2287, October.

Sharaf M (1983) *Fury on Earth: A Biography of Wilhelm Reich.* St. Martin's-Marek, NY.

Silberstein L (1921) The Propagation of Light in Rotating Systems. *J. Optical Society of America*, 5(4) 291-307.

Silberstein L (1925) The Relativity Theory and Ether Drift", *Science Supplement - Science News*, 62(1596) viii, 31 July.

Silvertooth EW (1986) Experimental Detection of the Ether. *Speculations in Science and Technology*, 10(1) 3-7.

Silvertooth EW, Whitney CK (1992) A New Michelson-Morley Experiment. *Physics Essays*, 5(1) 82-89.

Shnoll S (1979) *Physico-Chemical Factors of Biological Evolution*, Nauka, Moscow. See Beloussov, et al, 2000.

Shnoll S (2009) *Cosmophysical Factors in Random Processes*. Svenska Fysikarkivet. Stockholm. English translation 2012, see WebRef.4

Stearns RL (1952) *A Statistical Analysis of Interferometer Data,* Thesis, Case Institute of Technology, Physics Dept.

Stecchini L (1966) The Inconstant Heavens, in *The Velikovsky Affair: Warfare of Science and Scientism*, A. deGrazia, Ed., University Books.

Stokes G (1845) On the Aberration of Light, *Philosophical Mag.* 27: 9-15, 1845.

Stokes G (1893) The Luminiferous Aether. *Smithsonian Report #931.*

Sullivan, JWN (1938) *Isaac Newton: 1642-1727.* Macmillan, NY p.118-124.

Swenson L (1970) The Michelson-Morley-Miller Experiments Before and After 1905. *Journal, History of Astronomy*, p.56-78.

Swenson L (1972) *The Ethereal Aether: A History of the Michelson-Morley-Miller Aether-Drift Experiments, 1880-1930.* Univ. Texas Press, Austin.

Thompson C see WebRef.17.

Tromp, SW (1949) *Psychical Physics*. Elsevier, NY.

Thompson DW (1942) *On Growth and Form.* Cambridge Univ Press, England.

Walkey OR (1946) An Abstraction on the Solar Apex. *Royal Astronomical Society*, 6:274-279.

Wheeler RH (1943) *The Effect of Climate on Human Behavior in History,* Kansas Academy of Sciences.

Web References:

1. List of Reich's publications, with supporting works.
   www.orgonelab.org/bibliog.htm
2. Mainstream media, "skeptic" and Wikipedia distortions of Reich.
   www.orgonelab.org/wikipedia.htm
   www.orgonelab.org/bibliogPLAGUE.htm
3. Lay Enthusiast Distortions of Reich's research.
   www.orgonelab.org/orgonenonsense.htm
4. English translation of Shnoll 2009.
   www.shnoll.ptep-online.com/publications.html
5. Planet 9 and Kuiper belt planetoids; Preprint by Batygin et al 2019.
   www.findplanetnine.com/2019/01/
   arXiv:1902.10103v
6. Milky Way gamma jet tilting
   insider.si.edu/2012/05/ghostly-gamma-ray-beams-blast-from-milky-ways-center/
7. The Plasmasphere and IMAGE Satellite
   www.plasmasphere.nasa.gov/regions.html
8. Interstellar Winds
   www.nasa.gov/content/goddard/interstellar-wind-changed-direction-over-40-years
   www.sciencemag.org/news/2013/09/change-interstellar-wind
9. Cosmic Ray Winds
   Díaz-Vélez, et al. arXiv:1708.03005v1
   Erylkin, et al. arXiv:1901.00160v4
10. Bernabei on DAMA: Dark Matter Wind
    people.roma2.infn.it/%7Edama/pdf/cris2010_bernabei_webs.ppt
    people.roma2.infn.it/~dama/web/home.html
11. Ligo/Aligo www.ligo.caltech.edu/page/ligos-ifo
12. Ligo Gravitational Wave Transient Catalog 1
    www.ligo.caltech.edu/page/four-new-detections-o1-o2-catalog
13, New Scientist: Grave Doubts over LIGOs Discovery
    www.newscientist.com/article/mg24032022-600-exclusive-grave-doubts-over-ligos-discovery-of-gravitational-waves
14. Faraday disk generator plans
    www.jnaudin.free.fr/html/farhom.htm

15. Inside Einstein's Mind, The 100th Anniversary of "The Perfect Theory" (2015) Segment on Diverging Atomic Clocks, NOVA/WGBH/PBS video, timecode 39:28 - 41:12.
    www.pbs.org/wgbh/nova/video/inside-einsteins-mind
16. Louis Essen rejects Einstein's Relativity
    www.naturalphilosophy.org/site/arryricker/2015/05/25/dr-louis-essen-inventor-of-atomic-clock-rejects-einsteins-relativity-theory
17. Carolyn Thompson's archived website
    web.archive.org/web/20160418014858/http://freespace.virgin.net/ch.thompson1
18. Nicos Hamaus website
    www.usm.lmu.de/people/hamaus/index.php
19. Milky Way, Sagittarius-A stars in motion
    earthsky.org/space/star-s2-s0-2-single-milky-way-monster-black-hole
20. M87 claimed black hole publicity image
    www.cnn.com/2019/04/10/world/black-hole-photo-scn/index.html
21. Katie Bouman web photo with star-ring on her computer
    nypost.com/2019/04/10/meet-katie-bouman-woman-behind-first-black-hole-photo
22. Grusenick experiment:
    www.youtube.com/watch?v=7T0d7o8X2-E
23. aftar.uaa.alaska.edu/gallery/img-165.html
    www.noao.edu/image_gallery/html/im0685.html
24. Superluminal velocity of the M87 Jet
    www.stsci.edu/ftp/science/m87/m87.html
    www.stsci.edu/ftp/science/m87/press.txt
25. Cohen, on Superluminal Motion
    ned.ipac.caltech.edu/level5/ESSAYS/Cohen/cohen.html
26. Historical ether-drift references organized chronologically, with PDFs
    www.orgonelab.org/energyinspace.htm
27. Newton's Letter to Boyle on the Ether, 1679
    www.orgonelab.org/NewtonLetter.htm

Newspaper Clippings From Various Archives, often without full citation info (organized chronologically)

Buel W (1925) Clevelander Bombs Einstein's Theory, Hurls Bomb, *Cleveland Plain Dealer*, 29 April.
Morse PM (1925) Local Man Proves Ether Drifts, Refuting Einstein"
anon (1925) Scientists Debate Recent Tests of Einstein Theory, *Washington Post* 29 April.
Serviss G (1925) Earth's Flight Through the Ether", *Cleveland News*, 1 Dec.

anon (1925) Case School Scientific Discovery, *Jefferson Ohio Gazette*, 19 Dec.

Charles F (1925) Einstein Theory Born on Heights, *Cleveland Plain Dealer*, 24 Dec.

Dietz D (1925) Einstein Theory Blast, US Scientist After 6 Years Tells of Work, *Citizen Cairo*, 29 Dec.

Dietz D (1925) Theory Pushed by Einstein is Given Setback. 29 Dec.

anon (1926) Doubt Einstein, Clevelander's Tests may Upset Relativity Theory

anon (1926) "Scientists to Argue Theory of Einstein",

anon (1926) "Einstein is Ready to Bet on His Theory",

anon (1926) "Einstein Bets Miller is Wrong, Willing to Wager Case Professor's Experiments are Faulty",

anon (1926) "Says Einstein's Theory of Relativity Stands, Laboratory Expert Denies It is Overthrown by Experiments Recently Made", unidentified newspaper articles, early January 1926, Case Western Reserve University Archives.

anon (1926) Goes to Disprove Einstein Theory: Case Scientist Will Conduct Further Studies in Ether Drift", *Cleveland Plain Dealer*, 27 Jan.

anon (1926) Leaves Nut for Einstein to Crack: Dr Miller Puts Next Move in Ether Controversity up to Opponent, *Cleveland Plain Dealer*, 26 Oct.

Deitz D (1926) With Use of Interferometer Prof. Miller Says... *Cleveland Press,* 6 April.

Deitz D (1926) Einstein's Relativity Theory Refuted by American Expert, *Port Huron Herald*, 26 Jan.

anon (1929) Is there an Ether? *Science News-Letter,* 9 November.

anon (1929) Solar System Rushing Through Space, *Science News-Letter,* 30 November.

anon (1929) Miller Challenges Einstein: Explains Ether Drift Research and Function of Interferometer. *The Case Alumnus,* December, p.10-12.

anon (1930) These Two Disagree on Scientific Precept: Case Savant Continues Attack on Einstein Theory. *Cleveland News*. 26 Jan.

anon (1931) May Dethrone Einstein: Dr. Miller of Case to Launch New Attack, Prepares 6-Month Test to Prove Ether Drift Exists. *Cleveland News,* 2 Dec.

Dietz D (1933) Case's Miller Seen Hero of 'Revolution'. New Revelations on Speed of Light Hint Change in Einstein Theory. *Cleveland Press*, 30 Dec.

Dietz D (1934) French Savants Fight Einstein. *Cleveland Press*. 25 July.

Laurence W (1936) New Evidence Held to Upset Einstein...Existence of Ether Seen. *New York Times,* 23 Feb. p.1.

Randall E (1940) Honoring Cleveland's Distinguished Scientist. Tribute Next Thursday to Case's Dr. Miller Recalls His Renowned Experiments with Ether-Drift. *Cleveland Plain Dealer*, 10 March.

Glossary

aberration, stellar - The angular shift in the apparent direction of a star caused by the orbital motion of the Earth.

aether - an older archaic spelling for the cosmic ether.

anisotropy - A variation in an otherwise smooth distribution of a given parameter, such as the anisotropy of the CMBR (below).

aphelion - The location of a planet at its farthest distance from the Sun. The Earth's aphelion is in early July.

astronomical unit, or AU - a method of measuring distance in the universe. One AU is equal to the distance from the Sun to the Earth, slightly less than 93 million miles.

azimuth - A fixed direction or location in the sky, defining the position of a star or galaxy as seen from Earth.

celestial coordinates - A grid system for locating things in the sky. The celestial poles lie above the Earth's north and south poles, and the celestial equator lies directly above the Earth's equator. Right ascension and declination (below) use this coordinate system.

CMBR - Cosmic Microwave Background Radiation. The temperature in open cosmic space, extremely cold, about 3° Kelvin (-270°C). A slight variation exists within the CMBR, called the CMBR anisotropy.

declination, or Dec - The celestial equivalent of latitude, denoting how far (in degrees) an object in the sky lies north or south of the celestial equator.

eccentricity - the measure of how much an orbiting planet or moon deviates from being exactly circular.

ecliptic, or plane of the ecliptic - A general flat plane defining the orbits of planets in our solar system as they move around the Sun.

ether - The cosmic medium filling all space, between all objects and penetrating inside of matter, and into the atoms, also filling the space between planets, stars and galaxies. Ether is present in the hard vacuum of space and inside vacuum tubes. It is the medium for transmission of light and electromagnetic waves, and in dynamic form, the prime mover and gravitational force. (Context easily separates the cosmic ether from the ether gas, used in surgery to numb the nerves.)

ether drag /ether entrainment - The coupling of cosmic ether with the surface of a planet or star, such that ether velocity is greatly reduced. It is similar to how the velocity of water inside a pipe is slower close to the interior walls of the pipe, and faster in the center of the pipe.

ether drift/ether wind - These terms are interchangeable, referencing the motion of the ether as measured on the Earth's surface. Ether drift can be created by the Earth's motion through a static ether, or ether wind by a dynamic motional ether pushing and blowing across the surface of the Earth, and similarly for any other planet or star.

km/sec - Kilometers per second, a velocity. One km/sec is equal approximately to 0.62 miles per second. Or one mile per second is equal to 1.61 km/sec.

luminiferous - Possessing light, able to transmit light, and also to glow.

parallax - the apparent offset of a star in the foreground against the background, as you move in relation to it. Stellar parallax is measured six months apart, using the Earth's orbital diameter to yield two different observing perspectives, as for measuring a star's distance.

perihelion - The location of a planet at its closest distance from the Sun. The Earth's perihelion is in early January..

PMT - The photomultiplier tube, used to amplify dim light flashes so that they can be recorded and counted.

right ascension, or RA - The celestial equivalent of longitude, denoting how far (in 15° segments or "hours") an object lies east of the sun's location during the March equinox.

scientism - The act of replacing objective observed or experimentally proven facts and evidence with mystical ad-hoc proclamations or never demonstrated imaginary things or claims.

solar corona - the outermost layer of the Sun, extending outwards by several solar diameters, or farther.

solar day - the time required for the Earth to rotate on its axis over a full 24 hours, to face the Sun at the same identical location.

sidereal day, or sidereal hour - A timekeeping system used by astronomers to locate celestial objects according to their position against the fixed background of stars. Sidereal hour is defined by right ascension, and one sidereal day is marked by the Earth's alignment towards the same identical location in the background of stars. The sidereal day is 23 hours, 56 minutes, 4 seconds, being 3 minutes and 56 seconds shorter than a solar day, and a sidereal year is one Earth day shorter than the solar year.

tilde, or " ~ " - A symbol placed before a number to indicate the meaning of "approximate". Such as ~10 km/sec.

unequivocal - a theory or idea having only one explanation, as opposed to being *equivocal,* having several different explanations.

zenith - the point in the sky directly overhead.

Additional astronomical glossaries are found on-line. The Northern Virginia Astronomy Club has a very good one, found here:
www.novac.com/wp/fp/resources/glossary

Index

Other Books by James DeMeo
Available from most on-line bookstores

Saharasia: The 4000 BCE Origins of Child Abuse, Sex-Repression, Warfare and Social Violence In the Deserts of the Old World, Revised Second Edition.

The Orgone Accumulator Handbook: Wilhelm Reich's Life-Energy Discoveries and Healing Tools for the 21st Century, with Construction Plans, Third Revised Edition.

In Defense of Wilhelm Reich: Opposing the 80-Years' War of Mainstream Defamatory Slander Against One of the 20th Century's Most Brilliant Physicians and Natural Scientists.

Marx Engels Lenin Trotsky: Genocide Quotes. The Hidden History of Communism's Founding Tyrants, in their Own Words.

Preliminary Analysis of Changes in Kansas Weather Coincidental to Experimental Operations with a Reich Cloudbuster: From a 1979 Research Project, reprinted 2010.

(as Editor) *Heretic's Notebook: Emotions, Protocells, Ether-Drift and Cosmic Life-Energy, with New Research Supporting Wilhelm Reich.*

(as Co-Editor) *Nach Reich: Neue Forschungen zur Orgonomie: Sexualökonomie, Die Entdeckung der Orgonenergie.*

(as Editor) *On Wilhelm Reich and Orgonomy.*

About the Author

James DeMeo, PhD, formally studied the Earth, Atmospheric, and Environmental Sciences at Florida International University and later at the University of Kansas, where he earned his PhD in 1986. His research has been interdisciplinary, ranging across the social and natural sciences. He undertook milestone research on the role of harsh desert climates and famine on the origins of human violence, with on-going drought-abatement field work in North America, Africa and Israel. Dr. DeMeo also has a lifelong interest in astronomy. His study of the ether-drift subject and related works on cosmic energy began in high school, with in-depth study and experimental work on those subjects in the universities. He served on the faculties of Geography at the University of Kansas, Illinois State University, the University of Miami and the University of Northern Iowa, teaching courses on Physical and Social Geography, Climatology and Renewable Energy. Since 1989, DeMeo has been Director of the private, non-profit *Orgone Biophysical Research Laboratory* which he founded, a high-altitude research institute and observatory in the Siskiyou Mountains of Southern Oregon. Seminars are periodically offered to serious students, and he occasionally lectures at scientific conferences and universities.

A full list of his published works and lectures is found at:
    www.orgonelab.academia.edu/JamesDeMeoPhD
    www.researchgate.net/profile/James_DeMeo
    www.orgonelab.org/demeopubs.htm

www.ingramcontent.com/pod-product-compliance
Lightning Source LLC
Chambersburg PA
CBHW021547210326
41599CB00010B/344